Cancer Drug Discovery and Development

Series editor
Beverly A. Teicher

More information about this series at http://www.springer.com/series/7625

Paolo A. Ascierto • David F. Stroncek
Ena Wang
Editors

Developments in T Cell Based Cancer Immunotherapies

Humana Press

Editors
Paolo A. Ascierto
Unit of Melanoma & Cancer Immunotherapy
National Tumor Institute Fondazione G.
Naples, Italy

David F. Stroncek
Department of Transfusion Medicine
National Institutes of Health
Bethesda, MD, USA

Ena Wang
Sidra Medical and Research Center
Research Branch
Doha, Qatar

ISSN 2196-9906 ISSN 2196-9914 (electronic)
Cancer Drug Discovery and Development
ISBN 978-3-319-79330-6 ISBN 978-3-319-21167-1 (eBook)
DOI 10.1007/978-3-319-21167-1

Springer Cham Heidelberg New York Dordrecht London

Softcover re-print of the Hardcover 1st edition 2015

Printed on acid-free paper

Humana Press is a brand of Springer
Springer International Publishing AG Switzerland is part of Springer Science+Business Media (www.springer.com)

Preface

After decades of disappointing results resilient to extensive efforts to improve the efficacy of immunotherapy against cancer, patients and scientists are witnessing a revolution. A rapid translation of concepts from the bench to the bedside is finally making a difference in overall survival of patients with different types of cancers, including those traditionally considered non-responsive to immunotherapy. Clinical studies have proven unequivocally the effectiveness of T cell-based therapies that can induce regression of late stage cancers otherwise resistant to standard therapy. Regressions are associated with prolonged patients' survival, achieving, in some cases, durable disease-free survival.

Many written accounts on large studies that validate the clinical usefulness of immunotherapy have appeared monthly in high-impact journals. This is leading to a rapid inflation of the field characterized by the rapid expansion of tumor immunotherapy clinical programs and participation of oncologists to meetings focused on this discipline. In the last 3 years, for instance, the Society for Immunotherapy of Cancer (SITC) has more than doubled participation to its annual meeting with nearly 2000 attendees.

This long-awaited success is giving both clinicians and scientists new opportunities. The high frequency of objective responses allows for a more efficient study of mechanisms of responsiveness and identification of biomarkers as a smaller number of patients must be accrued to observe a sufficient number of responding cases. The shortened length of time necessary to perform informative clinical studies expedites the feedback loop stimulating research based on clinical evidence while simultaneously helping the design of second-generation clinical studies. In addition, the expansion of clinical protocols to larger patients cohorts in phase three or even post-licensing studies allows for a less fragmented approach to the understanding of human cancer biology by evaluating more homogenous patient populations in better controlled settings. This provides grounds for prospective validation of concepts developed during monitoring of early phase trials.

The clinical success has led to unprecedented nimbleness of regulatory agencies in approving novel therapeutics. This, in turn, has allowed a more flexible off-label use of therapeutics in combination. Combination therapy trials demonstrated that the therapeutic potential of immunotherapy agents is complementary and not overlapping. Thus, the tremendous success of combining conceptually similar approaches such as anti-CTLA-4 with anti-PD1/PD1L as checkpoint inhibitor agents, which have shown synergistic enhancement of unleash T cell activation. Considering the central role that checkpoint inhibitors are taking in the treatment of several cancers, their relatively limited toxicity, and simplicity of administration, it can be anticipated that future combinatorial approaches will be centered around the addition of other therapeutics such as pathway inhibitors, anti-cancer vaccines, agonistic antibodies, cytokines, adoptive cellular therapies, anti-angiogenesis agents, chemotherapy, epigenetic therapy, and radiotherapy to checkpoint inhibitors. In particular, radiotherapy is taking a novel role in the treatment of cancer as a facilitator of anti-cancer immune effects through the demonstration of its abscopal effects: concurrent not-irradiated tumors regress in the presence of checkpoint inhibitor therapy after radiation. The abscopal effect is revolutionizing our understanding of the role played by radiation in modulating the biology of human cancers.

Several new concepts have also emerged throughout the implementation of clinical trials: a salient one is derived by the observation that, contrary to other anti-cancer therapies, responses to immunotherapeutic agents are of long duration and linked to long-term survival. It has also become clear that immune responses follow a distinct dynamic pattern diverging from that of classical responses to standard chemotherapy. The latter is characterized, when successful, by immediate although often ephemeral reduction in tumor burden. Tumors that respond to immunotherapy often increase in size before a reduction can be observed. This phenomenon is believed to be due to the inflammatory process induced by immunotherapy that leads to recruitment of immune cells within the tumor microenvironment. Another pattern peculiar to immunotherapy is the observation that several patients seem to benefit from long-term stable disease although the biology of this "halting" of tumor growth is currently poorly understood.

Challenges remain. The cost of immunotherapy treatments is quite significant. Therefore, several therapies are not readily available to all potential beneficiaries. Interestingly, a cost-effectiveness analysis of these treatments is not, to our knowledge, reported. Although the price for the individual treatment may be costly, its effectiveness, short duration, and limited toxicity may mitigate the overall cost of care compared to traditional approaches.

The optimal way to simultaneously avoid unnecessary exposure of patients to ineffective therapies while relieving the society from wasteful spending would be to predict a priori likelihood of response. The identification of predictive biomarkers will, therefore, take a leading role in the next future. We and others have shown that the functional orientation of immune cells toward a Th1 polarization is a harbinger of likelihood or response. Interestingly, the same functional orientation has been associated with good prognostic connotation in most cancers. Lack of immune activation is likely to correspond to resistance to immunotherapy, while the presence of a Th1-polarized immune phenotype may indicate a microenvironment pre-conditioned to respond. Immunotherapy further enhances the otherwise lingering immune response leading to a full-blown activation of an acute inflammatory process similar to that observed during acute flares of autoimmunity or during acute transplant rejection. We referred to this phenomenon as "the Immunologic Constant of Rejection." An intermediate condition occurs when the same Th1 polarization is observed in association with improved prognosis. In that case, the immune response is not sufficient to completely eradicate the growth of cancer but can slow its progression. These observations will need further validation in the future before such signatures could be used for patient selection. A comprehensive discussion about the revolutionary role played by signatures of Th1 polarization in reshaping cancer staging or prediction of its responsiveness to therapy is beyond the scope of this volume. However, these findings clearly emphasize the central role played by T cells in controlling tumor growth.

Another limitation to the broad utilization of immunotherapy is the resistance to treatment peculiar to some cancer types. While novel immune therapeutics have greatly increased the range of immunotherapy expanding its proven efficacy to cancers previously judged to be immune-resistant, several cancers such as breast cancer remain quite unresponsive. Further work will need to be done to understand how ontogeny, together with genetic background of the host and somatic alterations, may affect immune responsiveness.

Thus, in conclusion, the progress of immunotherapy has been exponential and the unprecedented clinical outcomes are promising for the years to come. However, several challenges remain. Moreover, as the mechanism leading to tumor rejection has not been fully investigated nor completely understood using integrated system biology approaches, a better understanding will likely lead to further outcomes improvement.

This volume illustrates salient aspects of cancer biology relevant to the successful implementation of immunotherapy. Coverage includes the enhancement of antigen-specific immune responses by anti-cancer vaccines, modulation of the function of T cells within the tumor microenvironment, and the effect of genetic, epigenetic, developmental, and environmental determinants on T cell function. Also covered is the ex vivo expansion of T or other immune cells and their genetic modification or reprogramming to increase their ability to survive and expand when adoptively transferred back to the patients. Specific attention is devoted to the genetic manipulation of T cells through the introduction of re-directed T cell receptors, chimeric antibody receptors, and other genetic manipulation aimed at improving the effectiveness of anti-cancer agents. Furthermore, the revolutionary role of checkpoint inhibitors and their potential in combination with other immunotherapeutic approaches or with standard chemo and radiation therapy is extensively discussed.

We hope that the readers will find this volume useful and we would like to conclude with the famous quote from Winston Churchill: "This is not the end, it is not even the beginning of the end but, perhaps it is the end of the beginning".

Doha, Qatar — Ena Wang
Doha, Qatar — Francesco M. Marincola

The original version of this book was revised. An erratum to this book can be found at DOI 10.1007/978-3-319-21167-1_14

Contents

1 **Insights on Peptide Vaccines in Cancer Immunotherapy** 1
Kwong Y. Tsang, Caroline Jochems, and Jeffrey Schlom

2 **Strategies for Improving Vaccines to Elicit T Cells to Treat Cancer** 29
Jay A. Berzofsky, Lauren V. Wood, and Masaki Terabe

3 **T Cell Fate in the Tumor Microenvironment** 53
Joel Crespo, Ilona Kryczek, Theodore Welling, Shuang Wei, and Weiping Zou

4 **Influence of Antigen Receptor Avidity, Affinity, and Specificity on Genetically Engineered T Cells** 75
Kaoru Nagato, Timothy T. Spear, and Michael I. Nishimura

5 **Toward the Identification of Genetic Determinants of Responsiveness to Cancer Immunotherapy** 99
Davide Bedognetti, Sara Tomei, Wouter Hendrickx, Francesco M. Marincola, and Ena Wang

6 **Production of Clinical T Cell Therapies** 129
David Stroncek, Jianjian Jin, Virginia David-Ocampo, Vicki Fellowes, Larry Moses, and Marianna Sabatino

7 **Clinical Success of Adoptive Cell Transfer Therapy Using Tumor Infiltrating Lymphocytes** 151
Jessica Chacon, Krit Ritthipichai, Geok Sim, Michiko Harao, Jie Qing Chen, Caitlin Creasy, Chantale Bernatchez, Patrick Hwu, and Laszlo Radvanyi

8 **Harnessing Stem Cell-Like Memory T Cells for Adoptive Cell Transfer Therapy of Cancer** 183
Enrico Lugli and Luca Gattinoni

9 T Cell Blockade Immunotherapy Against Cancer and Abscopal Effect in Combination Therapy 211
Giuseppe V. Masucci, Luigi De Petris, Andreas Lundqvist, Rolf Kiessling, and Rolf Lewensohn

10 T Cell Modulation: Anti-PD-1 Antibodies for the Treatment of Cancer 231
Patrick A. Ott and F. Stephen Hodi

11 Enhancing T Cell Performance Against Cancer in Combination Treatment Strategies 245
Maria Libera Ascierto, Salvador Martin Algarra, Ignacio Melero, and Paolo A. Ascierto

12 Chimeric Antigen Receptor (CAR) T Cells 259
Daniel W. Lee and Alan S. Wayne

13 Rapamycin-Resistant T Cells and Pentostatin-Based Immuno-Selective Conditioning for the Allogeneic T Cell Therapy of Cancer 281
Daniel H. Fowler and David C. Halverson

Erratum E1

Index 303

Contributors

Salvador Martin Algarra Department of Oncology, University of Navarra, Pamplona, Spain

Maria Libera Ascierto Department of Oncology, Johns Hopkins University, Baltimore, MD, USA

Paolo A. Ascierto Unit of Melanoma, Cancer Immunotherapy and Innovative Therapy, Istituto Nazionale Tumori Fondazione "G. Pascale", Naples, Italy

Davide Bedognetti Tumor Biology, Immunology and Therapy Section, Division of Translational Medicine, Research Branch, Sidra Medical and Research Center, Doha, Qatar

Chantale Bernatchez Department of Melanoma Medical Oncology, The University of Texas M. D. Anderson Cancer Center, Houston, TX, USA

Jay A. Berzofsky, M.D., Ph.D. Vaccine Branch, Center for Cancer Research, National Cancer Institute, National Institutes of Health, Bethesda, MD, USA

Jessica Chacon Department of Melanoma Medical Oncology, The University of Texas M. D. Anderson Cancer Center, Houston, TX, USA

Jie Qing Chen Department of Melanoma Medical Oncology, The University of Texas M. D. Anderson Cancer Center, Houston, TX, USA

Caitlin Creasy Department of Melanoma Medical Oncology, The University of Texas M. D. Anderson Cancer Center, Houston, TX, USA

Joel Crespo Department of Surgery, University of Michigan School of Medicine, Ann Arbor, MI, USA

Virginia David-Ocampo Cell Processing Section, Department of Transfusion Medicine, Clinical Center, National Institutes of Health, Bethesda, MD, USA

Vicki Fellowes Cell Processing Section, Department of Transfusion Medicine, Clinical Center, National Institutes of Health, Bethesda, MD, USA

Daniel H. Fowler, M.D. Experimental Transplantation and Immunology Branch, National Institutes of Health, National Cancer Institute, Bethesda, MD, USA

Luca Gattinoni, M.D. Experimental Transplantation and Immunology Branch, Center for Cancer Research, National Cancer Institute, Bethesda, MD, USA

David C. Halverson, M.D. Experimental Transplantation and Immunology Branch, National Institutes of Health, National Cancer Institute, Bethesda, MD, USA

Michiko Harao Department of Melanoma Medical Oncology, The University of Texas M. D. Anderson Cancer Center, Houston, TX, USA

Wouter Hendrickx Tumor Biology, Immunology and Therapy Section, Division of Translational Medicine, Research Branch, Sidra Medical and Research Center, Doha, Qatar

F. Stephen Hodi Dana Farber Cancer Institute, Harvard Medical School, Melanoma Disease Center and Center for Immuno-Oncology, Boston, MA, USA

Patrick Hwu Department of Melanoma Medical Oncology, The University of Texas M. D. Anderson Cancer Center, Houston, TX, USA

Jianjian Jin Cell Processing Section, Department of Transfusion Medicine, Clinical Center, National Institutes of Health, Bethesda, MD, USA

Caroline Jochems, M.D., Ph.D. Laboratory of Tumor Immunology and Biology, Center for Cancer Research, National Cancer Institute, National Institutes of Health, Bethesda, MD, USA

Rolf Kiessling Department of Oncology and Pathology, Karolinska Institutet, Stockholm, Sweden

Ilona Kryczek Department of Surgery, University of Michigan School of Medicine, Ann Arbor, MI, USA

Daniel W. Lee Center for Cancer Research, National Cancer Institute, Bethesda, MD, USA

Rolf Lewensohn Department of Oncology and Pathology, Karolinska Institutet, Stockholm, Sweden

Enrico Lugli, Ph.D. Principal Investigator, Laboratory of Translational Immunology, Humanitas Clinical and Research Center, Rozzano, Milan, Italy

Andreas Lundqvist Department of Oncology and Pathology, Karolinska Institutet, Stockholm, Sweden

Francesco M. Marincola Research Branch, Sidra Medical and Research Center, Doha, Qatar

Giuseppe V. Masucci Department of Oncology and Pathology, Karolinska Institutet, Stockholm, Sweden

Ignacio Melero Department of Oncology, Centro de Investigación Médica Aplicada (CIMA), Clinica Universidad de Navarra, Pamplona, Spain

Larry Moses Cell Processing Section, Department of Transfusion Medicine, Clinical Center, National Institutes of Health, Bethesda, MD, USA

Kaoru Nagato Department of Surgery, Cardinal Bernardin Cancer Center, Loyola University Medical Center, Maywood, IL, USA

Michael I. Nishimura Department of Surgery, Cardinal Bernardin Cancer Center, Loyola University Medical Center, Maywood, IL, USA

Patrick A. Ott, M.D., Ph.D. Dana Farber Cancer Institute, Harvard Medical School, Melanoma Disease Center and Center for Immuno-Oncology, Boston, MA, USA

Luigi De Petris Department of Oncology and Pathology, Karolinska Institutet, Stockholm, Sweden

Laszlo Radvanyi, Ph.D. Immuno-Oncology Translational Innovation Platform, EMD Serono Research & Development Institute, Billerica, MA, USA

Krit Ritthipichai Department of Melanoma Medical Oncology, The University of Texas M. D. Anderson Cancer Center, Houston, TX, USA

Marianna Sabatino Kite Pharma, Inc., Santa Monica, CA, USA

Jeffrey Schlom, Ph.D. Laboratory of Tumor Immunology and Biology, Center for Cancer Research, National Cancer Institute, National Institutes of Health, Bethesda, MD, USA

Geok Sim Department of Melanoma Medical Oncology, The University of Texas M. D. Anderson Cancer Center, Houston, TX, USA

Timothy T. Spear Department of Surgery, Cardinal Bernardin Cancer Center, Loyola University Medical Center, Maywood, IL, USA

David Stroncek, M.D. Cell Processing Section, Department of Transfusion Medicine, Clinical Center, National Institutes of Health, Bethesda, MD, USA

Masaki Terabe, Ph.D. Molecular Immunogenetics & Vaccine Research Section, Vaccine Branch, Center for Cancer Research, National Cancer Institute, National Institutes of Health, Bethesda, MD, USA

Sara Tomei Omics Core and Biorepository, Division of Translational Medicine, Research Branch, Sidra Medical and Research Center, Doha, Qatar

Kwong Y. Tsang, Ph.D. Laboratory of Tumor Immunology and Biology, Center for Cancer Research, National Cancer Institute, National Institutes of Health, Bethesda, MD, USA

Ena Wang Sidra Medical and Research Center, Research Branch, Doha, Qatar

Alan S. Wayne Division of Hematology, Oncology and Blood & Marrow Transplantation, Children's Center for Cancer and Blood Diseases, Children's Hospital Los Angeles, Los Angeles, CA, USA

Shuang Wei Department of Surgery, University of Michigan School of Medicine, Ann Arbor, MI, USA

Theodore Welling Department of Surgery, University of Michigan School of Medicine, Ann Arbor, MI, USA

Lauren V. Wood, M.D. Clinical Trials Team, Vaccine Branch, Center for Cancer Research, National Cancer Institute, National Institutes of Health, Bethesda, MD, USA

Weiping Zou, M.D., Ph.D. The University of Michigan Comprehensive Cancer Center, University of Michigan School of Medicine, Ann Arbor, MI, USA

About the Editors

Ena Wang, M.D., is Director of Molecular Science, Infectious Diseases and Immunogenetics Section at the National Institutes of Health. During her career at NIH, Dr. Wang was granted the Investigator Travel Fellowship Award twice; she was recipient of an NIH Bench-to-Bedside Award in 2002, she was also recipient of the NCI Director's 2006 Intramural Innovation Award as co-investigator in 2006 and recipient of the Clinic Center Director's Award for scientific excellence in 2007. She has contributed 17 book chapters and published more than 130 peer-reviewed articles. The focus of her research is the identification of genetic traits in humans that could explain the relationship between pathogens and the host with particular interest in cancer and chronic infections. She has been cited over 6500 times in her scientific career.

Paolo Antonio Ascierto, M.D., is Vice Director of the Unit of Melanoma, Cancer Immunotherapy and Innovative Therapy at the National Institute Fondazione G. Pascale in Naples Italy. Dr. Ascierto has published over 117 papers in his scientific career and has been cited over 2100 times.

David F. Stroncek, M.D., is the Chief of the Cell Processing Section at the Department of Transfusion Medicine at the National Institutes of Health. Dr. Stroncek has published over 264 papers in his scientific career and has been cited over 5300 times.

Chapter 1
Insights on Peptide Vaccines in Cancer Immunotherapy

Kwong Y. Tsang, Caroline Jochems, and Jeffrey Schlom

Abstract Human tumor-associated antigens are generally weakly immunogenic and therefore able to escape detection by the immune system. Numerous studies have shown, however, that immune cells infiltrate many tumors, and that these cells are vital for keeping tumor burden in check. Immunotherapy can enhance this process by further stimulating tumor-recognizing cells while decreasing the function of immunosuppressive cells, such as regulatory T cells and myeloid-derived suppressor cells, thereby creating a more immune-activating tumor microenvironment.

Peptide vaccines can stimulate and activate T cells specific to tumor-associated antigens. Because peptides endogenously expressed by tumor cells are often weak immunogens, researchers are investigating various strategies for making them more immunogenic and more potent as vaccines. Here we review multiple strategies for enhancing peptide immunogenicity, including (a) peptides with amino acid substitutions at anchor residues and heteroclitic analogs, (b) multiple variance long peptides, (c) whole protein and 15-mer overlapping peptides, (d) multiple peptides recognizing different tumor-associated antigens, (e) class I and II epitope hybrid vaccines, (f) peptide-pulsed dendritic cells, and (g) combining peptide vaccines with other therapies. While it is unlikely that peptide vaccines alone could significantly affect progressive disease, the combination of these vaccines with the right adjuvants and/or immunomodulatory agents has shown promising results in clinical trials.

Keywords Peptide cancer vaccines • Immunotherapy • Combination therapy • Cytokines • Prime-boost regimen • Checkpoint inhibitors

---

K.Y. Tsang, Ph.D.
Laboratory of Tumor Immunology and Biology, Center for Cancer Research, National Cancer Institute, National Institutes of Health, 10 Center Drive, Building 10, Room 8B08, Bethesda, MD 20892, USA
e-mail: tsangkwo@mail.nih.gov

C. Jochems, M.D., Ph.D. • J. Schlom, Ph.D. (✉)
Laboratory of Tumor Immunology and Biology, Center for Cancer Research, National Cancer Institute, National Institutes of Health, 10 Center Drive, Building 10, Room 8B09, Bethesda, MD 20892, USA
e-mail: jochemscm@mail.nih.gov; js141c@nih.gov

P.A. Ascierto et al. (eds.), *Developments in T Cell Based Cancer Immunotherapies*, Cancer Drug Discovery and Development,
DOI 10.1007/978-3-319-21167-1_1

Introduction

Immune editing is an extrinsic mechanism of cancer suppression that initiates only after cellular transformation has occurred and intrinsic mechanisms of cancer suppression have failed [1]. The process of immune editing occurs in three phases: elimination, equilibrium, and escape. In the elimination phase, innate and adaptive immunity join forces to eliminate cancer cells before they become clinically apparent, rendering the host virtually cancer-free. In the equilibrium phase, cancer cells not eliminated in the elimination phase are prevented from proliferating by host immunity, which maintains the cancer cells in a state of functional dormancy. Equilibrium is a function of adaptive immunity, which may restrain cancer cell growth in the host for a lifetime. In the escape phase, cancer cells once held in equilibrium may escape recognition by adaptive immunity due to insensitivity to immune effector mechanisms and induction of immune suppression in the tumor microenvironment. Cancer cells that escape immune recognition proliferate and become clinically apparent. Therapies such as peptide vaccines have the potential to keep cancer cells in the elimination and/or equilibrium phase.

This review describes studies employing peptide-based cancer vaccines and prospects for improving their efficacy through the use of peptides with amino acid substitutions at anchor residues and heteroclitic analogs, multiple variance long peptides, whole protein and 15-mer overlapping peptides, multiple peptide epitopes from different tumor-associated antigens (TAAs), class I and II hybrid peptide vaccines, peptide-pulsed dendritic cells (DCs), adjuvants including toll-like receptor (TLR) agonists and cytokines, and combinations of peptide vaccines with various other therapies.

Peptide vaccines have several advantages over other cancer vaccine approaches. Short peptides (9 or 10 amino acid residues) that bind to major histocompatibility complex (MHC) class I molecules can induce specific $CD8^+$ T-cell responses that can lyse tumor cells expressing the cognate MHC class I and peptide [2, 3]. The quality of the immune response depends on the peptides and adjuvants used in the vaccine. Immune response rates approaching 100 % have been reported in some cases using multipeptide melanoma vaccines [4–6]. A mixture of a dozen peptides restricted to human leukocyte antigen (HLA)-A1, -A2, -A3, -A11, and -A24 can be a stable platform for a vaccine that can be used in 85 % of cancer patients, thus overcoming the limitation of peptide restriction. It has been demonstrated that this type of peptide mixture can induce immune responses in vaccinated patients with no negative effects from antigenic competition among the peptides in the mixture [4, 7, 8]. Other advantages of peptide vaccines include low production costs, stability, safety, their use as an off-the-shelf reagent, and their effectiveness as booster vaccines. On the other hand, peptide vaccines have some considerable limitations. In vivo, when a peptide vaccine is delivered into subcutaneous (s.c.) tissue, short peptides may bind to MHC on nonprofessional antigen-presenting cells (APCs) without optimal costimulation, which may induce tolerance. In addition, peptides in human plasma are rapidly degraded by exopeptidases and endopeptidases, and have a short half-life. In vaccinated patients, short peptides have no tertiary structure and thus may rapidly degrade before they can reach APCs. For example, the estimated half-life of MelanA/MART-1 peptide in fresh human plasma is about 22 s [9]. These issues can be overcome by combining the peptide with the proper adjuvant, which may not only emulsify it for better delivery, but also increase the half-life and stimulate the immune system more efficiently to avoid possible induction of tolerance.

There are numerous TAAs being used as vaccine targets. Below is a description of TAAs employed in vaccines developed at the National Cancer Institute.

Tumor-Associated Antigens

Carcinoembryonic Antigen

Carcinoembryonic antigen (CEA; CD66) is a 180-kDa immunoglobulin-like oncofetal glycoprotein that is expressed on the cell surface of normal colonic mucosa and primarily functions in cellular adhesion [10]. CEA is also commonly overexpressed on adenocarcinomas arising from the breast, cervix, lung, and gastrointestinal tract [11, 12].

Mucin 1

Mucin 1 (MUC1; CD227) is a large transmembrane glycoprotein normally expressed at the apical surface of glandular epithelial cells [13]. In adenocarcinomas (i.e., breast, prostate, colorectal, ovarian, lung, bladder, and pancreatic) it is overexpressed and aberrantly glycosylated [14, 15]. Loss of epithelial-cell polarization also results in MUC1 expression throughout the cell surface. These characteristics make MUC1 a potential target for immunotherapy [16]. MUC1 is also expressed in hematologic malignancies such as B-cell lymphoma, chronic myelogenous leukemia, and multiple myeloma [17–19]. The N-terminal (MUC1-N) is the large extracellular domain that consists of a variable number of tandem repeats (VNTR) region and a non-VNTR region. MUC1-N is shed from cells, is present in the circulation of patients with advanced cancer, and is used as a tumor marker (CA15.3) in breast cancer patients [20]. The C-terminal of MUC1 (MUC1-C) has been shown by several groups to be extremely important in the initiation and progression of a range of human neoplasms [21–23]. Overexpression of MUC1-C makes it possible for malignant cells of epithelial or hematopoietic origin to exploit the physiologic stress response, and thus stimulate their expansion and survival [24].

Prostate-Specific Antigen

Prostate-specific antigen (PSA) is a 34-kDa glycoprotein that is expressed in normal prostate tissue and prostate cancer [25]. PSA is also expressed at very low levels in the paraurethral and perianal glands, placenta, breast (including breast cancer), and thyroid. However, except for breast cancer, these tissues do not secrete a significant amount of PSA into the serum. Normally, PSA is secreted into the prostatic ducts. However, in prostate cancer the disordered glandular architecture causes increased amounts of PSA to diffuse into the serum, allowing PSA measurements to serve as screening and prognostic markers for prostate cancer. The immunogenicity of PSA has been demonstrated in multiple studies. Because PSA is secreted, it is not a good target for an antibody response. However, T cells can recognize any proteins made by cells once fragments of these proteins (peptides) are processed and presented on MHC molecules. It has been demonstrated that human cytotoxic T lymphocytes (CTLs) specific for PSA can be generated in vitro [26], and that some patients with advanced prostate cancer have naturally occurring PSA-specific T-cell responses [27]. Furthermore, Gulley et al. demonstrated that in patients with prostate cancer, a PSA vaccine could generate PSA-specific T cells that secrete interferon gamma (IFN-γ) and lyse PSA-expressing tumor cells in an MHC-restricted manner [28].

Brachyury

The transcription factor brachyury was initially identified as a molecule relevant to the formation of the mesoderm during murine embryonic development, which involves conversion of epithelial cells into mesenchymal cells [29]. Brachyury is thus a mediator of normal physiologic epithelial-mesenchymal transition (EMT) and metastasis. Subsequent studies revealed brachyury to be expressed in a range of human tumors, with limited levels in human adult testes and thyroid, and little or no expression in other normal adult tissues [30–33], making it an ideal target for cancer immunotherapy. Transcription factors such as brachyury, however, are generally believed to be difficult to target with small molecule targeted therapies due to their nuclear location and lack of a specific groove for the tight binding of a small molecule inhibitor [34]. An alternative approach to targeting transcription factors is vaccine-mediated T-cell therapy. Recent studies have identified an HLA-A2 class I brachyury peptide that is capable of inducing human $CD8^+$ CTLs in vitro [30]; these T cells were shown to be capable of selectively lysing a range of brachyury-expressing human carcinoma cell lines [30]. Two clinical trials are ongoing employing recombinant vectors expressing brachyury (NCT01519817 and NCT02179515).

Peptides/Proteins as TAAs

Many different TAAs have been used in cancer vaccines, either as the whole native protein, 9-mer peptide epitopes, 15-mer peptide epitopes, or after making changes in the endogenously expressed proteins in order to make them more immunogenic.

Peptides with Amino Acid Substitutions

Preclinical studies have shown that inducing strong T-cell responses requires repeated in vitro stimulation with TAA epitopes [35, 36], and that native peptide vaccines may have limited immunogenicity when used as vaccines for cancer patients [37, 38]. One method for improving the immunogenicity of self-antigenic peptides is to alter the amino acid residues at either the MHC binding site or the T-cell receptor (TCR) binding site. Modifying amino acids at the MHC binding site may enhance MHC binding affinity and significantly improve interaction with the TCR, thus enhancing antitumor immunity [39–45]. For instance, Terasawa et al. [42] demonstrated that changing an isoleucine to a leucine at position 155 on PSA (within the MHC binding region) made the peptide bind with higher affinity to the MHC class I molecule, and the generated PSA-specific T cells produced higher levels of IFN-γ. T-cell lines generated with this agonist epitope efficiently lysed tumor cells expressing native PSA in an MHC-restricted manner [42]. In addition, Zaremba et al. [46] demonstrated that an enhancer agonist epitope of CAP1, an immunogenic HLA-A2 allele binding peptide derived from CEA, could generate specific T cells from nonimmunized individuals, whereas the native epitope could not. These T cells recognized cells expressing both the native and agonist CEA sequence and lysed tumor cells endogenously expressing native CEA [46]. Seven novel CTL epitopes in the MUC1-C region of MUC1 have recently been identified, along with enhancer agonists for each of these epitopes [44]. This was demonstrated by the greater ability of the agonist peptides, compared to their corresponding native peptides, to generate MUC1-C-specific T-cell lines, enhance IFN-γ production by T cells, and lyse human tumor cell targets endogenously expressing the native epitopes in an MHC-restricted manner. The MUC1-C agonist epitopes span class I MHC HLA-A2, -A3, and -A24, which encompass the majority of the population. The studies provide the rationale for clinical immunotherapy studies employing a range of vaccines that target the C-terminus of MUC1.

An agonist epitope of brachyury has also recently been identified. This agonist epitope, which has enhanced binding to MHC class I, (a) increased IFN-γ production from brachyury-specific T cells, (b) generated brachyury-specific T cells with (i) greater levels of perforin, (ii) increased proliferation, and (iii) greater proficiency at lysing human carcinoma cells endogenously expressing the native epitope, and (c) achieved greater brachyury-specific T-cell responses in vivo in HLA-A2-transgenic mice [47]. Furthermore, CTL agonist epitopes of the tumor-associated proteins POTE [48] and TARP [49] have recently been identified. POTE was found to be expressed in a wide variety of human cancers, including prostate, colon, lung, breast, ovary, and pancreas [48]. TARP was found to be a prostate and breast tumor-associated antigen [49].

Another method for increasing the immunogenicity of self-antigenic peptides is by using heteroclitic analogs, which are peptides with substitutions outside the MHC anchor residues. These substitutions are usually made in the residues that contact the TCR, and the peptides are called "altered peptide ligands". It has been demonstrated that such peptides can strongly stimulate T cells and therefore achieve more potent immune responses than native peptide epitopes [46, 50, 51]. These peptides can also act as agonists [52] and modulate phosphorylation patterns and intracellular signaling involved in T-cell activation [53]. In clinical trials in melanoma and colorectal carcinoma, altered peptide ligands have been shown to enhance CD8$^+$ T-cell responses as measured by ELISPOT, tetramers, and chromium-release assays [54, 55]. It has been reported that vaccination with an altered peptide ligand of CEA (CAP1-6D) with Flt3 ligand was able to expand DCs for immunotherapy of colon and non-small cell lung cancer (NSCLC) [56]. These DCs were able to induce CD8$^+$ cytotoxic T lymphocytes that recognized tumor cells expressing endogenous CEA. After vaccination, 2 of 12 patients experienced dramatic tumor regression, 1 patient had a mixed response, and 2 had stable disease. It was also demonstrated that clinical responses correlated with the expansion of CD8$^+$ tetramer positive T cells [56].

The recently published results [57] of a randomized pilot study of CAP1-6D with Montanide and GM-CSF in patients with pancreatic adenocarcinoma showed a clinical signal of benefit, with 7/19 patients alive at a minimum of 32 months after study initiation, including 3 with unresectable disease. Other studies have shown that treatment with heteroclitic peptides did not significantly improve clinical responses in cancer patients [58, 59].

Multiple Variance Long Peptides

An alternative approach to peptide vaccines consists of multivalent synthetic long peptides. These synthetic long-peptide vaccines, which incorporate both class I CTL and class II helper epitopes, are predominantly taken up by APCs and processed for presentation by both class I and class II molecules. A phase I clinical trial using a 20-mer NY-ESO-1f peptide (NY-ESO-91-110) that includes multiple epitopes recognized by CD4$^+$ and CD8$^+$ T cells was conducted in patients with tumors expressing NY-ESO-1 antigen [60]. Ten patients were immunized with NY-ESO-1f peptide mixed with picibanil (OK-432) and Montanide ISA 51. Increased CD4$^+$ and CD8$^+$ T-cell responses were observed in 9/10 patients, demonstrating that NY-ESO-1f peptide vaccine was well tolerated and elicited humoral CD4$^+$- and CD8$^+$-specific T-cell responses in immunized patients [60]. To determine whether the addition of helper peptides would increase CD8$^+$ T-cell responses in a multipeptide vaccine, 167 patients with resected stage IIB-IV melanoma were vaccinated with 12 MHC class I-restricted melanoma peptides (12MP) to stimulate CD8$^+$ T cells, and randomized to receive either a tetanus helper peptide or a mixture of 6 melanoma helper peptides (6MHP) to stimulate CD4$^+$ T cells [6]. T-cell responses were assessed by ex vivo IFN-γ ELISPOT assay. Vaccination with 12MP plus tetanus peptide induced CD8$^+$ T-cell responses in 78 % of patients, and CD4$^+$ T-cell responses to tetanus peptides in 93 % of patients. Vaccination with 12MP plus 6MHP induced CD8$^+$ T-cell responses in only 19 % of patients, and CD4$^+$ T-cell responses to 6MHP in 48 % of patients. Thus, in this study setting, melanoma vaccine-associated helper peptides paradoxically decreased the $CD8^+$ T-cell response to the melanoma vaccine. Similar negative effects of combining helper peptides with class I peptides in metastatic melanoma were also seen in an Eastern Cooperative Oncology Group trial (E1602) [61].

Whole Protein Vaccines and 15-Mer Overlapping Peptides Covering the Whole Protein

$CD4^+$ T cells play an important role in the generation of an antitumor immune response by initiating and maintaining $CD8^+$ T-cell responses [62]. To this end, there are TAA protein vaccines composed of whole protein or protein subunits able to induce both CTL and helper T-cell responses. However, whole protein vaccines have a disadvantage in terms of cost of manufacturing; alternatively, synthetic long-peptide vaccines have been developed. These can be taken up by APCs and then processed for presentation by both MHC class I and class II molecules. A phase I study in end-stage cervical cancer used long overlapping peptides that covered the entire length of the oncogenic proteins E6 and E7 of human papilloma virus type 16 (HPV16), which are likely to contain multiple HLA class I and class II T-cell epitopes [63, 64]. The vaccine, given s.c. with Montanide ISA 51 4 times at 3-week intervals, displayed low toxicity and robust immunogenicity, inducing both $CD4^+$ and $CD8^+$ T-cell responses. The treatment resulted in partial or complete regression of premalignant lesions in a majority of women with $HPV16^+$ grade 3 vulvar intraepithelial neoplasia [65]. Further evidence for the clinical activity of long-peptide vaccines was reported in patients with metastatic colorectal cancer [66] and ovarian cancer [67]. Treatment with a p53 long-peptide vaccine consisting of 10 synthetic long overlapping peptides of wild-type p53 protein induced p53 T-cell responses in these patients [66, 67].

Vaccines Consisting of Multiple Peptides from Different Tumor-Associated Antigens

Helper T cells play a crucial role in the generation of CTLs. Some peptide vaccines consist of MHC class II-restricted helper epitopes recognized by $CD4^+$ T cells in addition to class I epitopes. Numerous helper epitopes have been identified and used with CTL epitopes in cancer vaccines [4, 68]. A helper epitope peptide capable of binding to pan HLA-DR (pan-DR epitope/PADRE) [69] has been used in peptide vaccine clinical trials in combination with Wilms tumor gene, proteinase 3, MUC1-derived peptides, Montanide ISA 51 and CpG7909 in patients with acute myeloid leukemia and multiple myeloma [70]. An increase in PADRE-specific $CD4^+$ T helper cells was observed. Different types of multipeptide vaccines have been developed to generate broad CTL responses against multiple epitopes and several TAAs at the same time.

The number of component peptides in the vaccine can vary from a few to more than 10. Some multipeptide vaccines consist of only CTL epitopes from different TAAs [7, 71] or CTL epitopes combined with helper epitopes [72–74]. Induction of immune responses and clinical efficacy were observed in a phase II clinical trial in metastatic NSCLC using IDM-2101, a 9-CTL epitope vaccine consisting of 2 native and 7 analog epitopes with enhanced TCR binding covering 5 different TAAs [72]. The study enrolled 63 HLA-A2$^+$ patients, none of whom experienced significant adverse events. One-year survival among treated patients was 60 %, and median overall survival was 17.3 months. One complete and 1 partial response were identified, and survival was longer in patients demonstrating an immune response to the peptide epitopes [72]. Walter et al. treated 96 HLA-A2$^+$ patients with advanced renal cell carcinoma (RCC) with IMA901 vaccine after single-dose cyclophosphamide in 2 consecutive studies (phase I and II). IMA901 consists of 10 tumor-associated peptides naturally present in human RCC [75]. In the phase I study, patients' T-cell responses to multiple TAAs were associated with improved disease control and reduced regulatory T cells (Tregs). In the phase II trial, a single dose of cyclophosphamide reduced the number of Tregs and confirmed that immune responses to tumor-associated peptides were associated with longer overall survival [75, 76]. The same group also conducted a phase I/II clinical trial in 19 HLA-A2$^+$ patients with biochemical relapse of hormone-sensitive prostate cancer treated with a multipeptide vaccine consisting of 11 HLA-A2-restricted peptides and 2 HLA class II peptides derived from prostate cancer [73]. The vaccine was used in combination with Montanide ISA 51, imiquimod, GM-CSF, MUC1-mRNA/protamine complex, local hyperthermia, or no adjuvant. The vaccine was well tolerated, and PSA doubling time increased from 4.9 to 25.8 months in 4/19 patients during treatment. Three of the responding patients received imiquimod and 1 received MUC1-mRNA/protamine complex as adjuvant; both are TLR-7 agonists [73].

Class I and II Epitope Hybrid Vaccines

Peptide sequences of most single-epitope and multi-epitope vaccines consist of native sequences with or without modification of the anchor amino acids, and are directed against either MHC class I or class II. Hybrid peptide vaccines are fusion peptides of one epitope specific for MHC class I and one epitope specific for MHC class II.

CD4$^+$ helper T-cell epitopes for HER-2/*neu* have been reported and tested in patients with breast cancer [77]. Thirteen of 18 patients responded by proliferation to at least 1 of the HER-2 peptides tested. Using similar peptides for HER-2/*neu* (777–790) (AE37) to treat patients with breast and ovarian cancer, Disis et al. also demonstrated generation of immunologic responses and antigen spreading [78]. A novel MHC class II epitope hybrid peptide of HER-2/*neu* (AE37) has been developed [79] and used in a clinical trial in patients with breast cancer that overexpressed HER-2/*neu* [80]. AE37 is the II-Key hybrid of HER-2/*neu* (777–790). Coupling the

II-Key segment of the II protein to the MHC class II epitope substantially increases the potency of epitope presentation. The II-Key peptide (4-amino-acid sequence (LRMK)) added to T-helper peptides facilitates direct antigenic epitope charging of MHC class II molecules at the cell surface. Enhanced epitope charging and increased antigen presentation can exponentially increase the potency of epitope presentation compared with unmodified class II epitopes in vitro [79, 81]. Fifteen breast cancer patients completed the HER-2/*neu* (AE37) dose-escalation clinical trial, including 5 dose groups with and without GM-CSF. No grade 3–5 toxicities were observed. The hybrid AE37 vaccine was safe and well tolerated, and is capable of eliciting HER-2/*neu*-specific immune responses even without adjuvant [80]. Perez et al. [82] reported a phase I clinical study using the II-Key/HER-2/*neu* (777–790) hybrid peptide vaccine with recombinant GM-CSF as adjuvant in 32 patients with castration-sensitive and castration-resistant prostate cancer expressing HER-2/*neu*. Their results showed that the AE37 vaccine was safe; grade ≥2 toxicities were not observed. A majority of patients (75 %) developed enhanced immunity to the AE37 vaccine, while 65 % developed enhanced immunity to the unmodified vaccine as detected by IFN-γ-based ELISPOT assays. In addition, significant decreases were seen in the frequency of circulating Tregs, plasma HER2/neu and serum TGF-β levels. These data suggest the potential of using HER-2/*neu* (777–790) vaccine for immunotherapy of HER-2/*neu*-expressing prostate cancer.

T_H1 immunity is essential for the induction of fully activated CTLs and immunologic memory [83, 84]. It has been demonstrated that synthetic long peptides derived from the naturally occurring sequence of human HPV16 were superior to short peptides in terms of generating an immune response in vulvar intraepithelial neoplasia [65]. A long-peptide vaccine containing both helper and CTL epitopes is a rational strategy for activating T_H1-dependent antitumor immunity. Takahashi et al. reported the use of a synthetic helper/killer-hybrid long-peptide epitope of MAGE-A4 combined with OK432 and Montanide ISA 51 for the treatment of pulmonary metastasis of colon cancer [85]. The vaccine induced MAGE-A4-specific T_H1 and $CD8^+$ CTL immune responses. The only side effects were injection-site reactions. Both tumor growth and CEA tumor markers significantly decreased.

Peptide-Pulsed Dendritic-Cell Vaccines

Another method of peptide vaccine treatment involves pulsing DCs with peptides and injecting patients with the peptide-pulsed DCs [86–90]. It has been demonstrated that vaccination with ex vivo-generated and peptide-pulsed DCs can be more effective than direct injection of peptides because the endogenous DCs may be functionally compromised in cancer patients [91, 92]. In a Gynecologic Oncology Group phase II trial, 25 $HLA\text{-}A2^+$ patients with stage III, IV, or recurrent ovarian cancer overexpressing the p53 protein with no evidence of disease were treated in 2 cohorts. Patients in cohort A received s.c. injections of wild-type p53 peptide (264–272) admixed with Montanide ISA 51 and GM-CSF. Patients in cohort B received wild-type p53 peptide

(264–272)-pulsed DCs by i.v. infusion. IL-2 was administered to both cohorts in alternate cycles. Results showed that 9/13 patients (69 %) in arm A and 5/6 patients (83 %) in arm B developed an immunologic response to wild-type p53 as determined by ELISPOT and tetramer assays. Median overall survival was 40.8 and 29.6 months for arms A and B, respectively. Median progression-free survival was 4.2 and 8.7 months for arms A and B, respectively. Both vaccination approaches generated comparable specific immune responses [93]. A phase I pilot study of vaccination with epitope-enhanced TARP peptide and TARP peptide-pulsed DCs in the treatment of stage D0 prostate cancer has been initiated at the National Cancer Institute by Berzofsky et al. (NCT00972309). In addition, a phase I/II trial employing peptide-pulsed autologous CD40L-matured DCs with codon 12 and codon 13 mutant ras peptides with or without IL-2 was recently completed in patients with locally advanced or metastatic colorectal cancer (NCT00019591).

Adjuvants

Peptide vaccines have been used with many different kinds of adjuvants, which are added both in order to emulsify the peptide so that it can be administered and to help induce a strong immune response. Different adjuvants have different properties and can induce a response that is more skewed towards either a T_H1 or T_H2 profile. Inducing a strong immune response while administering the peptide also decreases the risk of developing tolerance.

Incomplete Freund's Adjuvant

Incomplete Freund's Adjuvant (IFA) is the most commonly used adjuvant in peptide vaccines. A recent study investigated the type of immune response achieved in human skin after vaccination with a multipeptide vaccine in IFA [94]. One week after a single vaccination with peptides in IFA, T_H2 cells (GATA-3$^+$), but not T_H1 cells (T-bet$^+$), increased in number, suggesting a T_H2-dominant vaccination-site microenvironment. This was reversed after repeated vaccinations. However, repeated vaccinations may also increase the number of Tregs and eosinophils in the skin [94].

Cytokines

Cytokines such as GM-CSF, IL-2, IL-7, IL-12, IL-15, IL-21, and type 1 interferons have been used as biologic adjuvants in peptide vaccines. GM-CSF acts as an immune stimulant by enhancing antigen processing and presentation by DCs, increasing expression of MHC class II molecules, augmenting the primary antibody response, and inducing localized inflammation when administered by injection [95, 96]. When administered intradermally with vaccine, GM-CSF elicits strong delayed-type hypersensitivity reactions to peptide antigens [97]. A number of small, uncontrolled clinical vaccine trials showed encouraging preliminary results with GM-CSF in the treatment of solid tumors, including melanoma, breast carcinoma, pancreatic cancer, RCC, NSCLC, and prostate cancer [78, 98–105]. Other reports, however, challenge the benefits of GM-CSF and suggest that it may induce a weaker immune response [106]. Recently, a multicenter, randomized phase II trial in 119 patients with resected stage IIB to IV melanoma was conducted with 12 MHC class I-restricted melanoma peptides given alone or in combination with GM-CSF [4]. The multipeptide vaccine achieved high immune response rates, but $CD8^+$ and $CD4^+$ T-cell responses were lower when vaccine was administered with GM-CSF, a finding that directly challenges the utility of GM-CSF as an adjuvant. There is also some evidence that systemic use of GM-CSF can increase tumor-mediated immune suppression by acting as a growth factor for myeloid-derived suppressor cells (MDSCs) [107, 108].

IL-12 can stimulate IFN-γ-producing T cells and has been used to improve antitumor immunity in cancer patients [109, 110]. However, systemic administration of IL-12 has been shown to be highly toxic [111], so many researchers are working on finding new ways of administering the agent. One such modality is NHS-IL-12, which is a novel immunocytokine consisting of 2 IL-12 molecules and a tumor necrosis-targeting human IgG1 antibody [112]. Other γ-chain cytokines such as IL-7, IL-15, and IL-21 also have properties that make them potential adjuvants for peptide vaccines. IL-7 is essential for the development and survival of T cells [113]. It has immunorestorative properties and enhances the activity of naïve T-cell populations. IL-15 is a non-IL-2 growth factor that signals through the IL-2 receptor β and γ chains [114]. IL-15 preferentially augments CD8 memory and NK cells. ALT-803 is a complex of an IL-15 superagonist mutant and a dimeric IL-15 receptor fusion protein, which has been found to have significantly stronger in vivo biological activity on T cells and NK cells, as well as increasing the survival of myeloma-bearing mice [115, 116]. In addition, ALT-803 induces expansion and activation of $CD8^+$ memory cells with robust antitumor activity [117]. IL-21, also in the γ-chain cytokine family [118, 119], is structurally similar to IL-2 and has similar properties. IL-21 preferentially augments the inflammatory T_H17 population [120, 121]. Carefully designed clinical trials are needed to evaluate the potential of these cytokines to amplify the effects of peptide vaccines against cancers.

CD40

CD40, a member of the tumor necrosis factor (TNF) receptor family expressed on B cells, DCs, and a small subset of activated T cells, plays a crucial role in cellular and humoral immunity. Interaction of CD40 with its ligand (CD40L, CD154, expressed by activated T-helper cells) promotes DC activation and initiates $CD8^+$ T-cell immune responses. In cancer patients, antibody to CD40 can induce CD40 ligation. An agonist antibody was recently investigated in a phase I trial in 29 patients with different carcinomas [122]. Four patients with melanoma displayed an objective partial response on day 43, and the treatment was found to be safe. CD40 can be expressed in hematologic malignancies, melanomas, and carcinomas, and CD40 targeting can also induce direct antiproliferative effects on the tumor [123]. In contrast, another study in melanoma showed no direct effects on the tumor with anti-CD40 antibody treatment, either in $CD40^+$ or $CD40^{neg}$ melanoma cell lines [124].

Keyhole Limpet Hemocyanin Conjugates

Keyhole limpet hemocyanin (KLH), a large immunogenic protein from the keyhole limpet, induces an antibody response. Synthetic peptides conjugated to KLH have been used in combination with the adjuvant QS-21 to induce a robust antitumor immune response in prostate cancer [125]. KLH has also been used with DC vaccines for melanoma [126] and RCC [127].

Toll-Like Receptors

TLRs, early mediators of innate immune responses to pathogens, are a family of pattern recognition receptors mainly expressed on APCs. TLR agonists can enhance antitumor T-cell responses and may improve the efficacy of peptide vaccines. Several TLR agonists have the potential to be effective vaccine adjuvants, including agonists for TLR3, 4, 7, and 9, which are likely to enhance the T_H1 environment [128, 129]. It has also been demonstrated that stimulation of TLR8 mediates reversal of $CD4^+$ Treg function [130]. In addition, a combination of several TLR agonists may act synergistically to further enhance vaccine efficacy.

CpG Oligodeoxynucleotides

Oligodeoxynucleotides (ODNs) that contain unmethylated CpG motifs can activate B cells [131], DCs [132], and natural killer (NK) cells [133], and can act as TLR9 agonists and induce a T_H1 cytokine response [134], as well as stimulate $CD8^+$ CTL activity [135]. A clinical trial in patients with NY-ESO-1- or LAGE-1-expressing tumors recently found detectable $CD8^+$ T-cell responses in 9/14 patients, 6 of whom had a favorable clinical outcome after vaccination with synthetic CpG 7909 ODN mixed with NY-ESO-1 peptide p157-165 and IFA [136]. A combination of MUC1 and HER-2/*neu* peptide vaccines with CpG ODN, GM-CSF, or both, is currently being investigated in patients with previously treated stage II or III breast adenocarcinoma (NCT00640861).

Poly-ICLC

Polyinosinic-polycytidylic acid stabilized by lysine and carboxymethylcellulose (poly-ICLC) is a TLR3 agonist currently being investigated in several clinical trials. A robust induction of antigen-specific $CD8^+$ T cells was found in glioma patients treated in a phase I trial of multiple peptides combined with intramuscular poly-ICLC [137, 138]. Both $CD8^+$ and $CD4^+$ T-cell responses were also seen in patients with ovarian carcinoma vaccinated with overlapping long peptides from NY-ESO-1 in combination with poly-ICLC [139]. A clinical study to investigate the efficacy of a MUC1 peptide/poly-ICLC adjuvant vaccine in preventing recurrence of polyps in patients with advanced colorectal adenoma was recently concluded (NCT00773097).

Liposomes

Liposomes, an established drug delivery system, must be combined with a potent immune activator in order to produce an adjuvant effect. There are currently several ongoing or recently finished clinical trials employing different liposomal vaccines in different cancers: L-BLP25 (Stimuvax, EMD Serono), ONT-10 (Oncothyreon Inc.), and Lipovaxin-MM (Lipotek Pty Ltd.). BLP-25, a liposomal vaccine containing the VNTR region of MUC1, has been investigated in completed trials in NSCLC (NCT00157209, NCT00157196 and NCT00960115) and an ongoing trial of combination therapy with bevacizumab in NSCLC (NCT00828009). L-BLP25 recently failed to meet the primary endpoint of increased overall survival in a phase III trial in NSCLC. ONT-10 is a liposomal MUC1 cancer vaccine currently being investigated in a phase I trial in patients with solid tumors (NCT01556789) and a phase Ib maintenance therapy trial (NCT01978964). Lipovaxin-MM, a liposomal vaccine loaded with melanoma-associated antigens, is currently being investigated in a phase I trial (NCT01052142).

SB-AS2

SB-AS2 is composed of monophosphoryl lipid A and QS-21 in an oil-in-water emulsion. It can induce high titers of IgG antibodies, mostly the T_H1-dependent subclasses, as well as $CD8^+$ and $CD4^+$ T-cell responses [140]. Three clinical trials employing SB-AS2 as an adjuvant have recently been completed with (a) MUC1 100-mer peptide in pancreatic caner (NCT00008099), which was found to be safe and to increase MUC1-specific antibodies and T-cell responses in some patients [141]; (b) MAGE-A10 peptide in melanoma (NCT00112216); and (c) MAGE-3-His fusion protein in stage IV melanoma (NCT00042783). One trial with the MAGE-3-His fusion protein in unresectable melanoma is still ongoing (NCT00086866).

Peptide Vaccines Combined with Different Modalities

It is possible that peptide vaccines used in combination with other treatments will enhance tumor lysis and be of clinical benefit to a large number of cancer patients.

Chemotherapy

Combining peptide vaccines with chemotherapy makes sense, since it has been shown that some such treatments can target Tregs and MDSCs (which inhibit the immune response to vaccine), thus producing a more responsive tumor microenvironment with restored T-cell and NK-cell functions [142–145]. In addition, paclitaxel, doxorubicin, mitomycin C, and methotrexate at very low doses have been found to up-regulate antigen presentation by DCs [146], which is also beneficial for combination with peptide vaccines. Moreover, chemotherapy can lead to immunogenic tumor-cell death, which can activate antigen-specific T cells [147–149].

Immunomodulatory Agents

Studies of the mechanisms involved in the generation of immune responses against tumors and the immune suppression achieved by tumors have demonstrated that several mechanisms may simultaneously prevent effective antitumor immunity. These mechanisms include the induction of negative immune regulation by Tregs and MDSCs, and inefficient presentation of tumor antigens by DCs. Therapies designed to simultaneously enhance antitumor immunity and prevent immune evasion by tumors are most likely to have clinical efficacy, but it is unlikely that a single immunotherapeutic strategy can meet all of these requirements. Immunotherapeutic approaches using a combination of vaccines to elicit antitumor T-cell responses and immunomodulatory agents to activate the immune system, either directly or indirectly by antagonizing immune suppression, have a very high potential for antitumor effects.

Immune Checkpoint Inhibitors

Immune checkpoint proteins have become increasingly important targets for pharmacologic blockade. Utilizing the immune system to eliminate cancer has great potential, and several classes of new agents have shown positive clinical activity for the treatment of cancers. Current immune checkpoint inhibitors include anti-CTLA4 (ipilimumab, tremelimumab) [150, 151], anti-PD-1 (nivolumab, pembrolizumab) [152, 153], anti-PD-L1 (MPDL-3280A, MED14736, AMP-224) [154, 155], KIR inhibitor (lirilumab) [156], anti-41BB (urelumab) [157], LAG-3 inhibitor [158], phosphatidyl-serine inhibitor (bavituximab) [159], and indoleamine 2,3-dioxygenase (IDO) inhibitor (INCB024360) [160].

Anti-CTLA4, Anti-PD-1, Anti-PD-L1

A promising therapeutic combination is the use of a peptide vaccine with anti-CTLA4, an immune checkpoint inhibitor. CTLA4 (CD152) is not found in resting T cells but is up-regulated for 2–3 days after T-cell activation. It inhibits CD28-mediated T-cell costimulation by binding with higher avidity to B7.1 (CD80) and B7.2 (CD86) than the natural ligand does [161, 162]. In a clinical trial in patients with metastatic melanoma, a peptide vaccine combined with IFN-α and anti-CTLA4 significantly down-regulated MDSCs and resulted in significant clinical activity [163]. A phase III study in metastatic melanoma using a combination of gp100 peptide vaccine and ipilimumab showed that ipilimumab with or without gp100 peptide vaccine improved overall survival compared to gp100 alone [164]. Grade 3 or 4 immune-related adverse events occurred in 10–15 % of patients treated with ipilimumab, but most resolved with appropriate treatment [164]. In another study, the combination of extended-dose ipilimumab with multipeptide vaccine (tyrosine$_{368-376}$, gp100$_{209-217}$, MART-1$_{26-35}$) emulsified in Montanide ISA 51 appeared to be associated with improved outcome in patients with resected high-risk stage IIIc/IV melanoma [165]. Immune monitoring of melanoma patients treated with ipilimumab and vaccinated with gp100 DNS (IMF-24), gp100$_{209-217}$, tyrosinase peptides plus GM-CSF DNA (IMF-32), or NY-ESO-1 plus imiquimod (IMF-11) indicated that ipilimumab could increase antigen-specific CD8$^+$ T-cell responses, as well as effector memory T-cell responses and cytokine profiles [166].

Another immune checkpoint inhibitor consists of antibodies that block the interactions between programmed cell death protein 1 (PD-1) and its ligand (PD-L1). Similar to CTLA-4, the PD-1 receptor is a T-cell coinhibitory molecule that binds to the PD-1 and PD-2 ligands (PD-L1 and PD-L2). PD-1 is expressed on activated T cells, B cells, and some myeloid cells. Its ligand, PD-L1, is expressed on some peripheral tissues and many types of tumors. PD-L2 is expressed on DCs. PD-L1 expression on peripheral tissues appears essential to maintaining peripheral tolerance. The interaction of PD-1 with PD-L1/PD-L2 can suppress T-cell activation, while blocking this interaction can restore immune function [167, 168]. In a phase I clinical trial, 296 patients with a variety of cancers (advanced melanoma, NSCLC, castration-resistant prostate cancer, RCC, and colorectal cancer) were treated with anti-PD-1 antibody (BMS-936558 or MDX-1106) [169]. Objective responses were seen in 26/94 patients with melanoma and 14/76 patients with NSCLC, but not in the 19 patients with metastatic colorectal cancer [169]. In September 2014, the FDA announced accelerated approval of pembrolizumab [170] for advanced melanoma. This is the first anti-PD-1 antibody to receive approval.

In another study, a monoclonal antibody to PD-L1 (BMS-936559) was used to treat 207 patients with NSCLC (n=75), melanoma (n=55), colorectal cancer (n = 18), RCC (n = 17), ovarian cancer (n = 17), pancreatic cancer (n = 14), gastrointestinal cancer (n = 7), and breast cancer (n = 4) [154]. Complete or partial responses were observed in 9/52 evaluable patients with melanoma, 2/17 with RCC, 5/49 with NSCLC, and 1/17 with ovarian cancer. Again, patients with colorectal cancer had no response to treatment [154]. The level of PD-L1 expression, which varies on tumor cells, was found to correlate with clinical response to anti-PD-L1 therapy, suggesting that PD-L1 expression on tumor cells may be a predictive biomarker for response to treatment with anti-PD-1 and anti-PD-L1 antibodies. The combination of peptide vaccines and immune checkpoint inhibitors is likely to result in clinical benefit for many patients.

Anti-CD137

CD137 (4-1BB), a member of the TNF-receptor family, is expressed on activated T- and B-cells, follicular DCs, monocytes, and epithelium, and when it binds to its ligand (4-1BB ligand), it functions as a costimulatory molecule for T cells, affecting T-cell activation and survival as well as DC development. Recently, advances have been made in producing an agonist CD137 antibody (BMS-663513, Bristol-Myers Squibb) that mimics the natural ligand. It has been investigated in a phase II trial in melanoma (NCT00612664), and in vitro studies have shown promising results when using it to improve the quality of $CD8^+$ melanoma tumor-infiltrating lymphocytes for adoptive T-cell transfer [171].

IDO Inhibitors

Indoleamine 2,3-dioxygenase 1 (IDO) is a tryptophan-catabolizing enzyme [172, 173]. IDO mediates oxidative degradation of the essential amino acid tryptophan into kynurenine and other downstream metabolites that suppress effector T-cell function and enhance the differentiation of suppressor cells. Cells involved in the IDO-driven mechanisms of immunosuppression include Tregs [174], M2 macrophages [175], NKT cells [176], mast cells [177, 178], MDSCs [179, 180], and regulatory or tolerogenic DCs [181]. The suppressive effects of IDO contribute to the immune system's inability to respond effectively against tumor antigens. IDO inhibitors offer an opportunity to overcome these effects. Preclinical studies have demonstrated that treatment of tumor-bearing animals with IDO inhibitors can enhance antitumor immune responses. IDO inhibitors have been used in combination with a variety of chemotherapeutic drugs, antitumor vaccines, and other immunotherapeutic modalities [182, 183]. A number of IDO inhibitors have been reported in the literature [184], the most extensively studied being 1-methyl-tryptophan (1MT) [185]. 1MT is currently being investigated in phase II trials in combination with sipuleucel-T for refractory metastatic prostate cancer (NCT01560923), with ipilimumab for metastatic melanoma (NCT02073123), and with a DC vaccine for metastatic breast cancer (NCT01042535). Another IDO inhibitor, INCB024360

[186], is currently being investigated in combination with a peptide vaccine consisting of an emulsion of 12 MHC class I melanoma peptides and 1 MHC class II tetanus toxoid helper peptide (MELITAC 12.1) [4] in patients with stage III-IV melanoma (NCT01961115). It is also being used in combination with DEC-205/NY-ESO-1 fusion protein CDX-1401 and poly-ICLC in patients with ovarian, fallopian tube, or peritoneal cancer (NCT02166905).

Prime-Boost Strategies

Generation of memory CTLs is essential to the development of long-lasting immune responses against tumors. It has been demonstrated that antigen kinetics determine immune reactivity; the strength of T cell responses is governed by antigen dose, localization and costimulatory signals. Exponentially increasing antigen doses stimulated stronger immune responses than constant stimulation with uniform doses or immunization with one large dose [187]. With peptide vaccines, dosage, route of administration, and scheduling of boosts are essential to the induction of proper antigen-specific T-cell responses. Vector-based vaccines have been used to treat many types of cancers [188–192]. Sometimes, pre-existing immunity to the vector can accelerate its clearance after booster injections, limiting immune responses by impairing antigen presentation and production of inflammatory cytokines. In these cases, a prime-boost regimen employing different vectors or peptide vaccines can effectively generate memory CTLs [188–190]. For example, when mice were given recombinant vaccinia CEA as a prime and recombinant CEA protein as a boost, greater T-cell responses to CEA were achieved than with the use of either immunogen alone [193]. This concept of heterologous prime-boost immunization has been validated in other models employing vaccines for infectious agents and cancer, and presents a strong rationale for designing vaccine regimens using peptide vaccines (peptides with agonist epitopes in particular) [44] as boosts.

Conclusions

Identification of TAA peptides expressed by different human tumors has provided the basis for antigen-specific active immunotherapy and facilitated the development of clinical cancer immunotherapy trials. A major limitation of all immunotherapy is the negative effects of the tumor microenvironment on the function and survival of effector cells that could potentially eliminate the tumor. It is unlikely that a single immunotherapeutic strategy will be able to address all the requirements for successful treatment of cancers, but combination therapies are more likely to succeed. Prospects for improved peptide vaccines include the use of long peptides, modified and optimized peptides (with agonist epitopes), peptides with multiple epitopes, modification of adjuvants, and combination therapy with other immunologically active agents. It is conceivable that improvements in vaccine immunogenicity and T-cell persistence, as well as successful countermeasures against tumor-associated immune dysfunction, will lead to peptide vaccines with greater therapeutic value.

References

1. Dunn GP, Old LJ, Schreiber RD (2004) The three Es of cancer immunoediting. Annu Rev Immunol 22:329–360
2. Yamshchikov GV, Barnd DL, Eastham S et al (2001) Evaluation of peptide vaccine immunogenicity in draining lymph nodes and peripheral blood of melanoma patients. Int J Cancer 92:703–711
3. Chianese-Bullock KA, Pressley J, Garbee C et al (2005) MAGE-A1-, MAGE-A10-, and gp100-derived peptides are immunogenic when combined with granulocyte-macrophage colony-stimulating factor and Montanide ISA-51 adjuvant and administered as part of a multipeptide vaccine for melanoma. J Immunol 174:3080–3086
4. Slingluff CL Jr, Petroni GR, Olson WC et al (2009) Effect of granulocyte/macrophage colony-stimulating factor on circulating CD8+ and CD4+ T-cell responses to a multipeptide melanoma vaccine: outcome of a multicenter randomized trial. Clin Cancer Res 15:7036–7044
5. Hamid O, Solomon JC, Scotland R et al (2007) Alum with interleukin-12 augments immunity to a melanoma peptide vaccine: correlation with time to relapse in patients with resected high-risk disease. Clin Cancer Res 13:215–222
6. Slingluff CL Jr, Petroni GR, Chianese-Bullock KA et al (2011) Randomized multicenter trial of the effects of melanoma-associated helper peptides and cyclophosphamide on the immunogenicity of a multipeptide melanoma vaccine. J Clin Oncol 29:2924–2932
7. Slingluff CL Jr, Petroni GR, Chianese-Bullock KA et al (2007) Immunologic and clinical outcomes of a randomized phase II trial of two multipeptide vaccines for melanoma in the adjuvant setting. Clin Cancer Res 13:6386–6395
8. Baumgaertner P, Rufer N, Devevre E et al (2006) Ex vivo detectable human CD8 T-cell responses to cancer-testis antigens. Cancer Res 66:1912–1916
9. Tuettenberg A, Becker C, Huter E et al (2006) Induction of strong and persistent MelanA/MART-1-specific immune responses by adjuvant dendritic cell-based vaccination of stage II melanoma patients. Int J Cancer 118:2617–2627
10. Benchimol S, Fuks A, Jothy S et al (1989) Carcinoembryonic antigen, a human tumor marker, functions as an intercellular adhesion molecule. Cell 57:327–334
11. Robbins PF, Eggensperger D, Qi CF et al (1993) Definition of the expression of the human carcinoembryonic antigen and non-specific cross-reacting antigen in human breast and lung carcinomas. Int J Cancer 53:892–897
12. Tendler A, Kaufman HL, Kadish AS (2000) Increased carcinoembryonic antigen expression in cervical intraepithelial neoplasia grade 3 and in cervical squamous cell carcinoma. Hum Pathol 31:1357–1362
13. Gendler SJ, Lancaster CA, Taylor-Papadimitriou J et al (1990) Molecular cloning and expression of human tumor-associated polymorphic epithelial mucin. J Biol Chem 265:15286–15293
14. Hollingsworth MA, Swanson BJ (2004) Mucins in cancer: protection and control of the cell surface. Nat Rev Cancer 4:45–60
15. Kufe D, Inghirami G, Abe M et al (1984) Differential reactivity of a novel monoclonal antibody (DF3) with human malignant versus benign breast tumors. Hybridoma 3:223–232

16. Kufe DW (2009) Functional targeting of the MUC1 oncogene in human cancers. Cancer Biol Ther 8:1197–1203
17. Kawano T, Ito M, Raina D et al (2007) MUC1 oncoprotein regulates Bcr-Abl stability and pathogenesis in chronic myelogenous leukemia cells. Cancer Res 67:11576–11584
18. Yin L, Ahmad R, Kosugi M et al (2010) Survival of human multiple myeloma cells is dependent on MUC1 C-terminal transmembrane subunit oncoprotein function. Mol Pharmacol 78:166–174
19. Yin L, Kufe D (2011) MUC1-C oncoprotein blocks terminal differentiation of chronic myelogenous leukemia cells by a ROS-mediated mechanism. Genes Cancer 2:56–64
20. Hayes DF, Zurawski VR Jr, Kufe DW (1986) Comparison of circulating CA15-3 and carcinoembryonic antigen levels in patients with breast cancer. J Clin Oncol 4:1542–1550
21. Li Y, Liu D, Chen D et al (2003) Human DF3/MUC1 carcinoma-associated protein functions as an oncogene. Oncogene 22:6107–6110
22. Raina D, Ahmad R, Joshi MD et al (2009) Direct targeting of the mucin 1 oncoprotein blocks survival and tumorigenicity of human breast carcinoma cells. Cancer Res 69:5133–5141
23. Wei X, Xu H, Kufe D (2007) Human mucin 1 oncoprotein represses transcription of the p53 tumor suppressor gene. Cancer Res 67:1853–1858
24. Uchida Y, Raina D, Kharbanda S et al (2013) Inhibition of the MUC1-C oncoprotein is synergistic with cytotoxic agents in the treatment of breast cancer cells. Cancer Biol Ther 14:127–134
25. Balk SP, Ko YJ, Bubley GJ (2003) Biology of prostate-specific antigen. J Clin Oncol 21:383–391
26. Correale P, Walmsley K, Nieroda C et al (1997) In vitro generation of human cytotoxic T lymphocytes specific for peptides derived from prostate-specific antigen. J Natl Cancer Inst 89:293–300
27. McNeel DG, Nguyen LD, Ellis WJ et al (2001) Naturally occurring prostate cancer antigen-specific T cell responses of a Th1 phenotype can be detected in patients with prostate cancer. Prostate 47:222–229
28. Gulley JL, Arlen PM, Bastian A et al (2005) Combining a recombinant cancer vaccine with standard definitive radiotherapy in patients with localized prostate cancer. Clin Cancer Res 11:3353–3362
29. Kispert A, Koschorz B, Herrmann BG (1995) The T protein encoded by brachyury is a tissue-specific transcription factor. EMBO J 14:4763–4772
30. Palena C, Polev DE, Tsang KY et al (2007) The human T-box mesodermal transcription factor brachyury is a candidate target for T-cell-mediated cancer immunotherapy. Clin Cancer Res 13:2471–2478
31. Hamilton DH, Litzinger MT, Fernando RI et al (2012) Cancer vaccines targeting the epithelial-mesenchymal transition: tissue distribution of brachyury and other drivers of the mesenchymal-like phenotype of carcinomas. Semin Oncol 39:358–366
32. Fernando RI, Litzinger M, Trono P et al (2010) The T-box transcription factor brachyury promotes epithelial-mesenchymal transition in human tumor cells. J Clin Invest 120:533–544
33. Palena C, Roselli M, Litzinger MT, Ferroni P, Costarelli L, Spila A, Cavaliere F, Huang B, Fernando RI, Hamilton DH, Jochems C, Tsang KY, Cheng Q, Lyerly HK, Schlom J, Guadagni F (2014) Overexpression of the EMT driver brachyury in breast carcinomas: association with poor prognosis. J Natl Cancer Inst 9:106(5). doi:10.1093/jnci/dju054
34. Palena C, Fernando RI, Litzinger MT et al (2011) Strategies to target molecules that control the acquisition of a mesenchymal-like phenotype by carcinoma cells. Exp Biol Med (Maywood) 236:537–545
35. Rivoltini L, Kawakami Y, Sakaguchi K et al (1995) Induction of tumor-reactive CTL from peripheral blood and tumor-infiltrating lymphocytes of melanoma patients by in vitro stimulation with an immunodominant peptide of the human melanoma antigen MART-1. J Immunol 154:2257–2265

36. Salgaller ML, Afshar A, Marincola FM et al (1995) Recognition of multiple epitopes in the human melanoma antigen gp100 by peripheral blood lymphocytes stimulated in vitro with synthetic peptides. Cancer Res 55:4972–4979
37. Marchand M, Weynants P, Rankin E et al (1995) Tumor regression responses in melanoma patients treated with a peptide encoded by gene MAGE-3. Int J Cancer 63:883–885
38. Weber JS, Hua FL, Spears L et al (1999) A phase I trial of an HLA-A1 restricted MAGE-3 epitope peptide with incomplete Freund's adjuvant in patients with resected high-risk melanoma. J Immunother 22:431–440
39. Sette A, Vitiello A, Reherman B et al (1994) The relationship between class I binding affinity and immunogenicity of potential cytotoxic T cell epitopes. J Immunol 153:5586–5592
40. Borbulevych OY, Baxter TK, Yu Z et al (2005) Increased immunogenicity of an anchor-modified tumor-associated antigen is due to the enhanced stability of the peptide/MHC complex: implications for vaccine design. J Immunol 174:4812–4820
41. Tsang KY, Palena C, Gulley J et al (2004) A human cytotoxic T-lymphocyte epitope and its agonist epitope from the nonvariable number of tandem repeat sequence of MUC-1. Clin Cancer Res 10:2139–2149
42. Terasawa H, Tsang KY, Gulley J et al (2002) Identification and characterization of a human agonist cytotoxic T-lymphocyte epitope of human prostate-specific antigen. Clin Cancer Res 8:41–53
43. Yokokawa J, Bera TK, Palena C et al (2007) Identification of cytotoxic T-lymphocyte epitope(s) and its agonist epitope(s) of a novel target for vaccine therapy (PAGE4). Int J Cancer 121:595–605
44. Jochems C, Tucker JA, Vergati M et al (2014) Identification and characterization of agonist epitopes of the MUC1-C oncoprotein. Cancer Immunol Immunother 63:161–174
45. Yokokawa J, Palena C, Arlen P et al (2005) Identification of novel human CTL epitopes and their agonist epitopes of mesothelin. Clin Cancer Res 11:6342–6351
46. Zaremba S, Barzaga E, Zhu M et al (1997) Identification of an enhancer agonist cytotoxic T lymphocyte peptide from human carcinoembryonic antigen. Cancer Res 57:4570–4577
47. Tucker JA, Jochems C, Boyerinas B et al (2014) Identification and characterization of a cytotoxic T-lymphocyte agonist epitope of brachyury, a transcription factor involved in epithelial to mesenchymal transition and metastasis. Cancer Immunol Immunother 63:1307–1317
48. Huang YH, Terabe M, Pendleton CD et al (2013) Identification and enhancement of HLA-A2.1-restricted CTL epitopes in a new human cancer antigen-POTE. PLoS One 8, e64365
49. Oh S, Terabe M, Pendleton CD et al (2004) Human CTLs to wild-type and enhanced epitopes of a novel prostate and breast tumor-associated protein, TARP, lyse human breast cancer cells. Cancer Res 64:2610–2618
50. Pinilla-Ibarz J, May RJ, Korontsvit T et al (2006) Improved human T-cell responses against synthetic HLA-0201 analog peptides derived from the WT1 oncoprotein. Leukemia 20:2025–2033
51. Zirlik KM, Zahrieh D, Neuberg D et al (2006) Cytotoxic T cells generated against heteroclitic peptides kill primary tumor cells independent of the binding affinity of the native tumor antigen peptide. Blood 108:3865–3870
52. Palena C, Schlom J, Tsang KY (2003) Differential gene expression profiles in a human T-cell line stimulated with a tumor-associated self-peptide versus an enhancer agonist peptide. Clin Cancer Res 9:1616–1627
53. Sette A, Newman M, Livingston B et al (2002) Optimizing vaccine design for cellular processing, MHC binding and TCR recognition. Tissue Antigens 59:443–451
54. Chiong B, Wong R, Lee P et al (2004) Characterization of long-term effector-memory T-cell responses in patients with resected high-risk melanoma receiving a melanoma Peptide vaccine. J Immunother 27:368–379
55. Hou Y, Kavanagh B, Fong L (2008) Distinct CD8+ T cell repertoires primed with agonist and native peptides derived from a tumor-associated antigen. J Immunol 180:1526–1534
56. Fong L, Hou Y, Rivas A et al (2001) Altered peptide ligand vaccination with Flt3 ligand expanded dendritic cells for tumor immunotherapy. Proc Natl Acad Sci U S A 98:8809–8814

57. Geynisman DM, Zha Y, Kunnavakkam R et al (2013) A randomized pilot phase I study of modified carcinoembryonic antigen (CEA) peptide (CAP1-6D)/Montanide/GM-CSF-vaccine in patients with pancreatic adenocarcinoma. J Immunother Cancer 1:8
58. Frank C, Hundemer M, Ho AD et al (2008) Cellular immune responses against the cancer-testis antigen SPAN-XB in healthy donors and patients with multiple myeloma. Leuk Lymphoma 49:779–785
59. Brown ME, Miao H, McKee MD (2007) Recognition of carcinoembryonic antigen peptide and heteroclitic peptide by peripheral blood T lymphocytes. J Immunother 30:350–358
60. Kakimi K, Isobe M, Uenaka A et al (2011) A phase I study of vaccination with NY-ESO-1f peptide mixed with Picibanil OK-432 and Montanide ISA-51 in patients with cancers expressing the NY-ESO-1 antigen. Int J Cancer 129:2836–2846
61. Slingluff CL Jr, Lee S, Zhao F et al (2013) A randomized phase II trial of multiepitope vaccination with melanoma peptides for cytotoxic T cells and helper T cells for patients with metastatic melanoma (E1602). Clin Cancer Res 19:4228–4238
62. Kalams SA, Walker BD (1998) The critical need for CD4 help in maintaining effective cytotoxic T lymphocyte responses. J Exp Med 188:2199–2204
63. Kenter GG, Welters MJ, Valentijn AR et al (2008) Phase I immunotherapeutic trial with long peptides spanning the E6 and E7 sequences of high-risk human papillomavirus 16 in end-stage cervical cancer patients shows low toxicity and robust immunogenicity. Clin Cancer Res 14:169–177
64. Welters MJ, Kenter GG, Piersma SJ et al (2008) Induction of tumor-specific CD4+ and CD8+ T-cell immunity in cervical cancer patients by a human papillomavirus type 16 E6 and E7 long peptides vaccine. Clin Cancer Res 14:178–187
65. Kenter GG, Welters MJ, Valentijn AR et al (2009) Vaccination against HPV-16 oncoproteins for vulvar intraepithelial neoplasia. N Engl J Med 361:1838–1847
66. Speetjens FM, Kuppen PJ, Welters MJ et al (2009) Induction of p53-specific immunity by a p53 synthetic long peptide vaccine in patients treated for metastatic colorectal cancer. Clin Cancer Res 15:1086–1095
67. Leffers N, Vermeij R, Hoogeboom BN et al (2012) Long-term clinical and immunological effects of p53-SLP(R) vaccine in patients with ovarian cancer. Int J Cancer 130:105–112
68. Krug LM, Dao T, Brown AB et al (2010) WT1 peptide vaccinations induce CD4 and CD8 T cell immune responses in patients with mesothelioma and non-small cell lung cancer. Cancer Immunol Immunother 59:1467–1479
69. Alexander J, Sidney J, Southwood S et al (1994) Development of high potency universal DR-restricted helper epitopes by modification of high affinity DR-blocking peptides. Immunity 1:751–761
70. Kuball J, de Boer K, Wagner E et al (2011) Pitfalls of vaccinations with WT1-, Proteinase3- and MUC1-derived peptides in combination with MontanideISA51 and CpG7909. Cancer Immunol Immunother 60:161–171
71. Meyer RG, Korn S, Micke P et al (2007) An open-label, prospective phase I/II study evaluating the immunogenicity and safety of a ras peptide vaccine plus GM-CSF in patients with non-small cell lung cancer. Lung Cancer 58:88–94
72. Barve M, Bender J, Senzer N et al (2008) Induction of immune responses and clinical efficacy in a phase II trial of IDM-2101, a 10-epitope cytotoxic T-lymphocyte vaccine, in metastatic non-small-cell lung cancer. J Clin Oncol 26:4418–4425
73. Feyerabend S, Stevanovic S, Gouttefangeas C et al (2009) Novel multi-peptide vaccination in Hla-A2+ hormone sensitive patients with biochemical relapse of prostate cancer. Prostate 69:917–927
74. Maslak PG, Dao T, Krug LM et al (2010) Vaccination with synthetic analog peptides derived from WT1 oncoprotein induces T-cell responses in patients with complete remission from acute myeloid leukemia. Blood 116:171–179
75. Walter S, Weinschenk T, Stenzl A et al (2012) Multipeptide immune response to cancer vaccine IMA901 after single-dose cyclophosphamide associates with longer patient survival. Nat Med 18:1254–1261

76. Bedke J, Stenzl A (2013) IMA901: a peptide vaccine in renal cell carcinoma. Expert Opin Investig Drugs 22:1329–1336
77. Tuttle TM, Anderson BW, Thompson WE et al (1998) Proliferative and cytokine responses to class II HER-2/neu-associated peptides in breast cancer patients. Clin Cancer Res 4:2015–2024
78. Disis ML, Grabstein KH, Sleath PR et al (1999) Generation of immunity to the HER-2/neu oncogenic protein in patients with breast and ovarian cancer using a peptide-based vaccine. Clin Cancer Res 5:1289–1297
79. Gillogly ME, Kallinteris NL, Xu M et al (2004) Ii-Key/HER-2/neu MHC class-II antigenic epitope vaccine peptide for breast cancer. Cancer Immunol Immunother 53:490–496
80. Holmes JP, Benavides LC, Gates JD et al (2008) Results of the first phase I clinical trial of the novel II-key hybrid preventive HER-2/neu peptide (AE37) vaccine. J Clin Oncol 26:3426–3433
81. Humphreys RE, Adams S, Koldzic G et al (2000) Increasing the potency of MHC class II-presented epitopes by linkage to Ii-Key peptide. Vaccine 18:2693–2697
82. Perez SA, Kallinteris NL, Bisias S et al (2010) Results from a phase I clinical study of the novel Ii-Key/HER-2/neu(776–790) hybrid peptide vaccine in patients with prostate cancer. Clin Cancer Res 16:3495–3506
83. Takeshima T, Chamoto K, Wakita D et al (2010) Local radiation therapy inhibits tumor growth through the generation of tumor-specific CTL: its potentiation by combination with Th1 cell therapy. Cancer Res 70:2697–2706
84. Kobayashi H, Celis E (2008) Peptide epitope identification for tumor-reactive CD4 T cells. Curr Opin Immunol 20:221–227
85. Takahashi N, Ohkuri T, Homma S et al (2012) First clinical trial of cancer vaccine therapy with artificially synthesized helper/killer-hybrid epitope long peptide of MAGE-A4 cancer antigen. Cancer Sci 103:150–153
86. Schadendorf D, Ugurel S, Schuler-Thurner B et al (2006) Dacarbazine (DTIC) versus vaccination with autologous peptide-pulsed dendritic cells (DC) in first-line treatment of patients with metastatic melanoma: a randomized phase III trial of the DC study group of the DeCOG. Ann Oncol 17:563–570
87. Slingluff CL Jr, Petroni GR, Yamshchikov GV et al (2003) Clinical and immunologic results of a randomized phase II trial of vaccination using four melanoma peptides either administered in granulocyte-macrophage colony-stimulating factor in adjuvant or pulsed on dendritic cells. J Clin Oncol 21:4016–4026
88. Su Z, Dannull J, Heiser A et al (2003) Immunological and clinical responses in metastatic renal cancer patients vaccinated with tumor RNA-transfected dendritic cells. Cancer Res 63:2127–2133
89. Wierecky J, Muller MR, Wirths S et al (2006) Immunologic and clinical responses after vaccinations with peptide-pulsed dendritic cells in metastatic renal cancer patients. Cancer Res 66:5910–5918
90. Lesterhuis WJ, Schreibelt G, Scharenborg NM et al (2011) Wild-type and modified gp100 peptide-pulsed dendritic cell vaccination of advanced melanoma patients can lead to long-term clinical responses independent of the peptide used. Cancer Immunol Immunother 60:249–260
91. Gabrilovich DI, Nadaf S, Corak J et al (1996) Dendritic cells in antitumor immune responses. II. Dendritic cells grown from bone marrow precursors, but not mature DC from tumor-bearing mice, are effective antigen carriers in the therapy of established tumors. Cell Immunol 170:111–119
92. Kichler-Lakomy C, Budinsky AC, Wolfram R et al (2006) Deficiences in phenotype expression and function of dentritic cells from patients with early breast cancer. Eur J Med Res 11:7–12
93. Rahma OE, Ashtar E, Czystowska M et al (2012) A gynecologic oncology group phase II trial of two p53 peptide vaccine approaches: subcutaneous injection and intravenous pulsed dendritic cells in high recurrence risk ovarian cancer patients. Cancer Immunol Immunother 61:373–384

94. Schaefer JT, Patterson JW, Deacon DH et al (2010) Dynamic changes in cellular infiltrates with repeated cutaneous vaccination: a histologic and immunophenotypic analysis. J Transl Med 8:79
95. Warren TL, Weiner GJ (2000) Uses of granulocyte-macrophage colony-stimulating factor in vaccine development. Curr Opin Hematol 7:168–173
96. Chang DZ, Lomazow W, Joy Somberg C et al (2004) Granulocyte-macrophage colony stimulating factor: an adjuvant for cancer vaccines. Hematology 9:207–215
97. Disis ML, Bernhard H, Shiota FM et al (1996) Granulocyte-macrophage colony-stimulating factor: an effective adjuvant for protein and peptide-based vaccines. Blood 88:202–210
98. Clancy-Thompson E, King LK, Nunnley LD et al (2013) Peptide vaccination in Montanide adjuvant induces and GM-CSF increases CXCR3 and cutaneous lymphocyte antigen expression by tumor antigen-specific CD8 T cells. Cancer Immunol Res 1:332–339
99. Gaudernack G, Gjertsen MK (1999) Combination of GM-CSF with antitumour vaccine strategies. Eur J Cancer 35(Suppl 3):S33–S35
100. Hung K, Hayashi R, Lafond-Walker A et al (1998) The central role of CD4(+) T cells in the antitumor immune response. J Exp Med 188:2357–2368
101. Leong SP, Enders-Zohr P, Zhou YM et al (1999) Recombinant human granulocyte macrophage-colony stimulating factor (rhGM-CSF) and autologous melanoma vaccine mediate tumor regression in patients with metastatic melanoma. J Immunother 22:166–174
102. Schmidt W, Maass G, Buschle M et al (1997) Generation of effective cancer vaccines genetically engineered to secrete cytokines using adenovirus-enhanced transferrinfection (AVET). Gene 190:211–216
103. Simons JW, Jaffee EM, Weber CE et al (1997) Bioactivity of autologous irradiated renal cell carcinoma vaccines generated by ex vivo granulocyte-macrophage colony-stimulating factor gene transfer. Cancer Res 57:1537–1546
104. Soiffer R, Lynch T, Mihm M et al (1998) Vaccination with irradiated autologous melanoma cells engineered to secrete human granulocyte-macrophage colony-stimulating factor generates potent antitumor immunity in patients with metastatic melanoma. Proc Natl Acad Sci U S A 95:13141–13146
105. Weber J, Sondak VK, Scotland R et al (2003) Granulocyte-macrophage-colony-stimulating factor added to a multipeptide vaccine for resected Stage II melanoma. Cancer 97:186–200
106. Parmiani G, Castelli C, Pilla L et al (2007) Opposite immune functions of GM-CSF administered as vaccine adjuvant in cancer patients. Ann Oncol 18:226–232
107. Talmadge JE (2007) Pathways mediating the expansion and immunosuppressive activity of myeloid-derived suppressor cells and their relevance to cancer therapy. Clin Cancer Res 13:5243–5248
108. Serafini P, Carbley R, Noonan KA et al (2004) High-dose granulocyte-macrophage colony-stimulating factor-producing vaccines impair the immune response through the recruitment of myeloid suppressor cells. Cancer Res 64:6337–6343
109. Robertson MJ, Cameron C, Atkins MB et al (1999) Immunological effects of interleukin 12 administered by bolus intravenous injection to patients with cancer. Clin Cancer Res 5:9–16
110. Gollob JA, Mier JW, Veenstra K et al (2000) Phase I trial of twice-weekly intravenous interleukin 12 in patients with metastatic renal cell cancer or malignant melanoma: ability to maintain IFN-gamma induction is associated with clinical response. Clin Cancer Res 6:1678–1692
111. Leonard JP, Sherman ML, Fisher GL et al (1997) Effects of single-dose interleukin-12 exposure on interleukin-12-associated toxicity and interferon-gamma production. Blood 90:2541–2548
112. Fallon J, Tighe R, Kradjian G et al (2014) The immunocytokine NHS-IL12 as a potential cancer therapeutic. Oncotarget 5:1869–1884
113. Kaech SM, Tan JT, Wherry EJ et al (2003) Selective expression of the interleukin 7 receptor identifies effector CD8 T cells that give rise to long-lived memory cells. Nat Immunol 4:1191–1198
114. Bamford RN, Grant AJ, Burton JD et al (1994) The interleukin (IL) 2 receptor beta chain is shared by IL-2 and a cytokine, provisionally designated IL-T, that stimulates T-cell

proliferation and the induction of lymphokine-activated killer cells. Proc Natl Acad Sci U S A 91:4940–4944
115. Zhu X, Marcus WD, Xu W et al (2009) Novel human interleukin-15 agonists. J Immunol 183:3598–3607
116. Xu W, Jones M, Liu B et al (2013) Efficacy and mechanism-of-action of a novel superagonist interleukin-15: interleukin-15 receptor alphaSu/Fc fusion complex in syngeneic murine models of multiple myeloma. Cancer Res 73:3075–3086
117. Wong HC, Jeng EK, Rhode PR (2013) The IL-15-based superagonist ALT-803 promotes the antigen-independent conversion of memory CD8 T cells into innate-like effector cells with antitumor activity. Oncoimmunology 2, e26442
118. Habib T, Senadheera S, Weinberg K et al (2002) The common gamma chain (gamma c) is a required signaling component of the IL-21 receptor and supports IL-21-induced cell proliferation via JAK3. Biochemistry 41:8725–8731
119. Asao H, Okuyama C, Kumaki S et al (2001) Cutting edge: the common gamma-chain is an indispensable subunit of the IL-21 receptor complex. J Immunol 167:1–5
120. Korn T, Bettelli E, Gao W et al (2007) IL-21 initiates an alternative pathway to induce proinflammatory T(H)17 cells. Nature 448:484–487
121. Wei L, Laurence A, Elias KM et al (2007) IL-21 is produced by Th17 cells and drives IL-17 production in a STAT3-dependent manner. J Biol Chem 282:34605–34610
122. Vonderheide RH, Flaherty KT, Khalil M et al (2007) Clinical activity and immune modulation in cancer patients treated with CP-870,893, a novel CD40 agonist monoclonal antibody. J Clin Oncol 25:876–883
123. Wang ZQ, Wang J, Ling WH et al (2013) Effects of CD40 ligation combined with chemotherapy drugs on human breast cancer cell lines. J Int Med Res 41:1495–1504
124. Kalbasi A, Fonsatti E, Natali PG et al (2010) CD40 expression by human melanocytic lesions and melanoma cell lines and direct CD40 targeting with the therapeutic anti-CD40 antibody CP-870,893. J Immunother 33:810–816
125. Slovin SF, Ragupathi G, Fernandez C et al (2005) A bivalent conjugate vaccine in the treatment of biochemically relapsed prostate cancer: a study of glycosylated MUC-2-KLH and Globo H-KLH conjugate vaccines given with the new semi-synthetic saponin immunological adjuvant GPI-0100 OR QS-21. Vaccine 23:3114–3122
126. Reichardt VL, Okada CY, Liso A et al (1999) Idiotype vaccination using dendritic cells after autologous peripheral blood stem cell transplantation for multiple myeloma—a feasibility study. Blood 93:2411–2419
127. Holtl L, Rieser C, Papesh C et al (1998) CD83+ blood dendritic cells as a vaccine for immunotherapy of metastatic renal-cell cancer. Lancet 352:1358
128. Lore K, Betts MR, Brenchley JM et al (2003) Toll-like receptor ligands modulate dendritic cells to augment cytomegalovirus- and HIV-1-specific T cell responses. J Immunol 171:4320–4328
129. Mata-Haro V, Cekic C, Martin M et al (2007) The vaccine adjuvant monophosphoryl lipid A as a TRIF-biased agonist of TLR4. Science 316:1628–1632
130. Peng G, Guo Z, Kiniwa Y et al (2005) Toll-like receptor 8-mediated reversal of CD4+ regulatory T cell function. Science 309:1380–1384
131. Davis HL, Weeratna R, Waldschmidt TJ et al (1998) CpG DNA is a potent enhancer of specific immunity in mice immunized with recombinant hepatitis B surface antigen. J Immunol 160:870–876
132. Sparwasser T, Koch ES, Vabulas RM et al (1998) Bacterial DNA and immunostimulatory CpG oligonucleotides trigger maturation and activation of murine dendritic cells. Eur J Immunol 28:2045–2054
133. Ballas ZK, Rasmussen WL, Krieg AM (1996) Induction of NK activity in murine and human cells by CpG motifs in oligodeoxynucleotides and bacterial DNA. J Immunol 157:1840–1845
134. Bohle B, Jahn-Schmid B, Maurer D et al (1999) Oligodeoxynucleotides containing CpG motifs induce IL-12, IL-18 and IFN-gamma production in cells from allergic individuals and inhibit IgE synthesis in vitro. Eur J Immunol 29:2344–2353

135. Warren TL, Bhatia SK, Acosta AM et al (2000) APC stimulated by CpG oligodeoxynucleotide enhance activation of MHC class I-restricted T cells. J Immunol 165:6244–6251
136. Karbach J, Gnjatic S, Bender A et al (2010) Tumor-reactive CD8+ T-cell responses after vaccination with NY-ESO-1 peptide, CpG 7909 and Montanide ISA-51: association with survival. Int J Cancer 126:909–918
137. Okada H, Butterfield LH, Hamilton RL et al (2015) Induction of robust type-I CD8+ T-cell responses in WHO grade 2 low-grade glioma patients receiving peptide-based vaccines in combination with poly-ICLC. Clin Cancer Res 21(2):286–294
138. Pollack IF, Jakacki RI, Butterfield LH et al (2014) Antigen-specific immune responses and clinical outcome after vaccination with glioma-associated antigen peptides and polyinosinic-polycytidylic acid stabilized by lysine and carboxymethylcellulose in children with newly diagnosed malignant brainstem and nonbrainstem gliomas. J Clin Oncol 32:2050–2058
139. Tsuji T, Sabbatini P, Jungbluth AA et al (2013) Effect of Montanide and poly-ICLC adjuvant on human self/tumor antigen-specific CD4+ T cells in phase I overlapping long peptide vaccine trial. Cancer Immunol Res 1:340–350
140. Lalvani A, Moris P, Voss G et al (1999) Potent induction of focused Th1-type cellular and humoral immune responses by RTS, S/SBAS2, a recombinant Plasmodium falciparum malaria vaccine. J Infect Dis 180:1656–1664
141. Ramanathan RK, Lee KM, McKolanis J et al (2005) Phase I study of a MUC1 vaccine composed of different doses of MUC1 peptide with SB-AS2 adjuvant in resected and locally advanced pancreatic cancer. Cancer Immunol Immunother 54:254–264
142. Ghiringhelli F, Menard C, Puig PE et al (2007) Metronomic cyclophosphamide regimen selectively depletes CD4+CD25+ regulatory T cells and restores T and NK effector functions in end stage cancer patients. Cancer Immunol Immunother 56:641–648
143. Dudley ME, Yang JC, Sherry R et al (2008) Adoptive cell therapy for patients with metastatic melanoma: evaluation of intensive myeloablative chemoradiation preparative regimens. J Clin Oncol 26:5233–5239
144. Suzuki E, Kapoor V, Jassar AS et al (2005) Gemcitabine selectively eliminates splenic Gr-1+/CD11b+ myeloid suppressor cells in tumor-bearing animals and enhances antitumor immune activity. Clin Cancer Res 11:6713–6721
145. Roselli M, Cereda V, di Bari MG et al (2013) Effects of conventional therapeutic interventions on the number and function of regulatory T cells. Oncoimmunology 2, e27025
146. Shurin GV, Tourkova IL, Kaneno R et al (2009) Chemotherapeutic agents in noncytotoxic concentrations increase antigen presentation by dendritic cells via an IL-12-dependent mechanism. J Immunol 183:137–144
147. van der Most RG, Currie AJ, Robinson BW et al (2008) Decoding dangerous death: how cytotoxic chemotherapy invokes inflammation, immunity or nothing at all. Cell Death Differ 15:13–20
148. Hodge JW, Garnett CT, Farsaci B et al (2013) Chemotherapy-induced immunogenic modulation of tumor cells enhances killing by cytotoxic T lymphocytes and is distinct from immunogenic cell death. Int J Cancer 133:624–636
149. Hodge JW, Kwilas A, Ardiani A et al (2013) Attacking malignant cells that survive therapy: exploiting immunogenic modulation. Oncoimmunology 2, e26937
150. Maker AV, Phan GQ, Attia P et al (2005) Tumor regression and autoimmunity in patients treated with cytotoxic T lymphocyte-associated antigen 4 blockade and interleukin 2: a phase I/II study. Ann Surg Oncol 12:1005–1016
151. Calabro L, Morra A, Fonsatti E et al (2013) Tremelimumab for patients with chemotherapy-resistant advanced malignant mesothelioma: an open-label, single-arm, phase 2 trial. Lancet Oncol 14:1104–1111
152. Zhang Y, Huang S, Gong D et al (2010) Programmed death-1 upregulation is correlated with dysfunction of tumor-infiltrating CD8+ T lymphocytes in human non-small cell lung cancer. Cell Mol Immunol 7:389–395
153. Hamid O, Robert C, Daud A et al (2013) Safety and tumor responses with lambrolizumab (anti-PD-1) in melanoma. N Engl J Med 369:134–144

154. Brahmer JR, Tykodi SS, Chow LQ et al (2012) Safety and activity of anti-PD-L1 antibody in patients with advanced cancer. N Engl J Med 366:2455–2465
155. Muhlbauer M, Fleck M, Schutz C et al (2006) PD-L1 is induced in hepatocytes by viral infection and by interferon-alpha and -gamma and mediates T cell apoptosis. J Hepatol 45:520–528
156. Jacobs R, Hintzen G, Kemper A et al (2001) CD56bright cells differ in their KIR repertoire and cytotoxic features from CD56dim NK cells. Eur J Immunol 31:3121–3127
157. Vinay DS, Kwon BS (2012) Immunotherapy of cancer with 4-1BB. Mol Cancer Ther 11:1062–1070
158. Matsuzaki J, Gnjatic S, Mhawech-Fauceglia P et al (2010) Tumor-infiltrating NY-ESO-1-specific CD8+ T cells are negatively regulated by LAG-3 and PD-1 in human ovarian cancer. Proc Natl Acad Sci U S A 107:7875–7880
159. Gerber DE, Stopeck AT, Wong L et al (2011) Phase I safety and pharmacokinetic study of bavituximab, a chimeric phosphatidylserine-targeting monoclonal antibody, in patients with advanced solid tumors. Clin Cancer Res 17:6888–6896
160. Liu X, Shin N, Koblish HK et al (2010) Selective inhibition of IDO1 effectively regulates mediators of antitumor immunity. Blood 115:3520–3530
161. Greenwald RJ, Freeman GJ, Sharpe AH (2005) The B7 family revisited. Annu Rev Immunol 23:515–548
162. Subudhi SK, Alegre ML, Fu YX (2005) The balance of immune responses: costimulation verse coinhibition. J Mol Med (Berl) 83:193–202
163. Tarhini AA, Butterfield LH, Shuai Y et al (2012) Differing patterns of circulating regulatory T cells and myeloid-derived suppressor cells in metastatic melanoma patients receiving anti-CTLA4 antibody and interferon-alpha or TLR-9 agonist and GM-CSF with peptide vaccination. J Immunother 35:702–710
164. Hodi FS, O'Day SJ, McDermott DF et al (2010) Improved survival with ipilimumab in patients with metastatic melanoma. N Engl J Med 363:711–723
165. Sarnaik AA, Yu B, Yu D et al (2011) Extended dose ipilimumab with a peptide vaccine: immune correlates associated with clinical benefit in patients with resected high-risk stage IIIc/IV melanoma. Clin Cancer Res 17:896–906
166. Yuan J, Ginsberg B, Page D et al (2011) CTLA-4 blockade increases antigen-specific CD8(+) T cells in prevaccinated patients with melanoma: three cases. Cancer Immunol Immunother 60:1137–1146
167. Fife BT, Pauken KE (2011) The role of the PD-1 pathway in autoimmunity and peripheral tolerance. Ann N Y Acad Sci 1217:45–59
168. Barber DL, Wherry EJ, Masopust D et al (2006) Restoring function in exhausted CD8 T cells during chronic viral infection. Nature 439:682–687
169. Topalian SL, Hodi FS, Brahmer JR et al (2012) Safety, activity, and immune correlates of anti-PD-1 antibody in cancer. N Engl J Med 366:2443–2454
170. Robert C, Ribas A, Wolchok JD et al (2014) Anti-programmed-death-receptor-1 treatment with pembrolizumab in ipilimumab-refractory advanced melanoma: a randomised dose-comparison cohort of a phase 1 trial. Lancet 384:1109–1117
171. Chacon JA, Wu RC, Sukhumalchandra P et al (2013) Co-stimulation through 4-1BB/CD137 improves the expansion and function of CD8(+) melanoma tumor-infiltrating lymphocytes for adoptive T-cell therapy. PLoS One 8, e60031
172. Mellor AL, Munn DH (2004) IDO expression by dendritic cells: tolerance and tryptophan catabolism. Nat Rev Immunol 4:762–774
173. Munn DH (2011) Indoleamine 2,3-dioxygenase, Tregs and cancer. Curr Med Chem 18:2240–2246
174. Hill M, Tanguy-Royer S, Royer P et al (2007) IDO expands human CD4+CD25 high regulatory T cells by promoting maturation of LPS-treated dendritic cells. Eur J Immunol 37:3054–3062
175. Sica A, Bronte V (2007) Altered macrophage differentiation and immune dysfunction in tumor development. J Clin Invest 117:1155–1166

176. Terabe M, Berzofsky JA (2007) NKT cells in immunoregulation of tumor immunity: a new immunoregulatory axis. Trends Immunol 28:491–496
177. Lu LF, Lind EF, Gondek DC et al (2006) Mast cells are essential intermediaries in regulatory T-cell tolerance. Nature 442:997–1002
178. Kawasaki H, Chang HW, Tseng HC et al (2014) A tryptophan metabolite, kynurenine, promotes mast cell activation through aryl hydrocarbon receptor. Allergy 69:445–452
179. Marigo I, Dolcetti L, Serafini P et al (2008) Tumor-induced tolerance and immune suppression by myeloid derived suppressor cells. Immunol Rev 222:162–179
180. Zoso A, Mazza EM, Bicciato S et al (2014) Human fibrocytic myeloid-derived suppressor cells express IDO and promote tolerance via Treg-cell expansion. Eur J Immunol 44:3307–3319
181. Jurgens B, Hainz U, Fuchs D et al (2009) Interferon-gamma-triggered indoleamine 2,3-dioxygenase competence in human monocyte-derived dendritic cells induces regulatory activity in allogeneic T cells. Blood 114:3235–3243
182. Lob S, Konigsrainer A, Rammensee HG et al (2009) Inhibitors of indoleamine-2,3-dioxygenase for cancer therapy: can we see the wood for the trees? Nat Rev Cancer 9:445–452
183. Holmgaard RB, Zamarin D, Munn DH et al (2013) Indoleamine 2,3-dioxygenase is a critical resistance mechanism in antitumor T cell immunotherapy targeting CTLA-4. J Exp Med 210:1389–1402
184. Muller AJ, Scherle PA (2006) Targeting the mechanisms of tumoral immune tolerance with small-molecule inhibitors. Nat Rev Cancer 6:613–625
185. Muller AJ, DuHadaway JB, Donover PS et al (2005) Inhibition of indoleamine 2,3-dioxygenase, an immunoregulatory target of the cancer suppression gene Bin1, potentiates cancer chemotherapy. Nat Med 11:312–319
186. Liu X, Newton RC, Friedman SM et al (2009) Indoleamine 2,3-dioxygenase, an emerging target for anti-cancer therapy. Curr Cancer Drug Targets 9:938–952
187. Johansen P, Storni T, Rettig L et al (2008) Antigen kinetics determines immune reactivity. Proc Natl Acad Sci U S A 105:5189–5194
188. Gulley JL, Arlen PM, Madan RA et al (2010) Immunologic and prognostic factors associated with overall survival employing a poxviral-based PSA vaccine in metastatic castrate-resistant prostate cancer. Cancer Immunol Immunother 59:663–674
189. Gulley JL, Arlen PM, Tsang KY et al (2008) Pilot study of vaccination with recombinant CEA-MUC-1-TRICOM poxviral-based vaccines in patients with metastatic carcinoma. Clin Cancer Res 14:3060–3069
190. Mohebtash M, Tsang KY, Madan RA et al (2011) A pilot study of MUC-1/CEA/TRICOM poxviral-based vaccine in patients with metastatic breast and ovarian cancer. Clin Cancer Res 17:7164–7173
191. Bilusic M, Heery CR, Arlen PM et al (2014) Phase I trial of a recombinant yeast-CEA vaccine (GI-6207) in adults with metastatic CEA-expressing carcinoma. Cancer Immunol Immunother 63:225–234
192. Marshall JL, Gulley JL, Arlen PM et al (2005) Phase I study of sequential vaccinations with fowlpox-CEA(6D)-TRICOM alone and sequentially with vaccinia-CEA(6D)-TRICOM, with and without granulocyte-macrophage colony-stimulating factor, in patients with carcinoembryonic antigen-expressing carcinomas. J Clin Oncol 23:720–731
193. Hodge JW, McLaughlin JP, Kantor JA et al (1997) Diversified prime and boost protocols using recombinant vaccinia virus and recombinant non-replicating avian pox virus to enhance T-cell immunity and antitumor responses. Vaccine 15:759–768

Chapter 2
Strategies for Improving Vaccines to Elicit T Cells to Treat Cancer

Jay A. Berzofsky, Lauren V. Wood, and Masaki Terabe

Abstract Cancers have not evolved to be good vaccines. Indeed, most clinically evident cancers have escaped from the immune system. Thus, unlike most existing vaccines, one cannot simply mimic the disease agent to make a successful vaccine, but instead may need to combine several approaches. Our lab has developed a push-pull strategy in which we first improve the immunogenicity of the antigens themselves by modifying the amino acid sequence to improve binding of epitopes to MHC molecules (a process called epitope enhancement). The next step is to "push" the response to improve not only the quantity but also the quality of the immune response, to achieve better avidity, longevity, and type of response, by using combinations of defined molecular adjuvants such as cytokines like IL-15, TLR ligands, and NKT cell agonist antigens. We and others have identified synergistic combinations of these. Then, it is still necessary to overcome negative regulatory mechanisms that cancers elicit to suppress and evade the immune system. These include regulatory cells like Treg cells, myeloid-derived suppressor cells, M2-macrophages, regulatory type II NKT cells, and others, plus regulatory receptors on the T cells themselves such as CTLA-4 and PD-1, and regulatory cytokines like TGF-beta, IL-13 and IL-10. We call this the "pull" in vaccine strategy, removing the brakes to allow vaccines to achieve their maximal potential. Here, we describe preclinical

J.A. Berzofsky, M.D., Ph.D. (✉)
Vaccine Branch, Center for Cancer Research, National Cancer Institute,
National Institutes of Health, Bldg 41-Room D702D, 41 Medlars Drive,
Bethesda, MD, 20892-5062, USA
e-mail: berzofsj@mail.nih.gov

L.V. Wood, M.D.
Clinical Trials Team, Vaccine Branch, Center for Cancer Research,
National Cancer Institute, National Institutes of Health, Bldg 10-Room 3B37,
10 Center Drive, Bethesda, MD, 20892, USA
e-mail: lw51p@nih.gov

M. Terabe, Ph.D.
Molecular Immunogenetics & Vaccine Research Section, Vaccine Branch,
Center for Cancer Research, National Cancer Institute, National Institutes of Health,
Bldg 41-Room D702H, 41 Medlars Drive, Bethesda, MD, 20892-5062, USA
e-mail: terabe@mail.nih.gov

P.A. Ascierto et al. (eds.), *Developments in T Cell Based Cancer Immunotherapies*, Cancer Drug Discovery and Development,
DOI 10.1007/978-3-319-21167-1_2

studies to examine each of these elements in cancer vaccines and clinical trials to translate these into patients, including an epitope-enhanced vaccine and blockade of TGF-beta. Overall, the combination of these strategies may allow more effective vaccines to treat established human cancers.

Keywords Cancer vaccines • Cancer immunotherapy • Immune regulation • NKT cells • Epitope enhancement • Cytokines • Checkpoint inhibitors

Introduction

Most conventional vaccines target organisms that produce acute self-limited infections and/or that produce toxins. Because these organisms naturally induce protective immunity against subsequent exposure, the vaccine needs only to mimic natural infection to be effective. Thus, the emphasis has been on preparing attenuated or killed organisms or inactivated toxins. However, viruses causing chronic infection, such as HIV or Hepatitis C virus, or cancers, do not naturally induce sufficient immunity to prevent the chronic disease. Indeed, in the case of cancer, there is now substantial evidence that the cancers that emerge clinically are the ones that have already escaped natural immunosurveillance [1]. Thus, to make an effective cancer (or HIV) vaccine, one needs to do more than simply administer cancer antigens or killed cells, but rather must also make these more immunogenic and overcome the mechanisms that the cancer (or virus) has developed to evade the immune system.

Looked at another way, neither chronic viruses nor cancers evolved to be effective vaccines. Quite the contrary, they evolved to escape the immune system. Thus, in these cases, we should be able to rationally do better than nature to induce an immune response against these invaders [2].

We have developed a multi-pronged push-pull strategy to accomplish this goal [3–5]. First, the antigen itself can often be made more immunogenic by slightly modifying the sequence of epitopes to increase their affinity for common MHC (HLA in humans) molecules. We call this approach epitope enhancement [2, 6]. After all, the organism or cancer antigens may have already undergone some selection for antigens that are poor MHC binders. Second, we need to improve the quality of the immune response in addition to its quantity by using an appropriate combination of defined molecular adjuvants. Thus, we have previously seen that the quality of T cells, such as their avidity or longevity, or their cytokine profile, can be more important than their quantity for eliminating invaders. This is the push of the "push-pull" strategy [3]. Even with these improvements, we still need to overcome the negative regulatory mechanisms that cancers (and chronic viruses) exploit to evade the immune system. These include regulatory cells (such as Treg cells, MDSC, M2 macrophages, regulatory type II NKT cells, regulatory DCs, etc), inhibitory receptors on T cells (such as PD-1 and CTLA-4), and immunosuppressive cytokines (such as TGF-beta or IL-13 or IL-10). Overcoming these is the "pull" of the "push-pull" strategy, removing the brakes so that the vaccine can achieve its maximum potential [3]. By combining all these steps in a push-pull strategy, we hope to finally accomplish the difficult task of inducing an immune response to reject an established cancer.

The sections below will address each of these steps sequentially, both in mouse models and in translation to human clinical trials. Although for space reasons we must focus especially on examples from our own research, selected relevant related literature will be reviewed as well.

Epitope Enhancement

The concept of epitope enhancement derives from the observation that the binding of antigenic peptides corresponding to T cell epitopes to class I or class II MHC molecules can be improved by making selective amino acid substitutions. These can either remove residues that interfere with binding [7] or substitute higher affinity "anchor" residues that bind more tightly in the peptide binding pockets of these MHC molecules [8]. The latter approach was greatly facilitated by the discovery of MHC-binding motifs for peptides that bind to particular MHC molecules, initially defining primarily anchor residues [8–10], and then also defining secondary anchor positions or other positions that influenced binding [11]. The goal is to improve binding to the MHC molecule without interfering with or altering what the T-cell receptor (TCR) recognizes, because it is critical that the T cells elicited by the enhanced vaccine also crossreact with the wild-type sequence still present in the cancer or virus. This is facilitated by the fact that it is possible to map some residues of the epitope peptide that point down or sideways to interact primarily with residues of the MHC molecule in the MHC groove in which the peptide sits, and others that stick out of the groove and interact primarily with the TCR. These two types of amino acid residues correspond with the original concepts of "agretope" and "epitope" regions of the peptide defined by their effects on these functions, respectively, before the crystal structure of the MHC molecule was known [12]. A number of computer algorithms have been developed to predict binding of peptides to particular MHC molecules, many of which are based on these known primary and secondary anchors or more broadly on the frequency with which different amino acids appear at each position in a survey of peptides binding to a particular MHC molecule [13].

Our approach to epitope enhancement began with the observation that we could improve binding of an HIV peptide to a class II MHC molecule by removing a Glu residue that seemed to interfere with binding, replacing it with an Ala [7]. To apply this concept to other peptides, we carried out "Alanine scans" in which we substituted each amino acid residue sequentially with an Ala residue, based on the concept that the Ala side chain is only a small and uncharged methyl group and therefore we would be removing any side chain that might interfere with binding because of its bulk (steric hindrance) or charge [7, 14]. We also replaced poor anchor residues with more effective ones where we could. This epitope enhancement approach was successfully carried out for epitopes of Hepatitis C virus (HCV) [14], HIV [7, 15–17], and cancer [18, 19]. Others have used similar approaches for other epitopes from cancer and viruses [20], and have also found substitutions that improved binding to a dominant TCR [21]. Some of these have been translated into human clinical trials [22, 23].

The epitope-enhanced tumor antigen that we have translated into a clinical trial is TARP, discovered by Ira Pastan's lab [24, 25] and standing for "T-cell receptor gamma chain Alternative Reading frame Protein." Its expression was detected in about 95 % of prostate cancers and about 50 % of breast cancers, but it turned out that although the nucleic acid sequence derived from the TCR gamma chain gene, the TARP transcript was read in a different reading frame and the amino acid sequence had nothing in common with that of the TCR gamma chain. It was subsequently found to be expressed in all stages of prostate cancer of all Gleason types [26], making it a good target for immunotherapy. The amino acid sequence is only 58 amino acids long, and within that we mapped several HLA-A*0201-binding epitopes [18], using several algorithms to predict binding peptides, followed by a flow-based binding assay measuring stabilization of HLA-A*0201 on the surface of TAP-deficient cells. We compared the affinities of different peptides for this MHC molecule by determining the concentration required to increase HLA-A*0201 expression by 50 % above the level without peptide. We found one high affinity epitope residues 27-35, and an overlapping moderate affinity epitope residues 29-37. While the substitutions we tried did not increase binding of the 27-35 peptide, some did improve the binding affinity of the 29-37 peptide [18]. These were found to improve immunogenicity in vivo in HLA-A2 transgenic mice. We then used them to expand specific T cells from human HLA-A*0201^{+} patients, and found that the enhanced 29-37 peptide with a Valine substitution at position 9 induced human T cells that could kill human tumor cells expressing TARP and HLA-A*0201 more effectively than could the wild type peptide itself, and just as well as the high affinity peptide 27-35. On this basis, TARP peptides 27-35 and 29-37-9 V were chosen to carry forward into a human clinical trial in prostate cancer.

Epitope Enhanced TARP Clinical Trials

We conducted the clinical translation of our epitope enhanced TARP vaccine platform in HLA-A*0201 men with Stage D0 prostate cancer who had undergone primary treatment and had prostate specific antigen (PSA) biochemical recurrence without radiographic evidence of tumor. This is an ideal population for the study of therapeutic vaccination because the tumor burden is micrometastatic and immune function is presumed to be normal as a result of limited prior exposure to chemo, hormonal or radiation therapies. In addition, the rate of PSA rise, expressed as Slope Log (PSA) or as PSA doubling time (PSADT) in nomogram calculations [27], is a validated measure of tumor growth and disease outcomes (recurrence and survival) in men with D0 disease [28–32]. Since the optimal method of peptide vaccination remains to be defined, we carried out a prospective, randomized clinical trial (ClinicalTrials.gov identifier NCT00972309) investigating wild type (WT) 27-35 and 29-37-9 V epitope enhanced (EE) TARP peptides administered as a peptide emulsion with Montanide ISA 51 VG and granulocyte-macrophage colon-stimulating factor (GM-CSF) or as a peptide-pulsed autologous dendritic cell vaccine (also pulsed with keyhole limpet hemocyanin (KLH) as a source of help). A total of five vaccines were delivered every 3 weeks with the option for an additional sixth dose of vaccine at 36 weeks based on PSADT or immunologic response criteria, with subsequent booster doses of vaccine at 48 and 96 weeks.

The primary study outcomes were safety and immunogenicity in addition to assessing the impact of TARP vaccination on Slope Log (PSA)/PSADT and tumor growth rates. As the Slope Log (PSA) (the rate at which the PSA is rising over time) decreases, the PSADT, which is proportional to the reciprocal slope, increases. Hence a decrease in Slope Log (PSA) is equivalent to an increase and lengthening in PSADT that is the preferred outcome of vaccination [28–32]. As might be anticipated with a peptide vaccine, TARP vaccination was very well tolerated with adverse events limited to local injection site reactions of short duration [159]. Immunogenicity was assessed by TARP-specific IFN-γ ELISPOT responses to vaccine platform WT27 and EE29-37-9 V TARP peptides as well as the non-vaccine WT29-37 TARP peptide. TARP vaccination was highly immunogenic and associated with statistically significant increases in ELISPOT reactivity over baseline at 12, 18 and 24 weeks in the majority of subjects. Importantly, we documented reactivity to vaccine (WT27-35, EE29-37-9V) as well as *non-vaccine* (WT29-37) TARP peptide. Hence in humans, vaccine-induced TARP immune responses to the epitope enhanced peptide also cross-react with the wild-type version of the same peptide sequence, replicating our initial observations in pre-clinical animal studies [18]. In a pooled analysis of all subjects, 72 and 74 % of patients demonstrated a decrease in Slope Log (PSA) at 24 and 48 weeks respectively, compared to their pre-treatment baseline [159]. However there was no correlation between decreasing Slope Log (PSA) and vaccine-induced TARP ELISPOT responses or any other baseline variables examined. Using pre- and post-vaccine PSA measurements in a two-phase exponential growth and regression mathematical model [33], TARP vaccination was also associated with a 50 % reduction in tumor growth rates. Hence additional studies of this vaccine platform in patient populations with more clinically aggressive or advanced disease will allow further verification of its direct impact on primary clinical outcomes such as progression free survival or overall survival.

To confirm the effects of TARP vaccination on decreasing Slope Log (PSA) and slowing tumor growth rates observed in our initial study, a prospective, phase II randomized, placebo-controlled trial of a second generation multi-epitope (ME) TARP vaccine in an identical population of men with stage D0 prostate cancer is planned, except without the HLA restriction. This second generation vaccine contains five additional overlapping peptides that span the entire TARP protein as well as the original two WT27-35 and EE29-37-9V peptides, thus eliminating the need for HLA restriction and minimizing the risk of tumor immune escape following vaccination. The use of longer, overlapping peptides that include CD4⁺T cell helper epitopes may also allow the generation of anti-TARP antibodies as well as better TARP-specific CD8 T cell responses with improved functional avidity and longevity [34].

Use of Defined Molecular Adjuvants to Increase Immune Response Quantity and Quality

Tumors and chronic virus infections may not only induce an inadequate magnitude of immune response, but may also induce a less than optimal qualitative type of immune response. For example, a Th2 cytokine profile may be induced where a Th1 profile is really needed [35, 36]. Similarly, we had found that high functional avidity CD8 T cells were much more effective at clearing a virus infection in vivo than low avidity T cells specific for the same peptide-MHC complex [37]. Others then confirmed this for virus infections [38–41] and for tumors [42, 43], as we observed as well [44]. If high avidity T cells to tumors have been eliminated by self-tolerance through negative selection or other mechanisms, the residual T cells may not be adequate to control the tumor. Sometimes this can be overcome by using subdominant epitopes, for which tolerance has not been induced, and using epitope enhancement to make these more immunogenic [45]. However, we sought a more general method for selectively inducing higher avidity $CD8^+$ T cells.

Costimulatory Molecules as Molecular Adjuvants

A number of costimulatory molecules have been found to increase T cell responses. For example, CD40L, which is one of the main mediators by which helper T cells activate dendritic cells or B cells, has been produced in soluble form and we have found it to have adjuvant activity [3]. Indeed, we found synergy between GM-CSF that can recruit antigen-presenting cells (APCs) and CD40L that can mature them [3]. We asked whether costimulatory molecules might allow immunization with a lower concentration of antigen to selectively induce higher avidity $CD8^+$ T cells. Using a TRICOM vector expressing three costimulatory molecules, ICAM-1, CD80, and LFA-3 [46], we found that indeed higher functional avidity T cells could be elicited in mice [47]. This approach was further developed to induce considerably higher functional avidity T cells with TRICOM-based vaccines [48].

IL-15 as an Adjuvant to Induce Higher Avidity Longer-Lived CD8⁺ T Cells

Interleukin 15 (IL-15) was discovered as a cytokine that had T-cell stimulatory properties akin to those of IL-2 [49]. It was found to have a receptor that shared the beta and gamma chains with that of IL-2, but differed in the alpha chain [49]. However, in contrast to IL-2, IL-15 was not made by T cells, but rather by APCs, especially DCs. It was then found that unlike IL-2, IL-15 could be presented by the IL-15Ralpha chain on DCs to NK cells or T cells lacking the alpha chain but expressing only the IL-2Rbeta and gamma chains, so called "trans presentation" [50]. More recently, it has been suggested that IL-15 is shed as a heterodimer with its IL-15Ralpha chain and that the heterodimeric complex is more stable and is the physiological form of the cytokine [51].

We and others examined IL-15 as a molecular adjuvant [52–59]. Regarding avidity, we noticed that when mice were immunized with a recombinant vaccinia virus vector vaccine expressing IL-15, over time (after about 2 months), the avidity of the T cells was seen to be higher than if the vaccine did not express IL-15 [53]. This apparent avidity maturation over time occurred despite the fact that the TCR does not undergo somatic mutation as antibody variable regions do. We found two complementary mechanisms to account for this avidity maturation. It was known that IL-15 promoted homeostatic proliferation of $CD8^+$ T cells [60]. At a population level, we found that higher avidity T cells expressed higher levels of IL-15Ralpha and therefore were more responsive to IL-15 than low avidity T cells specific for the same epitope. Thus, over time, the high avidity T cells persisted or expanded while the low avidity T cells were dying by attrition, rendering the average population avidity higher over time [53]. The second mechanism was at the cellular level, in which we found that that IL-15 increased expression of CD8alpha and CD8beta, which contribute to apparent functional avidity even for the same TCR. This effect was greatest for the higher functional avidity T cells that expressed more IL-15Ra, so there was a positive feedback effect [53]. The net result was apparent avidity maturation over time, and IL-15 became a valuable molecular adjuvant to elicit a higher avidity $CD8^+$ T cell response.

We also found that IL-15, but not IL-2, expressed by a recombinant vaccinia vaccine vector, elicited longer-lived $CD8^+$T cells [52]. Kutzler et al. [54] had also found that IL-15 as an adjuvant expressed by a DNA vaccine could induce $CD8^+$ T cells in the partial absence of $CD4^+$ T cell help. Because $CD4^+$ T cell help for $CD8^+$ T cells had been shown to be mediated at least in part by activation of DCs, and DCs could be induced by $CD4^+$ helper cells to secrete IL-15 (and/or express it on their surface for trans presentation), we asked whether IL-15 might be a major mediator of $CD4^+$ T cell help. $CD4^+$ T cell help had been shown to be necessary to prevent $CD8^+$ T cells from undergoing apoptosis when re-exposed to antigen, mediated by TRAIL [61]. Therefore, we asked whether IL-15 could substitute for $CD4^+$ help and prevent such antigen-reexposure-induced apoptosis. Indeed, we found that priming $CD8^+$ T cells in the presence of IL-15, even in the absence of $CD4^+$ help, led to the same phenotype as $CD8^+$ T cells primed in the presence of help [58]. Thus, IL-15 was sufficient to substitute for $CD4^+$ help. To know if IL-15 was a key mediator of help, we had to know whether it was also necessary for help. To test this, we examined DCs from IL-15 knock-out mice that could not produce IL-15 even in the presence of adequate T cell help. When antigen-pulsed DCs were used as a vaccine, if the DCs could not produce IL-15, the resulting $CD8^+$ T cells did not show the long-lived phenotype [58]. Thus, IL-15 was both necessary and sufficient for $CD4^+$ help for $CD8^+$ T cells and was therefore a natural mediator of such help.

Because IL-15 as a vaccine adjuvant could induce $CD8^+$T cells that had both higher functional avidity and greater longevity, and were resistant to antigen-reexposure induced apoptosis, we believe that IL-15 is a powerful adjuvant to achieve at least two qualitative improvements in $CD8^+$ T cell responses in addition to a simply higher magnitude response.

IL-15 synergized with TLR ligands to increase polyfunctional $CD8^+$T cell immunity in rhesus macaques immunized with a peptide and MVA vaccine and to upregulate the expression of IL-15Ralpha on dendritic cells, promoting cross-presentation of IL-15 [59]. A plasmid DNA encoding IL-15 enhanced the T cell response and protective efficacy of a SHIV-based DNA vaccine in rhesus macaques [56]. DNA encoding IL-15 or IL-15 fused to an Ig-Fc domain also enhanced $CD8^+$ T cell responses to the gp100 melanoma antigen in mice [62]. Plasmid DNA encoding IL-15 plus IL-6 also enhanced both cellular and humoral responses to a DNA vaccine targeting the VP1 capsid protein of foot-and-mouth-disease virus [63]. Likewise, DNA encoding IL-15 enhanced the longevity of $CD8^+$ T cells induced by a hepatitis B vaccine [64]. A DNA vaccine expressing the IL-15/IL-15Ralpha heterodimer along with SIV antigens was effective at inducing SIV-specific T cells in rhesus macaques in a therapeutic setting under cover of anti-retroviral drugs (ART) [65]. Also, a plasmid encoding IL-15 increased effector memory $CD8^+$ T cell responses to an SIV DNA vaccine in macaques [66]. A plasmid encoding IL-15 also enhanced $CD8^+$ T cell responses and protection induced by a plasmid DNA vaccine against Brucella [67]. An IL-15 plasmid also enhanced T cell immunity and protection against vaginal challenge of rhesus macaques immunized with a vif-deleted SIV vaccine [68]. In a survivin-based cancer vaccine study in mice, a plasmid encoding IL-15 enhanced protective efficacy against the CT26 colon carcinoma [69]. An IL-15-encoding plasmid also enhanced central memory $CD8^+$ T cell and antibody responses in macaques to a replicating adenoviral vaccine against SIV, but was not sufficient to protect against mucosal challenge with SIV [70]. A cautionary note was raised by the finding that whereas a low dose of IL-15 DNA enhanced T and antibody responses of macaques to a flu vaccine, high doses actually inhibited both types of response [71]. A plasmid encoding IL-15Ralpha alone or in combination with one for IL-15 also enhanced immune responses [72]. However, in a human clinical trial, a DNA plasmid encoding IL-15 (or IL-12) failed to enhance immune responses to a vaccine, suggesting that the mode and level of expression may be critical in achieving the adjuvant effect of these cytokines [73].

Synergistic Combinations of TLR Ligands as Molecular Vaccine Adjuvants

Toll-like receptors (TLRs) are cellular "pattern recognition" receptors that recognize pathogen-associated molecular patterns or PAMPS, such as CpG-rich bacterial DNA (TLR9), or single-stranded (TLR 7, 8) or double-stranded RNA (TLR 3), or bacterial lipopolysaccharide (LPS) (TLR4), peptidoglycan (TLR2) or flagellin (TLR 5) (reviewed in [74]). TLR2 forms a heterodimer with TLR1 or TLR6, and these two heterodimers have somewhat different specificity. These receptors are expressed especially on professional antigen-presenting cells such as DCs, and triggering of the TLRs can lead to DC activation and/or maturation, promoting antigen presentation and cytokine secretion and thereby essentially alerting the immune system to the presence of such potential pathogens. In this way, TLR ligands can serve some of the same functions as helper T cells in activating or licensing DCs.

We reasoned that such PAMPs often occur in combinations in pathogens, and therefore, the immune system may have evolved to respond especially to certain combinations of TLR ligands, possibly in a synergistic fashion. Dual viral-bacterial infections are also common, and may require an even stronger alert to the immune system. Different TLR ligands can differentially affect the quality of $CD4^+$ and $CD8^+$ T cell responses in non-human primates [75]. TLR ligands can mediate vaccine adjuvant functions through multiple pathways [76], suggesting possible synergies. Indeed, several labs described synergistic combinations of TLR ligands in activating DCs [77, 78]. We examined TLR ligands and their combinations for their ability to serve as vaccine adjuvants [79]. We discovered that certain pairs, like TLR 2/6 and TLR 3 or TLR 9 and TLR 3 were synergistic to increase $CD8^+$ T cell responses. Because TLR3 signals through the adaptor protein TRIF, whereas TLR2/6 and TLR9 signal through the adaptor protein MyD88, we reasoned that perhaps these particular pairs were synergistic because they activated both pathways. Indeed, we found that the synergy was dependent on both MyD88 and TRIF, using knockout mice [79]. Interestingly, the enhancement turned out to be dependent on a unidirectional cross-talk between these two pathways, in which the TRIF pathway increased the activity through the MyD88 pathway, in an AP-1 dependent fashion, to upregulate IL-12 production by the DCs, which appeared to be the main correlate of increased $CD8^+$ T cell response [79].

In the course of these studies, we noted that a triple combination of TLR2/6, 3 and 9 ligands as a vaccine adjuvant did not increase the number of antigen-specific (tetramer-positive) $CD8^+$T cells beyond what the best double combinations could do. We reasoned that this was likely due to the fact that TLR2/6 and TLR9 both signaled through MyD88, so each could synergize with the TRIF-dependent TLR3, but they might not synergize with each other. However, when we studied induction of protective $CD8^+$ T cell immunity to reduce viral load of a recombinant vaccinia virus expressing the HIV envelope protein, we found that the triple combination was substantially more effective than the best of the double combinations [80]. This led to the enigma that the CD8-dependent protective efficacy was greater even though the induction of $CD8^+$ T cells was not greater. We reasoned that if there was no difference in quantity, the difference must be in quality. Indeed, the triple combination was found to induce on average higher functional avidity $CD8^+$ T cells than the best double combinations [80]. As we had previously found that high avidity $CD8^+$ T cells were more effective at clearing virus infections [37], it made sense that greater avidity could account for greater antiviral efficacy even with the same numbers of T cells. Thus, here as in the earlier study [37], it can be seen that quality may be more important than quality for $CD8^+$ T cell responses [80]. As at least a partial explanation, it was found that the triple combination of TLR ligands upregulated IL-15 production [80], which in turn could induce higher avidity T cells [53]. Thus, appropriate synergistic combinations of TLR ligands may be extremely effective as vaccine adjuvants. It may be that the best of the empirically-discovered adjuvants, such as complete Freund's adjuvant, are so effective because they naturally combine several synergistic TLR ligands. However, using experimentally optimized combinations of molecularly-defined TLR-ligand adjuvants may allow effective induction of immunity with fewer side effects.

NKT Cell Agonists as Molecular Vaccine Adjuvants: α-GalCer and β-ManCer

A major point of using TLR ligands as adjuvants is to induce rapid production of cytokines in the microenvironment of T cell priming to skew T cell responses toward a certain direction. Along the same lines, NKT cell agonists are potent adjuvants as NKT cells rapidly produce a large amount of cytokines upon stimulation as well as induce maturation of DCs to be licensed to activate CD8 T cells, CD4 T cells, or B cells [81, 82].

NKT cells are a T cell population that recognizes lipid antigens, not peptides, presented by CD1d [82–85]. There are two subsets of NKT cells based on the TCR that they express. Type I or invariant NKT cells express a semi-invariant TCRα chain with Vα14Jα18 (Vα24Jα18 in humans) with very limited TCRβ repertoire (Vβ8, 7, 2 in mice and Vβ11 in humans). Type II NKT cells are defined as non-type I NKT cells that do not express the TCR utilizes Vα14Jα18, and they have a diverse TCR repertoire (but still restricted by CD1d). As these two types of NKT cells are defined by their TCR, they recognize different sets of antigens. This feature of NKT cells allows us to manipulate specific NKT cell subsets in vivo. Surface markers expressed on NKT cells (at least type I NKT cells) at a steady state are distinct from those on conventional T cells where the NKT cells from naïve mice are positive for activation/memory markers CD44, CD69, CD122 (IL-2Rβ) and negative for naïve T cell marker, CD62L. Consistent with the surface marker phenotype, they rapidly (within 3 hr) produce a large amount of cytokines upon stimulation. Thus, they have a characteristic of innate immunity on the top of the characteristics of adaptive immunity.

The interaction of type I NKT cells with DCs to recognize agonistic antigen such as α-galactosylceramide (α-GalCer) induces maturation of DCs and licenses DCs.

The maturation is initiated by IFN-γ production by activated type I NKT cells and NK cells together with TNF-α produced by DCs, which subsequently induces expression of co-stimulatory molecules, CD40, CD80, CD86 and CD70 as well as cytokines such as IL-12 [86, 87]. CD40L on NKT cells can also signal through CD40 on DCs to induce DC maturation. It is believed that instruction given by $CD4^+$ T cells or by inflammatory stimuli to DCs is essential for optimal $CD8^+$T cell priming that allows the $CD8^+$ T cells to be able respond to a secondary stimulation [61]. As mentioned above, the interaction between type I NKT cells and DCs induces expression of surface molecules and cytokine production that are necessary for the activated DCs to be able to prime $CD8^+$ T cells.

Because the interaction between type I NKT cells and DCs allows DCs to be fully licensed to activate conventional T cells, type I NKT cell ligands have adjuvant properties to facilitate antigen specific T cell responses induced by a vaccine. The adjuvant effect of α-GalCer was originally reported with a malaria sporozoite vaccine [88]. Subsequently the adjuvant effect was reported in tumor settings [86, 89]. It is important to note that in these studies, soluble or cell-associated proteins, which usually do not induce $CD8^+$ T cell responses, were used as a vaccine to induce CD8-mediated protection against tumors. It was also demonstrated that α-GalCer allows protein antigen to induce $CD8^+$ T cell responses and anti-tumor efficacy when administered orally [90]. The adjuvant activity is not limited to α-GalCer but also to its analogues such as OCH and β-ManCer. Recently Tsuji et al. reported that an analogue of α-GalCer, 7DW8-5, which induces more Th1 skewed responses in human type I NKT cells, has adjuvant activity [91]. Therefore, one may consider type I NKT cell agonists to serve as promising vaccine adjuvants to "push" and steer the immune responses.

Overcoming Negative Regulation to Improve Vaccine Responses

Most types of immune responses usually spontaneously resolve after some time. This is a result of negative regulation of immune responses to prevent deleterious over reaction of the immune system to pathogens or tissue damage. The negative regulation is very important to prevent immunopathology, such as autoimmune responses or excessive inflammation; however, such regulation limits the magnitude of anti-tumor immune responses. Here we will discuss cells and molecules that are involved in the negative regulation.

Regulatory NKT Cells, Treg Cells and Other Regulatory Cells

There are multiple cell types that are involved in the negative regulation of tumor immunity. These include MDSC, TAM, TAN, M2 macrophages, N2 neutrophils, $Foxp3^+$ Treg cells, Th2 $CD4^+$ T cells, regulatory type II NKT cells, and suppressor

CD8$^+$T cells. Among them, MDSC and Treg cells have been heavily studied in the last decade and are reviewed elsewhere [92–94]. All of the cell types listed above have been shown to play critical roles in the suppression of tumor immunity in some mouse tumor models. However, it seems that the necessity of any single regulatory cell type is not always observed in all models. For example, anti-CD25 treatment to inhibit Treg induced rejection of six out of eight tumors tested in the original study [95]. This result suggests that there is no single type of cell that is necessary for the immune suppression to occur in all tumor-bearing individuals. In addition, the negative regulators may also interact with each other to form a network of negative regulation and complement each other's functions.

One example is the relationship between Treg cells and regulatory type II NKT cells where both types of cells can suppress tumor immunity. In an individual who has functional type I NKT cells, the suppressive activity of type II NKT cells is regulated by type I NKT cells. In those individuals, the blockade of natural Treg cells by anti-CD25 induced significant protection against tumor growth in both an s.c. CT26 colon carcinoma model and an s.c. R331 renal cell carcinoma model [96]. Although similar protection could be achieved in NKT cell-deficient CD1d$^{-/-}$ mice, which do not have any NKT cells, the protective effect of Treg cell blockade could not be seen in type I NKT cell-deficient Jα18$^{-/-}$ mice. In Jα18$^{-/-}$ mice, blockade of both Treg and CD1d-restricted NKT cells (type II NKT cells) was necessary to induce the protection. This piece of data suggests that in the absence of type I NKT cells, both Treg cells and type II NKT cells suppress tumor immunity probably simultaneously [96, 97]. Type I NKT cells, which are well documented to induce protective tumor immunity (but paradoxically can also support Treg cells) [98], have been shown to suppress the immunosuppressive activity of type II NKT cells [99–101]. Indeed adoptive transfer of type I NKT cells in Jα18$^{-/-}$ mice made Treg cell-blockade protective [96]. Furthermore, activation of type II NKT cells by sulfatide in vivo to tip the balance between these two types of NKT cells toward type II NKT cell dominance made Treg cell-blockade ineffective to protect wild-type mice. These data suggested that the balance between two types of NKT cells determines the outcome of Treg cell-blockade in tumor-bearing animals, and that the balance between two types of regulatory T cells, Treg and type II NKT cells, is controlled by a third type of T cell, type I NKT cells. Loss of function in type I NKT cells has been reported in patients with many types of cancers [102–109]. Thus, it is likely that type II NKT cell function is gained in the patients. In those patients, it may be necessary to block both Treg cells and type II NKT cells to remove immune suppression or recover the functions of type I NKT cells. These results also indicate that host immunological status will change the outcome of Treg targeted therapy and that this might be a reason for a very limited success of Treg-targeted therapies in patients.

This example describes only a small part of the complexity of the network formed by regulatory cells [83]. Since it may be very overwhelming to deal with all regulatory cells, it will be important to identify the right target patient population when regulatory cell targeted therapies will be translated.

Blocking Regulatory Molecules: CTLA-4, PD-1, IL-13, IL-10, and TGF-Beta

Another way to deal with negative regulation of tumor immunity in patients is to target molecules that mediate the functions of immune regulatory cells. One type of molecule recently drawing attention in the field of immunotherapy is the category of cell surface molecules mediating "Checkpoints" in T cell responses [110]. Recent approval of three monoclonal antibody checkpoint inhibitors (anti-CTLA4 and anti-PD1) by the FDA finally brought immunotherapy to the attention of non-immunologists as an option for cancer treatment. As these drugs facilitate tumor immunity in patients, it is possible that these antibodies will enhance the effect of vaccines. In fact, blockade of CTLA4 or PD-1/PD-L1 has been shown to increase the efficacy of tumor vaccines in mice [111–114], and anti-CTLA4 has been reported to be effective to improve the efficacy of a prostate cancer vaccine [115]. Furthermore, blockade of both molecules also can synergistically enhance vaccine efficacy in a mouse tumor model since the mechanisms of action of these two molecules are different [116]. As the synergy of two antibodies, anti-CTLA4 and anti-PD1, was reported to improve natural immunosurveillance without a vaccine in cancer patients, there is a very high likelihood that this combination can also facilitate vaccine efficacy.

Another type of target is cytokines that are involved in the functions of regulatory cells (e.g. TGF-β, IL-10, IL-13). TGF-β is a cytokine that has a strong suppressive effect on cell proliferation [117]. Thus, it can inhibit the growth of pre-malignant cells. However, cancers usually do not respond to TGF-β because of loss of function in the signaling pathway and they make large quantities of TGF-β. This cytokine has been reported to induce MDSC, Tregs (with IL-2) [118, 119] and IL-17-producing T cells (with IL-6) [120, 121]. TGF-β is also produced by regulatory cells and is used by them to suppress immune cells [122, 123]. Thus, inhibiting TGF-β can block multiple pathways of immune suppression. Blockade of TGF-β not only facilitates natural immunosurveillance, but also can enhance the efficacy of vaccines [124–128].

A cytokine that induces TGF-β production by myeloid cells is IL-13 [129–131]. As a cytokine produced by immunosuppressive type II NKT cells which is activated in a tumor-bearing individual, this cytokine together with TNF-α induces TGF-β [132]. IL-13 also induces arginase in MDSC [133] and can convert M1 macrophages to M2 phenotype [134]. Blockade of IL-13 has been shown to be able to enhance vaccine efficacy as well [3]. Thus, IL-13 can be another attractive target to inhibit.

Clinical Translation of Blockade of Negative Regulation (CTLA-4, PD-1, TGF-Beta, Treg)

Negative regulation of the immune system by cells (MDSCs, Tregs), cytokines and regulatory receptors (CTLA-4, PD-1, TGF-β) and their respective ligands (e.g. PDL-1) permits immune homeostasis through termination of effective immune responses and control of excessive inflammation in the normal healthy host. However in patients with advanced metastatic cancer, these negative regulatory elements are amplified, thereby allowing the development of resistance and immune escape [135, 136] that prevent the induction of optimal anti-tumor responses that might otherwise allow for control or eradication of disease. Hence targeting these elements alone or in combination with other treatment modalities that could result in potential mechanistic synergy and improved clinical outcomes is currently an intense focus in the field of cancer immunotherapy.

A pleiotropic cytokine that plays a pivotal role in tumor growth, disease progression and metastasis, TGF-β is elevated in many patients with advanced cancers [137, 138]. Based on our earlier work in preclinical mouse models [126, 127, 129, 139, 140], we conducted a phase I study of a human anti-TGF-β monoclonal antibody (GC1008, Genzyme Corp) in patients with advanced malignant melanoma and renal cell carcinoma [141]. One patient achieved a significant (89 %) partial response lasting about a year, while six others had stable disease or mixed responses associated with a median time to progression of 24 weeks in these patients. Although no dose-limiting toxicity was observed, the development of hyperkeratosis and cutaneous reversible keratoacanthomas/squamous cell carcinomas (resolving after cessation of antibody therapy) in four patients on high doses of antibody poses a potential challenge to further clinical development of this agent.

In contrast, significant progress has been realized in the successful translation of agents with unique features and the potential for inducing durable clinical responses. As of December 2014, there are now three different monoclonal antibodies addressing negative regulation that are approved for treatment of human cancer: the anti-CTLA4 T cellcheckpoint inhibitor ipilumumab (Yervoy™) and the new PD-1 inhibitors pembrolizumab (Keytruda™) and nivolumab (Opdivo™). While all are approved for the treatment of metastatic melanoma, they are also being studied in a plethora of clinical trials for other advanced solid tumors including non-small cell lung, bladder, brain, breast, colorectal, ovarian, pancreatic, prostate and renal cell cancers as well as lymphomas, multiple myeloma and acute and chronic leukemias. Since they "take the brakes off" the immune system, these agents are often associated with autoimmune phenomena that also seem to correlate with treatment response and improved survival [142–145]. Observed autoimmune adverse events commonly involve the colon (colitis), skin (dermatitis, vitiligo), liver (hepatitis), lung (pneumonitis) and endocrine system (thyroiditis, hypophysitis), and are more frequent with anti-CTLA-4 than with anti-PD-1. Hence it appears that breaking tolerance to self antigens may have a fundamental role in inducing anti-tumor effects that are associated with objective responses and positive clinical outcomes, including improved survival. Parallel to observations in preclinical animal models [116, 146], concurrent combinations of two checkpoint inhibitors with unique mechanisms of action such as nivolumab (an inhibitor of PD-1) and ipilumumab (an inhibitor of CTLA-4) has been shown to be associated with more pronounced anti-tumor activity including greater objective response rates, tumor regression and clinical activity that is distinct from that observed with use of either of these agents as monotherapy [147].

The preclinical push/pull strategy is fully captured in human clinical trials investigating a variety of therapeutic vaccines in combination with systemically delivered checkpoint inhibitors [115, 148–150]. In general, combination of these inhibitors with vaccines has been associated with immunologic activity and modest potential impact on signals of clinical outcomes [149, 150], without resulting in any observed intensification of known immune-related adverse events [148]. There remains a great need for continued examination of these dual approaches to immune modulation that should be built on a foundation, whenever possible, of well-designed preclinical animal studies to provide scientific justification for their combination and further insights into mechanisms of action and synergy.

T regulatory cells and MDSCs play a critical role in cancer development, progression, prognosis and maintenance of an immunosuppressive environment that interferes with generation of effective innate and adaptive antitumor immunity. Directly targeting immunosuppressive cell populations with low doses of cytotoxic agents to maximize immune responses to delivered tumor antigens is being investigated utilizing multiple vaccine platforms in both animals [151–154] and humans [155–157]. While the most commonly used agent is cyclophosphamide (cytoxan) other chemotherapy agents such as temazolamide [158] have also been utilized. In addition to its inhibition of Tregs, cytoxan also appears to have multiple immune modifying properties of its own. The use of these types of cellular depletion strategies followed by therapeutic vaccination reflects the "pull" preceding the "push" but has the potential to further augment anti-tumor responses, particularly in the local tumor microenvironment where effective tumor responses are most needed.

Unique features associated with immunotherapy include induction of autoimmune toxicities (e.g. anti-CTLA4, anti-PD-1), lack of typical drug resistance and the potential for durable clinical responses. Importantly, therapeutic cancervaccines constitute a diverse collection of complex biologics with kinetics and clinical profiles distinct from standard chemotherapy. Using traditional and combinatorial therapies, the field of immuno-oncology is well positioned to further transform clinical responses from durable to curable resulting in the complete eradication of cancer.

References

1. Dunn GP, Old LJ, Schreiber RD (2004) The three Es of cancer immunoediting. Annu Rev Immunol 22:329–360
2. Berzofsky JA (1993) Epitope selection and design of synthetic vaccines: molecular approaches to enhancing immunogenicity and crossreactivity of engineered vaccines. Ann N Y Acad Sci 690:256–264
3. Ahlers JD, Belyakov IM, Terabe M, Koka R, Donaldson DD, Thomas E, Berzofsky JA (2002) A push-pull approach to maximize vaccine efficacy: abrogating suppression with an IL-13 inhibitor while augmenting help with GM-CSF and CD40L. Proc Natl Acad Sci U S A 99(20):13020–13025
4. Berzofsky JA, Ahlers J, Janik J, Morris J, Oh S, Terabe M, Belyakov IM (2004) Progress on new vaccine strategies against chronic viral infections. J Clin Invest 114:450–462

5. Berzofsky JA, Terabe M, Oh S, Belyakov IM, Ahlers JD, Janik JE, Morris JC (2004) Progress on new vaccine strategies for the immunotherapy and prevention of cancer. J Clin Invest 113(11):1515–1525
6. Berzofsky JA, Ahlers JD, Belyakov IM (2001) Strategies for designing and optimizing new generation vaccines. Nat Rev Immunol 1(3):209–219
7. Boehncke W-H, Takeshita T, Pendleton CD, Sadegh-Nasseri S, Racioppi L, Houghten RA, Berzofsky JA, Germain RN (1993) The importance of dominant negative effects of amino acids side chain substitution in peptide-MHC molecule interactions and T cell recognition. J Immunol 150:331–341
8. Rammensee H-G, Friede T, Stevanovíc S (1995) MHC ligands and peptide motifs: first listing. Immunogenetics 41:178–228
9. Falk K, Rötzschke O, Rammensee H-G (1990) Cellular peptide composition governed by major histocompatibility complex class I molecules. Nature 348:248–251
10. Romero P, Corradin G, Luescher IF, Maryanski JL (1991) H-2Kd-restricted antigenic peptides share a simple binding motif. J Exp Med 174:603–612
11. Ruppert J, Sidney J, Celis E, Kubo RT, Grey HM, Sette A (1993) Prominent role of secondary anchor residues in peptide binding to HLA-A2.1 molecules. Cell 74:929–937
12. Heber-Katz E, Hansburg D, Schwartz RH (1983) The Ia molecule of the antigen-presenting cell plays a critical role in immune response gene regulation of T cell activation. J Mol Cell Immunol 1:3–14
13. De Groot AS, Cohen T, Ardito M, Moise L, Martin B, Berzofsky JA (2010) Use of bioinformatics to predict MHC ligands and T-cell epitopes: application to epitope-driven vaccine design. Methods in Microbiology 37:35–66
14. Sarobe P, Pendleton CD, Akatsuka T, Lau D, Engelhard VH, Feinstone SM, Berzofsky JA (1998) Enhanced in vitro potency and in vivo immunogenicity of a CTL epitope from hepatitis C virus core protein following amino acid replacement at secondary HLA-A2.1 binding positions. J Clin Invest 102:1239–1248
15. Ahlers JD, Takeshita T, Pendleton CD, Berzofsky JA (1997) Enhanced immunogenicity of HIV-1 vaccine construct by modification of the native peptide sequence. Proc Natl Acad Sci U S A 94:10856–10861
16. Okazaki T, Pendleton DC, Lemonnier F, Berzofsky JA (2003) Epitope-enhanced conserved HIV-1 peptide protects HLA-A2-transgenic mice against virus expressing HIV-1 antigen. J Immunol 171:2548–2555
17. Okazaki T, Pendleton CD, Sarobe P, Thomas EK, Harro C, Schwartz D, Iyengar S, Berzofsky JA (2006) Epitope-enhancement of a CD4 HIV epitope toward the development of the next generation HIV vaccine. Journal of Immunol 176(6):3753–3759
18. Oh S, Terabe M, Pendleton CD, Bhattacharyy A, Bera TK, Epel M, Reiter Y, Phillips J, Linehan WM, Kasten-Sportes C et al (2004) Human CTL to wild type and enhanced epitopes of a novel prostate and breast tumor-associated protein, TARP, lyse human breast cancer cells. Cancer Res 64:2610–2618
19. Huang Y, Terabe M, Pendleton CD, Khursigara DS, Bera TK, Pastan I, Berzofsky JA (2013) Identification and enhancement of HLA-A2.1-restricted CTL epitopes in a new human cancer antigen-POTE. PLoS One 8(6), e64365
20. Parkhurst MR, Salgaller ML, Southwood S, Robbins PF, Sette A, Rosenberg SA, Kawakami Y (1996) Improved induction of melanoma-reactive CTL with peptides from the melanoma antigen gp100 modified at HLA-A*0201-binding residues. J Immunol 157:2539–2548
21. Zaremba S, Barzaga E, Zhu M, Soares N, Tsang KY, Schlom J (1997) Identification of an enhancer agonist cytotoxic T lymphocyte peptide from human carcinoembryonic antigen. Cancer Res 57(20):4570–4577
22. Morse MA, Deng Y, Coleman D, Hull S, Kitrell-Fisher E, Nair S, Schlom J, Ryback ME, Lyerly HK (1999) A Phase I study of active immunotherapy with carcinoembryonic antigen peptide (CAP-1)-pulsed, autologous human cultured dendritic cells in patients with metastatic malignancies expressing carcinoembryonic antigen. Clin Cancer Res 5(6):1331–1338

23. Kammula US, Marincola FM, Rosenberg SA (2000) Real-time quantitative polymerase chain reaction assessment of immune reactivity in melanoma patients after tumor peptide vaccination. J Natl Cancer Inst 92(16):1336–1344
24. Wolfgang CD, Essand M, Vincent JJ, Lee B, Pastan I (2000) TARP: a nuclear protein expressed in prostate and breast cancer cells derived from an alternate reading frame of the T cell receptor gamma chain locus. Proc Natl Acad Sci U S A 97(17):9437–9442
25. Essand M, Vasmatzis G, Brinkmann U, Duray P, Lee B, Pastan I (1999) High expression of a specific T-cell receptor gamma transcript in epithelial cells of the prostate. Proc Natl Acad Sci U S A 96(16):9287–9292
26. True L, Coleman I, Hawley S, Huang CY, Gifford D, Coleman R, Beer TM, Gelmann E, Datta M, Mostaghel E et al (2006) A molecular correlate to the Gleason grading system for prostate adenocarcinoma. Proc Natl Acad Sci U S A 103(29):10991–10996
27. Arlen PM, Bianco F, Dahut WL, D'Amico A, Figg WD, Freedland SJ, Gulley JL, Kantoff PW, Kattan MW, Lee A et al (2008) Prostate Specific Antigen Working Group guidelines on prostate specific antigen doubling time. J Urol 179(6):2181–2185, discussion 5–6
28. Pound CR, Partin AW, Eisenberger MA, Chan DW, Pearson JD, Walsh PC (1999) Natural history of progression after PSA elevation following radical prostatectomy. JAMA 281(17):1591–1597
29. Freedland SJ, Humphreys EB, Mangold LA, Eisenberger M, Dorey FJ, Walsh PC, Partin AW (2005) Risk of prostate cancer-specific mortality following biochemical recurrence after radical prostatectomy. JAMA 294(4):433–439
30. Slovin SF, Wilton AS, Heller G, Scher HI (2005) Time to detectable metastatic disease in patients with rising prostate-specific antigen values following surgery or radiation therapy. Clin Cancer Res 11(24 Pt 1):8669–8673
31. Lee AK, Levy LB, Cheung R, Kuban D (2005) Prostate-specific antigen doubling time predicts clinical outcome and survival in prostate cancer patients treated with combined radiation and hormone therapy. Int J Radiat Oncol Biol Phys 63(2):456–462
32. Freedland SJ, Humphreys EB, Mangold LA, Eisenberger M, Dorey FJ, Walsh PC, Partin AW (2007) Death in patients with recurrent prostate cancer after radical prostatectomy: prostate-specific antigen doubling time subgroups and their associated contributions to all-cause mortality. J Clin Oncol 25(13):1765–1771
33. Stein WD, Gulley JL, Schlom J, Madan RA, Dahut W, Figg WD, Ning YM, Arlen PM, Price D, Bates SE et al (2011) Tumor regression and growth rates determined in five intramural NCI prostate cancer trials: the growth rate constant as an indicator of therapeutic efficacy. Clin Cancer Res 17(4):907–917
34. Welters MJ, Kenter GG, Piersma SJ, Vloon AP, Lowik MJ, Berends-van der Meer DM, Drijfhout JW, Valentijn AR, Wafelman AR, Oostendorp J et al (2008) Induction of tumor-specific CD4+ and CD8+ T-cell immunity in cervical cancer patients by a human papillomavirus type 16 E6 and E7 long peptides vaccine. Clin Cancer Res 14(1):178–187
35. Tsukui T, Hildesheim A, Schiffman MH, Lucci J III, Contois D, Lawler P, Rush BB, Lorincz AT, Corrigan A, Burk RD et al (1996) IL-2 production in vitro by peripheral lymphocytes in response to human papillomavirus-derived peptides: correlation with cervical pathology. Cancer Res 56:3967–3974
36. Lucey DR, Clerici M, Shearer GM (1996) Type 1 and type 2 cytokine dysregulation in human infectious, neoplastic, and inflammatory diseases. Clin Microbiol Rev 9(4):532–562
37. Alexander-Miller MA, Leggatt GR, Berzofsky JA (1996) Selective expansion of high or low avidity cytotoxic T lymphocytes and efficacy for adoptive immunotherapy. Proc Natl Acad Sci U S A 93:4102–4107
38. Gallimore A, Dumrese T, Hengartner H, Zinkernagel RM, Rammensee HG (1998) Protective immunity does not correlate with the hierarchy of virus-specific cytotoxic T cell responses to naturally processed peptides. J Exp Med 187:1647–1657
39. Almeida JR, Price DA, Papagno L, Arkoub ZA, Sauce D, Bornstein E, Asher TE, Samri A, Schnuriger A, Theodorou I et al (2007) Superior control of HIV-1 replication by CD8+ T cells is reflected by their avidity, polyfunctionality, and clonal turnover. J Exp Med 204(10):2473–2485

40. Almeida JR, Sauce D, Price DA, Papagno L, Shin SY, Moris A, Larsen M, Pancino G, Douek DC, Autran B et al (2009) Antigen sensitivity is a major determinant of CD8+ T-cell polyfunctionality and HIV-suppressive activity. Blood 113(25):6351–6360
41. Appay V, Iglesias MC (2011) Antigen sensitivity and T-cell receptor avidity as critical determinants of HIV control. Curr Opin HIV AIDS 6(3):157–162
42. Yee C, Savage PA, Lee PP, Davis MM, Greenberg PD (1999) Isolation of high avidity melanoma-reactive CTL from heterogeneous populations using peptide-MHC tetramers. J Immunol 162(4):2227–2234
43. Zeh HJ 3rd, Perry-Lalley D, Dudley ME, Rosenberg SA, Yang JC (1999) High avidity CTLs for two self-antigens demonstrate superior in vitro and in vivo antitumor efficacy. J Immunol 162(2):989–994
44. Derby MA, Alexander-Miller MA, Tse R, Berzofsky JA (2001) High avidity CTL exploit two complementary mechanisms to provide better protection against viral infection than low avidity CTL. J Immunol 166:1690–1697
45. Gross DA, Graff-Dubois S, Opolon P, Cornet S, Alves P, Bennaceur-Griscelli A, Faure O, Guillaume P, Firat H, Chouaib S et al (2004) High vaccination efficiency of low-affinity epitopes in antitumor immunotherapy. J Clin Invest 113(3):425–433
46. Hodge JW, Sabzevari H, Yafal AG, Gritz L, Lorenz MGO, Schlom J (1999) A triad of costimulatory molecules synergize to amplify T-cell activation. Cancer Res 59:5800–5807
47. Oh S, Hodge JW, Ahlers JD, Burke DS, Schlom J, Berzofsky JA (2003) Selective induction of high avidity CTL by altering the balance of signals from antigen presenting cells. J Immunol 170:2523–2530
48. Hodge JW, Chakraborty M, Kudo-Saito C, Garnett CT, Schlom J (2005) Multiple costimulatory modalities enhance CTL avidity. J Immunol 174(10):5994–6004
49. Waldmann TA (2006) The biology of interleukin-2 and interleukin-15: implications for cancer therapy and vaccine design. Nat Rev Immunol 6(8):595–601
50. Dubois S, Mariner J, Waldmann TA, Tagaya Y (2002) IL-15Ralpha recycles and presents IL-15 In trans to neighboring cells. Immunity 17(5):537–547
51. Bergamaschi C, Rosati M, Jalah R, Valentin A, Kulkarni V, Alicea C, Zhang GM, Patel V, Felber BK, Pavlakis GN (2008) Intracellular interaction of interleukin-15 with its receptor alpha during production leads to mutual stabilization and increased bioactivity. J Biol Chem 283(7):4189–4199
52. Oh S, Berzofsky JA, Burke DS, Waldmann TA, Perera LP (2003) Coadministration of HIV vaccine vectors with vaccinia viruses expressing IL-15 but not IL-2 induces long-lasting cellular immunity. Proc Natl Acad Sci U S A 100(6):3392–3397
53. Oh S, Perera LP, Burke DS, Waldmann TA, Berzofsky JA (2004) IL-15/IL-15R alpha-mediated avidity maturation of memory $CD8^{+}$ T cells. Proc Natl Acad Sci U S A 101(42):15154–15159
54. Kutzler MA, Robinson TM, Chattergoon MA, Choo DK, Choo AY, Choe PY, Ramanathan MP, Parkinson R, Kudchodkar S, Tamura Y et al (2005) Coimmunization with an optimized IL-15 plasmid results in enhanced function and longevity of CD8 T cells that are partially independent of CD4 T cell help. J Immunol 175(1):112–123
55. Perera PL, Mosca J, Oh S, Berzofsky JA, Waldmann TA (2007) Development of smallpox vaccine candidates with integrated IL-15 that demonstrate superior immunogenicity, efficacy and safety in mice. J Virol 81:8774–8783
56. Boyer JD, Robinson TM, Kutzler MA, Vansant G, Hokey DA, Kumar S, Parkinson R, Wu L, Sidhu MK, Pavlakis GN et al (2007) Protection against simian/human immunodeficiency virus (SHIV) 89.6P in macaques after coimmunization with SHIV antigen and IL-15 plasmid. Proc Natl Acad Sci U S A 104(47):18648–18653
57. Calarota SA, Dai A, Trocio JN, Weiner DB, Lori F, Lisziewicz J (2008) IL-15 as memory T-cell adjuvant for topical HIV-1 DermaVir vaccine. Vaccine 26(40):5188–5195
58. Oh S, Perera LP, Terabe M, Ni L, Waldmann TA, Berzofsky JA (2008) IL-15 as a mediator of CD4+ help for CD8+ T cell longevity and avoidance of TRAIL-mediated apoptosis. Proc Natl Acad Sci U S A 105(13):5201–5206

59. Sui Y, Zhu Q, Gagnon S, Dzutsev A, Terabe M, Vaccari M, Venzon D, Klinman D, Strober W, Kelsall B et al (2010) Innate and adaptive immune correlates of vaccine and adjuvant-induced control of mucosal transmission of SIV in macaques. Proc Natl Acad Sci U S A 107(21):9843–9848
60. Wherry EJ, Becker TC, Boone D, Kaja MK, Ma A, Ahmed R (2002) Homeostatic proliferation but not the generation of virus specific memory CD8 T cells is impaired in the absence of IL-15 or IL-15Ralpha. Adv Exp Med Biol 512:165–175
61. Janssen EM, Droin NM, Lemmens EE, Pinkoski MJ, Bensinger SJ, Ehst BD, Griffith TS, Green DR, Schoenberger SP (2005) CD4+ T-cell help controls CD8+ T-cell memory via TRAIL-mediated activation-induced cell death. Nature 434(7029):88–93
62. Ferrone CR, Perales MA, Goldberg SM, Somberg CJ, Hirschhorn-Cymerman D, Gregor PD, Turk MJ, Ramirez-Montagut T, Gold JS, Houghton AN et al (2006) Adjuvanticity of plasmid DNA encoding cytokines fused to immunoglobulin Fc domains. Clin Cancer Res 12(18):5511–5519
63. Su B, Wang J, Zhao G, Wang X, Li J, Wang B (2012) Sequential administration of cytokine genes to enhance cellular immune responses and CD4 (+) T memory cells during DNA vaccination. Hum Vaccin Immunother 8(11):1659–1667
64. Zhang W, Dong SF, Sun SH, Wang Y, Li GD, Qu D (2006) Coimmunization with IL-15 plasmid enhances the longevity of CD8 T cells induced by DNA encoding hepatitis B virus core antigen. World J Gastroenterol 12(29):4727–4735
65. Valentin A, von Gegerfelt A, Rosati M, Miteloudis G, Alicea C, Bergamaschi C, Jalah R, Patel V, Khan AS, Draghia-Akli R et al (2010) Repeated DNA therapeutic vaccination of chronically SIV-infected macaques provides additional virological benefit. Vaccine 28(8):1962–1974
66. Li S, Qi X, Gao Y, Hao Y, Cui L, Ruan L, He W (2010) IL-15 increases the frequency of effector memory CD8+ T cells in rhesus monkeys immunized with HIV vaccine. Cell Mol Immunol 7(6):491–494
67. Hu XD, Chen ST, Li JY, Yu DH, Yi Z, Cai H (2010) An IL-15 adjuvant enhances the efficacy of a combined DNA vaccine against Brucella by increasing the CD8+ cytotoxic T cell response. Vaccine 28(12):2408–2415
68. Dubie RA, Maksaereekul S, Shacklett BL, Lemongello D, Cole KS, Villinger F, Blozis SA, Luciw PA, Sparger EE (2009) Co-immunization with IL-15 enhances cellular immune responses induced by a vif-deleted simian immunodeficiency virus proviral DNA vaccine and confers partial protection against vaginal challenge with SIVmac251. Virology 386(1):109–121
69. Yang Z, Wang L, Wang H, Shang X, Niu W, Li J, Wu Y (2008) A novel mimovirus vaccine containing survivin epitope with adjuvant IL-15 induces long-lasting cellular immunity and high antitumor efficiency. Mol Immunol 45(6):1674–1681
70. Demberg T, Boyer JD, Malkevich N, Patterson LJ, Venzon D, Summers EL, Kalisz I, Kalyanaraman VS, Lee EM, Weiner DB et al (2008) Sequential priming with simian immunodeficiency virus (SIV) DNA vaccines, with or without encoded cytokines, and a replicating adenovirus-SIV recombinant followed by protein boosting does not control a pathogenic SIVmac251 mucosal challenge. J Virol 82(21):10911–10921
71. Yin J, Dai A, Laddy DJ, Yan J, Arango T, Khan AS, Lewis MG, Andersen H, Kutzler MA, Draghia-Akli R et al (2009) High dose of plasmid IL-15 inhibits immune responses in an influenza non-human primates immunogenicity model. Virology 393(1):49–55
72. Kraynyak KA, Kutzler MA, Cisper NJ, Laddy DJ, Morrow MP, Waldmann TA, Weiner DB (2009) Plasmid-encoded interleukin-15 receptor alpha enhances specific immune responses induced by a DNA vaccine in vivo. Hum Gene Ther 20(10):1143–1156
73. Kalams SA, Parker S, Jin X, Elizaga M, Metch B, Wang M, Hural J, Lubeck M, Eldridge J, Cardinali M et al (2012) Safety and immunogenicity of an HIV-1 gag DNA vaccine with or without IL-12 and/or IL-15 plasmid cytokine adjuvant in healthy, HIV-1 uninfected adults. PLoS One 7(1), e29231
74. O'Neill LA (2008) The interleukin-1 receptor/Toll-like receptor superfamily: 10 years of progress. Immunol Rev 226:10–18

75. Wille-Reece U, Flynn BJ, Lore K, Koup RA, Miles AP, Saul A, Kedl RM, Mattapallil JJ, Weiss WR, Roederer M et al (2006) Toll-like receptor agonists influence the magnitude and quality of memory T cell responses after prime-boost immunization in nonhuman primates. J Exp Med 203(5):1249–1258
76. Celis E (2007) Toll-like receptor ligands energize peptide vaccines through multiple paths. Cancer Res 67(17):7945–7947
77. Pulendran B (2004) Modulating vaccine responses with dendritic cells and Toll-like receptors. Immunol Rev 199:227–250
78. Napolitani G, Rinaldi A, Bertoni F, Sallusto F, Lanzavecchia A (2005) Selected Toll-like receptor agonist combinations synergistically trigger a T helper type 1-polarizing program in dendritic cells. Nat Immunol 6(8):769–776
79. Zhu Q, Egelston C, Vivekanandhan A, Uematsu S, Akira S, Klinman DM, Belyakov IM, Berzofsky JA (2008) Toll-like receptor ligands synergize through distinct dendritic cell pathways to induce T cell responses: Implications for vaccines. Proc Natl Acad Sci U S A 105(42):16260–16265
80. Zhu Q, Egelston C, Gagnon S, Sui Y, Belyakov IM, Klinman DM, Berzofsky JA (2010) Using 3 TLR ligands as a combination adjuvant induces qualitative changes in T cell responses needed for antiviral protection in mice. J Clin Invest 120(2):607–616
81. Terabe M, Berzofsky JA (2014) The immunoregulatory role of type I and type II NKT cells in cancer and other diseases. Cancer Immunol Immunother 63(3):199–213
82. Fujii S, Shimizu K, Okamoto Y, Kunii N, Nakayama T, Motohashi S, Taniguchi M (2013) NKT cells as an ideal anti-tumor immunotherapeutic. Front Immunol 4:409
83. Terabe M, Berzofsky JA (2008) The role of NKT cells in tumor immunity. Adv Cancer Res 101:277–348
84. Bendelac A, Savage PB, Teyton L (2007) The biology of NKT cells. Annu Rev Immunol 25:297–336
85. Kronenberg M (2005) Toward an understanding of NKT cell biology: progress and paradoxes. Annu Rev Immunol 23:877–900
86. Fujii S, Shimizu K, Smith C, Bonifaz L, Steinman RM (2003) Activation of natural killer T cells by alpha-galactosylceramide rapidly induces the full maturation of dendritic cells in vivo and thereby acts as an adjuvant for combined CD4 and CD8 T cell immunity to a coadministered protein. J Exp Med 198(2):267–279
87. Fujii S, Liu K, Smith C, Bonito AJ, Steinman RM (2004) The linkage of innate to adaptive immunity via maturing dendritic cells in vivo requires CD40 ligation in addition to antigen presentation and CD80/86 costimulation. J Exp Med 199(12):1607–1618
88. Gonzalez-Aseguinolaza G, Van Kaer L, Bergmann CC, Wilson JM, Schmieg J, Kronenberg M, Nakayama T, Taniguchi M, Koezuka Y, Tsuji M (2002) Natural killer T cell ligand alpha-galactosylceramide enhances protective immunity induced by malaria vaccines. J Exp Med 195(5):617–624
89. Hermans IF, Silk JD, Gileadi U, Salio M, Mathew B, Ritter G, Schmidt R, Harris AL, Old L, Cerundolo V (2003) NKT cells enhance CD4+ and CD8+ T cell responses to soluble antigen in vivo through direct interaction with dendritic cells. J Immunol 171(10):5140–5147
90. Silk JD, Hermans IF, Gileadi U, Chong TW, Shepherd D, Salio M, Mathew B, Schmidt RR, Lunt SJ, Williams KJ et al (2004) Utilizing the adjuvant properties of CD1d-dependent NK T cells in T cell-mediated immunotherapy. J Clin Invest 114(12):1800–1811
91. Li X, Fujio M, Imamura M, Wu D, Vasan S, Wong CH, Ho DD, Tsuji M (2010) Design of a potent CD1d-binding NKT cell ligand as a vaccine adjuvant. Proc Natl Acad Sci U S A 107(29):13010–13015
92. Sakaguchi S (2004) Naturally arising CD4+ regulatory t cells for immunologic self-tolerance and negative control of immune responses. Annu Rev Immunol 22:531–562
93. Gabrilovich DI, Nagaraj S (2009) Myeloid-derived suppressor cells as regulators of the immune system. Nat Rev Immunol 9(3):162–174

94. Pere H, Tanchot C, Bayry J, Terme M, Taieb J, Badoual C, Adotevi O, Merillon N, Marcheteau E, Quillien VR et al (2012) Comprehensive analysis of current approaches to inhibit regulatory T cells in cancer. Oncoimmunology 1(3):326–333
95. Onizuka S, Tawara I, Shimizu J, Sakaguchi S, Fujita T, Nakayama E (1999) Tumor rejection by in vivo administration of anti-CD25 (interleukin-2 receptor alpha) monoclonal antibody. Cancer Res 59(13):3128–3133
96. Izhak L, Ambrosino E, Kato S, Parish ST, O'Konek JJ, Weber H, Xia Z, Venzon D, Berzofsky JA, Terabe M (2013) Delicate balance among three types of T cells in concurrent regulation of tumor immunity. Cancer Res 73(5):1514–1523
97. Izhak L, Berzofsky JA, Terabe M (2013) Balance is a key for happiness. Oncoimmunology 2(5), e24211
98. Jiang S, Game DS, Davies D, Lombardi G, Lechler RI (2005) Activated CD1d-restricted natural killer T cells secrete IL-2: innate help for CD4+CD25+ regulatory T cells? Eur J Immunol 35(4):1193–1200
99. Ambrosino E, Terabe M, Halder RC, Peng J, Takaku S, Miyake S, Yamamura T, Kumar V, Berzofsky JA (2007) Cross-regulation between type I and type II NKT cells in regulating tumor immunity: a new immunoregulatory axis. J Immunol 179:5126–5136
100. Ambrosino E, Berzofsky JA, Terabe M (2008) Regulation of tumor immunity: the role of NKT cells. Expert Opin Biol Ther 8(6):725–734
101. Berzofsky JA, Terabe M (2008) NKT cells in tumor immunity: opposing subsets define a new immunoregulatory axis. J Immunol 180(6):3627–3635
102. Tahir SM, Cheng O, Shaulov A, Koezuka Y, Bubley GJ, Wilson SB, Balk SP, Exley MA (2001) Loss of IFN-gamma production by invariant NK T cells in advanced cancer. J Immunol 167(7):4046–4050
103. Yanagisawa K, Seino K, Ishikawa Y, Nozue M, Todoroki T, Fukao K (2002) Impaired proliferative response of V alpha 24 NKT cells from cancer patients against alpha-galactosylceramide. J Immunol 168(12):6494–6499
104. Giaccone G, Punt CJ, Ando Y, Ruijter R, Nishi N, Peters M, von Blomberg BM, Scheper RJ, van der Vliet HJ, van den Eertwegh AJ et al (2002) A phase I study of the natural killer T-cell ligand alpha-galactosylceramide (KRN7000) in patients with solid tumors. Clin Cancer Res 8(12):3702–3709
105. Dhodapkar MV, Geller MD, Chang DH, Shimizu K, Fujii S, Dhodapkar KM, Krasovsky J (2003) A reversible defect in natural killer T cell function characterizes the progression of premalignant to malignant multiple myeloma. J Exp Med 197(12):1667–1676
106. Molling JW, Kolgen W, van der Vliet HJ, Boomsma MF, Kruizenga H, Smorenburg CH, Molenkamp BG, Langendijk JA, Leemans CR, von Blomberg BM et al (2005) Peripheral blood IFN-gamma-secreting Valpha24+Vbeta11+ NKT cell numbers are decreased in cancer patients independent of tumor type or tumor load. Int J Cancer 116(1):87–93
107. Molling JW, Langius JA, Langendijk JA, Leemans CR, Bontkes HJ, van der Vliet HJ, von Blomberg BM, Scheper RJ, van den Eertwegh AJ (2007) Low levels of circulating invariant natural killer T cells predict poor clinical outcome in patients with head and neck squamous cell carcinoma. J Clin Oncol 25(7):862–868
108. Motohashi S, Nakayama T (2008) Clinical applications of natural killer T cell-based immunotherapy for cancer. Cancer Sci 99(4):638–645
109. Terabe M, Berzofsky JA (2004) Immunoregulatory T cells in tumor immunity. Curr Opin Immunol 16(2):157–162
110. Pardoll D, Allison J (2004) Cancer immunotherapy: breaking the barriers to harvest the crop. Nat Med 10(9):887–892
111. Geng H, Zhang GM, Xiao H, Yuan Y, Li D, Zhang H, Qiu H, He YF, Feng ZH (2006) HSP70 vaccine in combination with gene therapy with plasmid DNA encoding sPD-1 overcomes immune resistance and suppresses the progression of pulmonary metastatic melanoma. Int J Cancer 118(11):2657–2664
112. Sutmuller RPM, Van Duivenvoorde LM, Van Elsas A, Schumacher TNM, Wildenberg ME, Allison JP, Toes REM, Offringa R, Melief CJM (2001) Synergism of cytotoxic T

lymphocyte-associated antigen 4 blockade and depletion of CD25+ regulatory T cells in antitumor therapy reveals alternative cytotoxic T lymphocyte responses. J Exp Med 194(6):823–832
113. Hurwitz AA, Yu TF-Y, Leach DR, Allison JP (1998) CTLA-4 blockade synergizes with tumor-derived granulocyte-macrophage colony-stimulating factor for treatment of an experimental mammary carcinoma. Proc Natl Acad Sci 95:10067–10071
114. Curran MA, Allison JP (2009) Tumor vaccines expressing flt3 ligand synergize with ctla-4 blockade to reject preimplanted tumors. Cancer Res 69(19):7747–7755
115. Jochems C, Tucker JA, Tsang KY, Madan RA, Dahut WL, Liewehr DJ, Steinberg SM, Gulley JL, Schlom J (2014) A combination trial of vaccine plus ipilimumab in metastatic castration-resistant prostate cancer patients: immune correlates. Cancer Immunol Immunother 63(4):407–418
116. Curran MA, Montalvo W, Yagita H, Allison JP (2010) PD-1 and CTLA-4 combination blockade expands infiltrating T cells and reduces regulatory T and myeloid cells within B16 melanoma tumors. Proc Natl Acad Sci U S A 107(9):4275–4280
117. Letterio JJ, Roberts AB (1998) Regulation of immune responses by TGF-beta. Annu Rev Immunol 16:137–161
118. Peng Y, Laouar Y, Li MO, Green EA, Flavell RA (2004) TGF-{beta} regulates in vivo expansion of Foxp3-expressing CD4+CD25+ regulatory T cells responsible for protection against diabetes. Proc Natl Acad Sci U S A 101(13):4572–4577
119. Marie JC, Letterio JJ, Gavin M, Rudensky AY (2005) TGF-{beta}1 maintains suppressor function and Foxp3 expression in CD4+CD25+ regulatory T cells. J Exp Med 201(7): 1061–1067
120. McGeachy MJ, Bak-Jensen KS, Chen Y, Tato CM, Blumenschein W, McClanahan T, Cua DJ (2007) TGF-beta and IL-6 drive the production of IL-17 and IL-10 by T cells and restrain T(H)-17 cell-mediated pathology. Nat Immunol 8(12):1390–1397
121. Korn T, Bettelli E, Oukka M, Kuchroo VK (2009) IL-17 and Th17 Cells. Annu Rev Immunol 27:485–517
122. Chen W, Wahl SM (2003) TGF-beta: the missing link in CD4+CD25+ regulatory T cell-mediated immunosuppression. Cytokine Growth Factor Rev 14(2):85–89
123. Nakamura K, Kitani A, Fuss I, Pedersen A, Harada N, Nawata H, Strober W (2004) TGF-beta1 plays an important role in the mechanism of CD4+CD25+ regulatory T cell activity in both humans and mice. J Immunol 172(2):834–842
124. Kim S, Buchlis G, Fridlender ZG, Sun J, Kapoor V, Cheng G, Haas A, Cheung HK, Zhang X, Corbley M et al (2008) Systemic blockade of transforming growth factor-beta signaling augments the efficacy of immunogene therapy. Cancer Res 68(24):10247–10256
125. Gunn GR, Zubair A, Peters C, Pan ZK, Wu TC, Paterson Y (2001) Two Listeria monocytogenes vaccine vectors that express different molecular forms of human papilloma virus-16 (HPV-16) E7 induce qualitatively different T cell immunity that correlates with their ability to induce regression of established tumors immortalized by HPV-16. J Immunol 167(11):6471–6479
126. Terabe M, Ambrosino E, Takaku S, O'Konek JJ, Venzon D, Lonning S, McPherson JM, Berzofsky JA (2009) Synergistic enhancement of CD8+ T cell-mediated tumor vaccine efficacy by an anti-transforming growth factor-beta monoclonal antibody. Clin Cancer Res 15(21):6560–6569
127. Takaku S, Terabe M, Ambrosino E, Peng J, Lonning S, McPherson JM, Berzofsky JA (2010) Blockade of TGF-beta enhances tumor vaccine efficacy mediated by CD8(+) T cells. Int J Cancer 126(7):1666–1674
128. Ueda R, Fujita M, Zhu X, Sasaki K, Kastenhuber ER, Kohanbash G, McDonald HA, Harper J, Lonning S, Okada H (2009) Systemic inhibition of transforming growth factor-beta in glioma-bearing mice improves the therapeutic efficacy of glioma-associated antigen peptide vaccines. Clin Cancer Res 15(21):6551–6559
129. Terabe M, Matsui S, Park J-M, Mamura M, Noben-Trauth N, Donaldson DD, Chen W, Wahl SM, Ledbetter S, Pratt B et al (2003) Transforming growth factor-b production and myeloid

cells are an effector mechanism through which CD1d-restricted T cells block cytotoxic T lymphocyte-mediated tumor immunosurveillance: abrogation prevents tumor recurrence. J Exp Med 198(11):1741–1752
130. Park JM, Terabe M, van den Broeke LT, Donaldson DD, Berzofsky JA (2004) Unmasking immunosurveillance against a syngeneic colon cancer by elimination of CD4+ NKT regulatory cells and IL-13. Int J Cancer 114(1):80–87
131. Terabe M, Berzofsky JA (2007) NKT cells in immunoregulation of tumor immunity: a new immunoregulatory axis. Trends Immunol 28(11):491–496
132. Fichtner-Feigl S, Terabe M, Kitani A, Young CA, Fuss I, Geissler EK, Schlitt HJ, Berzofsky JA, Strober W (2008) Restoration of tumor immunosurveillance via targeting of interleukin-13 receptor-alpha 2. Cancer Res 68(9):3467–3475
133. Gallina G, Dolcetti L, Serafini P, De Santo C, Marigo I, Colombo MP, Basso G, Brombacher F, Borrello I, Zanovello P et al (2006) Tumors induce a subset of inflammatory monocytes with immunosuppressive activity on CD8+ T cells. J Clin Invest 116(10):2777–2790
134. Sinha P, Clements VK, Ostrand-Rosenberg S (2005) Interleukin-13-regulated M2 macrophages in combination with myeloid suppressor cells block immune surveillance against metastasis. Cancer Res 65(24):11743–11751
135. Mellman I, Coukos G, Dranoff G (2011) Cancer immunotherapy comes of age. Nature 480(7378):480–489
136. Drake CG, Jaffee E, Pardoll DM (2006) Mechanisms of immune evasion by tumors. Adv Immunol 90:51–81
137. Friedman E, Gold LI, Klimstra D, Zeng ZS, Winawer S, Cohen A (1995) High levels of transforming growth factor beta 1 correlate with disease progression in human colon cancer. Cancer Epidemiol Biomarkers Prev 4(5):549–554
138. Krasagakis K, Tholke D, Farthmann B, Eberle J, Mansmann U, Orfanos CE (1998) Elevated plasma levels of transforming growth factor (TGF)-beta1 and TGF-beta2 in patients with disseminated malignant melanoma. Br J Cancer 77(9):1492–1494
139. Nam JS, Terabe M, Kang MJ, Chae H, Voong N, Yang YA, Laurence A, Michalowska A, Mamura M, Lonning S et al (2008) Transforming growth factor {beta} subverts the immune system into directly promoting tumor growth through interleukin-17. Cancer Res 68(10):3915–3923
140. Nam JS, Terabe M, Mamura M, Kang MJ, Chae H, Stuelten C, Kohn E, Tang B, Sabzevari H, Anver MR et al (2008) An anti-transforming growth factor {beta} antibody suppresses metastasis via cooperative effects on multiple cell compartments. Cancer Res 68(10): 3835–3843
141. Morris JC, Tan AR, Olencki TE, Shapiro GI, Dezube BJ, Reiss M, Hsu FJ, Berzofsky JA, Lawrence DP (2014) Phase I study of GC1008 (fresolimumab): a human anti-transforming growth factor-beta (TGFbeta) monoclonal antibody in patients with advanced malignant melanoma or renal cell carcinoma. PLoS One 9(3), e90353
142. Yang JC, Hughes M, Kammula U, Royal R, Sherry RM, Topalian SL, Suri KB, Levy C, Allen T, Mavroukakis S et al (2007) Ipilimumab (anti-CTLA4 antibody) causes regression of metastatic renal cell cancer associated with enteritis and hypophysitis. J Immunother 30(8):825–830
143. Hodi FS, O'Day SJ, McDermott DF, Weber RW, Sosman JA, Haanen JB, Gonzalez R, Robert C, Schadendorf D, Hassel JC et al (2010) Improved survival with ipilimumab in patients with metastatic melanoma. N Engl J Med 363(8):711–723
144. Robert C, Thomas L, Bondarenko I, O'Day S, Weber J, Garbe C, Lebbe C, Baurain JF, Testori A, Grob JJ et al (2011) Ipilimumab plus dacarbazine for previously untreated metastatic melanoma. N Engl J Med 364(26):2517–2526
145. Topalian SL, Hodi FS, Brahmer JR, Gettinger SN, Smith DC, McDermott DF, Powderly JD, Carvajal RD, Sosman JA, Atkins MB et al (2012) Safety, activity, and immune correlates of anti-PD-1 antibody in cancer. N Engl J Med 366(26):2443–2454

146. Selby M, Engelhardt J, Lu L, Quigley M, Wang C, Chen B, Korman AJ (2013) Antitumor activity of concurrent blockade of immune checkpoint molecules CTLA-4 and PD-1 in preclinical models. J Clin Oncol 31(Suppl):3061
147. Wolchok JD, Kluger H, Callahan MK, Postow MA, Rizvi NA, Lesokhin AM, Segal NH, Ariyan CE, Gordon RA, Reed K et al (2013) Nivolumab plus ipilimumab in advanced melanoma. N Engl J Med 369:122–133
148. Madan RA, Mohebtash M, Arlen PM, Vergati M, Rauckhorst M, Steinberg SM, Tsang KY, Poole DJ, Parnes HL, Wright JJ et al (2012) Ipilimumab and a poxviral vaccine targeting prostate-specific antigen in metastatic castration-resistant prostate cancer: a phase 1 dose-escalation trial. Lancet Oncol 13(5):501–508
149. Le DT, Lutz E, Uram JN, Sugar EA, Onners B, Solt S, Zheng L, Diaz LA Jr, Donehower RC, Jaffee EM et al (2013) Evaluation of ipilimumab in combination with allogeneic pancreatic tumor cells transfected with a GM-CSF gene in previously treated pancreatic cancer. J Immunother 36(7):382–389
150. Gibney GT, Kudchadkar RR, DeConti RC, Thebeau MS, Czupryn MP, Tetteh LF, Eysmans C, Richards A, Schell MJ, Fisher K et al (2015) Safety, correlative markers and clinical results of adjuvant nivolumab in combination with vaccine in resected high-risk metastatic melanoma. Clin Cancer Res 21:712–720
151. Machiels JP, Reilly RT, Emens LA, Ercolini AM, Lei RY, Weintraub D, Okoye FI, Jaffee EM (2001) Cyclophosphamide, doxorubicin, and paclitaxel enhance the antitumor immune response of granulocyte/macrophage-colony stimulating factor-secreting whole-cell vaccines in HER-2/neu tolerized mice. Cancer Res 61(9):3689–3697
152. Wada S, Yoshimura K, Hipkiss EL, Harris TJ, Yen HR, Goldberg MV, Grosso JF, Getnet D, Demarzo AM, Netto GJ et al (2009) Cyclophosphamide augments antitumor immunity: studies in an autochthonous prostate cancer model. Cancer Res 69(10):4309–4318
153. Chen C, Hou J, Lin Z, Yao M, Jiang R, Wang Y, Gao Y, Shao Q, Deng L, Chen Y et al (2013) A bystander cell-based GM-CSF secreting vaccine synergized with a low dose of cyclophosphamide presents therapeutic immune responses against murine hepatocellular carcinoma. Cell Mol Immunol 10(4):349–359
154. Peng S, Lyford-Pike S, Akpeng B, Wu A, Hung CF, Hannaman D, Saunders JR, Wu TC, Pai SI (2013) Low-dose cyclophosphamide administered as daily or single dose enhances the antitumor effects of a therapeutic HPV vaccine. Cancer Immunol Immunother 62(1):171–182
155. Slingluff CL Jr, Petroni GR, Chianese-Bullock KA, Smolkin ME, Ross MI, Haas NB, von Mehren M, Grosh WW (2011) Randomized multicenter trial of the effects of melanoma-associated helper peptides and cyclophosphamide on the immunogenicity of a multipeptide melanoma vaccine. J Clin Oncol 29(21):2924–2932
156. Walter S, Weinschenk T, Stenzl A, Zdrojowy R, Pluzanska A, Szczylik C, Staehler M, Brugger W, Dietrich PY, Mendrzyk R et al (2012) Multipeptide immune response to cancer vaccine IMA901 after single-dose cyclophosphamide associates with longer patient survival. Nat Med 18:1254–1261
157. Le DT, Jaffee EM (2013) Harnessing immune responses in the tumor microenvironment: all signals needed. Clin Cancer Res 19(22):6061–6063
158. Ridolfi L, Petrini M, Granato AM, Gentilcore G, Simeone E, Ascierto PA, Pancisi E, Ancarani V, Fiammenghi L, Guidoboni M et al (2013) Low-dose temozolomide before dendritic-cell vaccination reduces (specifically) CD4+CD25++Foxp3+ regulatory T-cells in advanced melanoma patients. J Transl Med 11:135
159. Wood LV, Fojo T, Roberson D, Hughes MSB, Dahut W, Gulley JL, Madan RA, Arlen PM, Sabatino M, Stroncek DF et al. TARP Vaccination is associated with slowing in PSA velocity and decreasing tumor growth rates in patients with Stage D0 prostate cancer. Submitted., manuscript submitted)

Chapter 3
T Cell Fate in the Tumor Microenvironment

Joel Crespo, Ilona Kryczek, Theodore Welling, Shuang Wei, and Weiping Zou

Abstract Immune infiltration of tumors is a well-known phenomenon in cancer patients. Nevertheless, the tumor and immune cell coexisting scenario is often accompanied by efficient cancer progression indicating a compromised immune phenotype. As a matter of fact, it is well documented that a wealthy source of immune-suppressive molecular and cellular networks at the tumor site foster faulty T cell responses and ultimately redirect T cell fate and patient outcome. In this chapter, we summarize recent discoveries of the acquired dysfunctions of effector T cells in the tumor microenvironment due to the lack of proper activation networks and underlying enforcers regulating T cell unresponsiveness and their impact in new therapeutic development. Specifically, the advance in the Th17 balance, T cell stemness, and polyfunctionality of T cells which may improve clinic outcome.

Keywords Tumor • Stemness • Anergy • Exhaustion • Polyfunctionality • Memory T cell • Th17 • PD-1 • B7-H1 (PD-L1)

J. Crespo • I. Kryczek
Department of Surgery, University of Michigan School of Medicine, 1150 W. Medical Center Dr., Ann Arbor, Ann Arbor, MI 48109, USA

T. Welling • S. Wei
Department of Surgery, University of Michigan School of Medicine, 1150 W. Medical Center Dr., Ann Arbor, Ann Arbor, MI 48109, USA

W. Zou, M.D., Ph.D. (✉)
The University of Michigan Comprehensive Cancer Center, University of Michigan School of Medicine, Ann Arbor, MI USA
e-mail: wzou@med.umich.edu

P.A. Ascierto et al. (eds.), *Developments in T Cell Based Cancer Immunotherapies*, Cancer Drug Discovery and Development,
DOI 10.1007/978-3-319-21167-1_3

Introduction

Cancer immunotherapies aim to enhance the host's immune response against malignancies which have grown out of the body's control. Such therapies mainly target the immunoregulatory pathways involved in the suppression of effector T lymphocytes. The tumor microenvironment is considered to be rich in immunosuppressive pathways intrinsic to the tumor cells which shape the ultimate fate of T lymphocytes at the malignant site [1–4]. We will therefore focus on describing the immune imbalances that exist in cancer and its ultimate effect on T cell immune fate.

Immune-Suppressive Networks in Tumors

Over the years, effector T cell function and phenotype have been the focus of much study. In order to understand their immunological fate in cancer, studies have used diverse mouse models that allow the focus on the progression of cancer and the immune system. Further, tumor immunologists have explored interdisciplinary pathological similarity from infectious disease and chronic inflammation models to understand whether the same functional principles hold true in tumor settings. Supported by recent advances in tumor immunology, it is now well understood that the tumor microenvironment is rich in immune suppressive molecules which promote and maintain dysfunctional T cell responses [5].

Proper T cell activation relies on the co-engagement of T cell receptor-specific stimulatory antigen presented in the context of MHC molecules and engagement of CD28 by classic B7 stimulatory family members CD80 and CD86 (a.k.a. B7.1 and B7.2 respectively) [6–9]. Stimulatory co-engagement of these receptors is usually provided by activated antigen presenting cells (APC) which leads to the activation of intracellular signaling via transcription factors that promote T cell activation, mainly nuclear factor kappa-light-chain-enhancer of activated B cells (NF-κB) and Nuclear Factor of Activated T cells (NFAT) [8–12]. Following elimination of the pathogenic insult, effector T cells enter a contractory phase that is facilitated by various immune regulatory molecules expressed by APC and other immune cells that are found in their microenvironment.

T cell responses normally subside once the antigenic insult has been eliminated and is actively mediated by the expression of immune regulatory molecules from APC and their immediate effect on T cell responses. It also involves the actions of immune regulatory cells such as regulatory T cells which mediate potent suppression on effector T cells. Tumor malignancies are known to hijack these immune suppressive networks and maintain an active state of suppression by expression co-inhibitory molecules [1, 2]. We will first introduce important immune suppressive molecules which affect effector T cell functionality followed by how myeloid-derived suppressor cells affect T cell responses in the tumor.

B7-Family Members

CD80/CD86 and CTLA-4

Classic T cell activation requires concerted T cell receptor stimulation by specific antigen and CD28 ligation by co-stimulatory ligands CD80/CD86 from activated APC [6–9]. This stimulation is known to induce proliferation, cytokine expression, and overall function on the target T cell. Central to these functions are the activation of intracellular factors that lead to cell cycle progression and T cell activation (mTOR and Ras/MAPK signaling pathways) and the expression of genes associated with NF-κB and NFAT [13, 14]. On the other hand, T cells require inactivation in order for immune action to be resolved and homeostasis to be re-established. CTLA-4 is a co-inhibitory receptor expressed by T cells that is upregulated upon T cell activation. CTLA-4 promotes T cell inactivation upon interaction with CD80/86. The molecular pathways involved in CTLA-4 binding and its effect on the ultimate fate of T cells in the tumor microenvironment have been extensively studied. Based on the solid evidence of CTLA-4 inhibition in tumor microenvironment, therapeutic intervention by block CTLA-4 has achieved clinic benefits in promote tumor rejection and prolong patient survival. In some rare cases, patient could reach durable responsiveness and reach disease free status [2, 6, 15] (Fig. 3.1).

B7-H1 and PD-1

B7-H1 is part of the co-stimulatory B7-family protein superfamily and was first described to be involved in immune suppression by promoting IL-10 expression by T cells [16]. Since then, binding of B7-H1 by the T cell co-inhibitory receptor PD-1 is known to promote cellular death, expression of immune suppressive molecule IL-10 and loss of IL-2, T_{Reg} differentiation, anergy and exhaustion [1, 2, 6]. It is also hypothesized that other B7-H1 interacting proteins exist and have yet to be identified as blocking PD-1 does not completely relieve B7-H1-mediated immune suppression. Furthermore, B7-H1 is ubiquitously expressed by tumor cells and upon co-engagement promotes tumor survival [17]. B7-H1 became an interesting target in the tumor microenvironment and as such was tested in *in vivo* mouse models [18–20]. Blockade of B7-H1 in mice was observed to promote tumor regression and enhanced T cell anti-tumor responses [19, 20]. Moreover, recent clinical trials show how increased T cell functions are observed upon blockade of the B7-H1 pathway [21, 22]. We will focus on the establishment and maintaining of anergic and exhausted T cells below.

B7-H4

B7-H4 is another co-inhibitory receptor of the B7-family with known T cell suppressive functions [23]. Its expression by APC is induced by IL-6 and IL-10 [24]. B7-H4 has also been detected in various cancers such as ovarian, breast and lung

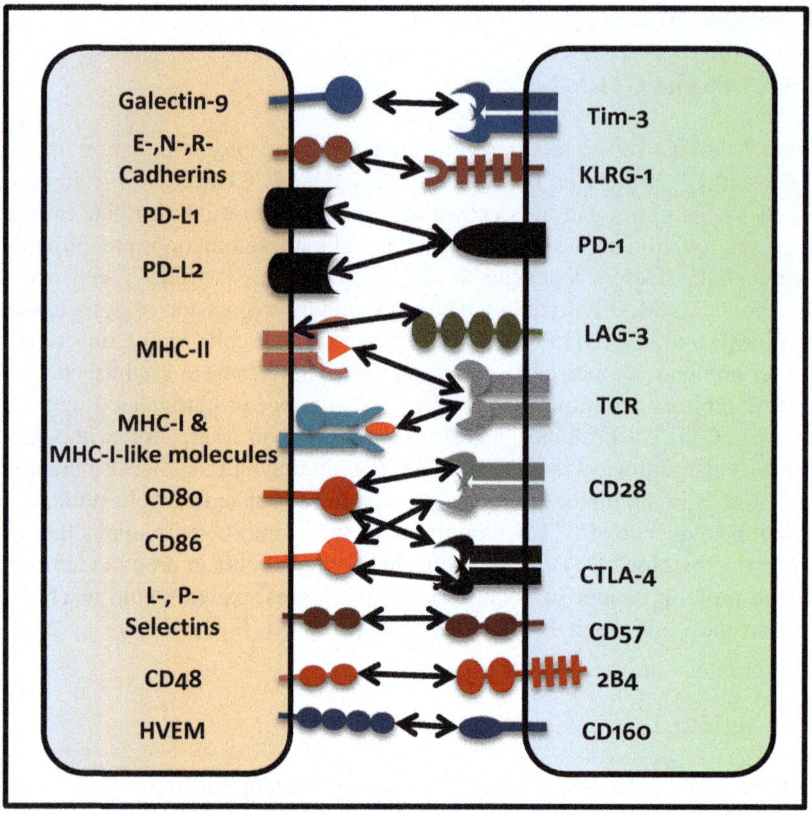

Fig. 3.1 Immunosuppressive molecules on T cells and their corresponding ligands. T cell activation relies on the efficient presentation of antigen in the context of T cell co-stimulatory CD28 engagement by CD80/86. Following activation, T cells gain expression of co-inhibitory molecules PD-1 and CTLA-4 which suppress and aims to resolve immune responses. Further expression of other co-inhibitory molecules such as Tim-3, KLRG-1, PD-1, LAG-3, CD57, 2B4, and CD160 promote T cell unresponsiveness. These immunoregulatory molecules are found in the tumor microenvironment and show phenotypic and therapeutic promise

cancers which suggests its importance in tumor progression through escape of immune surveillance [25–29]. Though the binding receptor for B7-H4 has not been discovered, contact between B7-H4-expressing APC with T cells induces cell cycle arrest and decreased cytokine expression on the target T cell [23, 24, 28, 30]. *In vivo* KO models of B7-H4 have also shown increased neutrophil infiltration to peripheral lymphoid tissues upon bacterial infection [31] and increased Th1 and Th17 responses in experimental autoimmune diseases [32]. B7-H4 has also been suggested to act intracellularly in tumor cells wherein it provides pro-survival signals [33]. We will also suggest B7-H4 is involved in T cell anergy.

Myeloid-Derived Suppressor Cells

Myeloid-derived suppressor cells (MDSC) are a population of immature myeloid cells and progenitor cells with potent immune suppressive activities. MDSC have been classically studied in the context of tumor progression. Identification of such cells in malignant tissues mainly rely on the common myeloid markers CD33, CD14, CD15 and HLA-DR for humans whereas in mice these have been classically defined using CD11b, Ly6G and Ly6C [34, 35]. Phenotypic and functional classifications have been discussed elsewhere [34–36]. This cell subset has been identified in various immunological contexts such as inflammation, infection, and cancer and are able to suppress T cell responses and macrophage functions [37–39]. MDSC are observed to be increased in frequency in peripheral blood of cancer patients [37]. It was hypothesized that this increase correlated with decreased DC function and consequently, T cell dysfunction. Further study into the functional impacts of this cell type showed co-injection of MDSC with tumor cells in mice led to increased tumor vascularization and decreased necrotic cell death [40]. It was also found that MDSC are able to promote tumor metastasis by inducing vascular endothelial growth factor-c through reactive oxygen (ROS) species production [34, 41]. Further, MDSC have been reported and implicated in various types of cancer including ovarian carcinoma, prostrate, hepatocellular carcinoma, and malignant melanoma which suggests their role in promoting tumor progression [42–45]. How MDSC are implicated in establishing T cell dysfunction will be considered in this chapter (Fig. 3.2).

Tumor-infiltrating effector T cell functions have been shown to be highly dampened due to MDSC activities in the tumor microenvironment in mice and human patients [34, 46]. MDSC suppressor functions on T cell may act in a cell-to-cell independent manner by modulating metabolites in the immediate environment and affecting T cell signaling pathways, and a cell-to-cell dependent manner through ligation of inhibitory receptors expressed by T cells in the tumor microenvironment. Cell-cell independent pathways mainly rely on MDSC depletion of L-arginine and production of ROS and peroxynitrite in the microenvironment. Specifically, MDSC-ARG-1 depletion of L-arginine from the environment promotes decreased cellular proliferation due to increased cell cycle regulators cyclin D3 and cyclin-dependent kinase 4 as well as decreased T cell responsiveness due to decreased expression of the CD3 ζ-chain [47, 48]. MDSC were also shown to expand regulatory T cells in cancers through increased production of arginase and IDO [49, 50]. Expansion of this highly immunosuppressive T cell subset allows for worse tumor outcome. Increased production of ROS by MDSC has also been shown to be involved in tumor progression and diminished effector T cell function as observed by increased ROS in tumor sites and anti-oxidants promote recovery of T cell functions *in vitro* [51–53]. Similarly, enhanced peroxynitrite-mediated oxidation promoted T cell unresponsiveness [54–57]. Furthermore, MDSC from human ovarian carcinomas and hepatocellular cancers have been shown to express various co-inhibitory molecules such as CTLA-4, B7-H1, B7-H4 [20, 24, 58, 59]. Therefore, contact between MDSC and T cells via inhibitory molecule interaction will then promote active suppression and will be discussed in depth in the context of various T cell dysfunction paradigms below.

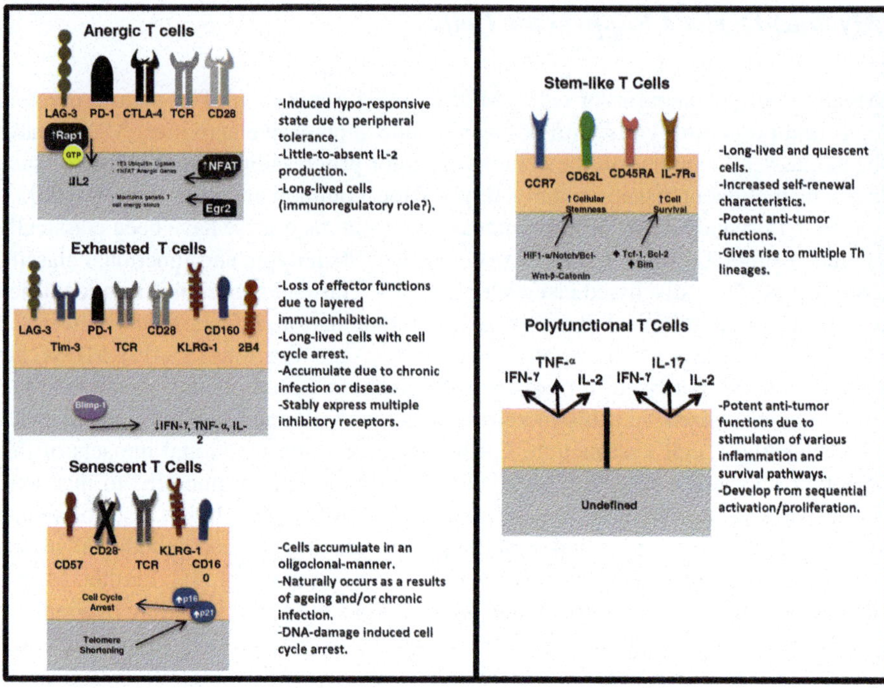

Fig. 3.2 Phenotypic and functional differences of T cell paradigms in tumors. T cell anergy is achieved upon poor stimulation of the cell or imbalanced high co-inhibitory molecule expression. Such cells are unresponsive to activation and have limited IL-2 expression. Exhausted T cells are the result of chronic activation which gain the expression of multiple co-inhibitory surface molecules. These cells lose the ability to express effector cytokines. Senescent T cells are terminally differentiated cells that remain unresponsive to activation and experience cell cycle arrest. These normally lose CD28 expression and may express multiple co-inhibitory receptors. Stem-like T cells are long-lived cells with the capacity to self-renew. These cells have potent anti-tumor functions and may give rise to distinct Th lineages. Polyfunctional T cells are potent effector cells with the capacity to promote anti-tumor responses by stimulating various arms of the immune system at the same time. Currently, their genetic, phenotypic, and differentiation is being studied

Immunological Fates of T Cells in Cancer

T Cell Anergy

T cellanergy has been described as an induced state of hyporesponsiveness naïve T cells fall to as part of suboptimal activation in the context of low co-stimulation and/or high co-inhibition. These cells remain unable to express high levels of classic effector cytokines IFN-γ and TNF-α, produce low levels of IL-2, and may become long-lived cells with low proliferation capabilities [60, 61]. A functional explanation has been proposed such that T cell anergy is a consequence of immune regulation and induced on peripheral naïve T cells to promote self-tolerance and protect the host from developing autoimmune diseases [61, 62]. Given its important role in regulating T cell responses in the periphery, there have been many studies which aim to understand the functional implications, phenotypic characteristics, and intrinsic enforcers this state requires in distinct disease settings including cancer. Here, we will consider central conditions that are required to promote T cell anergy in the tumor microenvironment and how these might affect effector T cell responses.

T cell anergy has been mainly defined based on its dysfunctional characteristics status since a specific cellular phenotype characterizing this induced state remains elusive. We will therefore consider those observations which describe T cell anergy is present in the context of tumors. Human tumors are surrounded by high concentration of immunosuppressive networks from both tumor cells and its associated antigen presenting cells [1–3, 18]. These make the ideal environment for T cells to be exposed to high co-inhibitory molecules which may lead to T cell anergy in the tumor. Specifically, T cell immune suppression status can be induced by interaction with co-inhibitory molecules B7-DC (CD273 or PD-L2), B7-H1 (CD274 or PD-L1), B7-H2 (CD275 or ICOS-L), B7-H3 (CD276), and B7-H4 (B7S1 or B7x), and very low/absent B7-1 (CD80) and B7-2 (CD86) co-stimulatory molecules being detected from tumor cells and infiltrating APC, including MDSC [2, 18, 20, 24, 58, 59, 63]. This active imbalance of low co-stimulation but high co-inhibition is conducive for T cells to achieve anergy. In theory, by tilting the expression patterns from co-inhibitory to co-stimulatory molecules in the tumor microenvironment, T cell anergy may be circumvented and T cell activation would occur. *In vivo* studies support such ideas. Specifically, B7-1 tumor transfections or functional blockade of B7 family members have shown reduced tumor growth which may result in tumor rejection [2, 18–20, 24, 64, 65]. Increasing B7-1 engagement in the tumor microenvironment led to increased anti-tumor responses. Furthermore, functional blockade of B7-H1 and B7-H4 co-inhibitory axis also showed increased tumor regression. There is further evidence that shows dysfunctional tumor-antigen-specific T cells are present in the tumor microenvironment which are actively being inhibited by MDSC networks of suppression discussed previously [40, 41, 51, 53, 56, 57]. Specifically, MDSC were shown to provide suboptimal T cell activation by ROS-mediated modifications to TCR of $CD8^+$ T cells [57]. MDSC are also a source of increased co-inhibitory molecules which disrupt proper T cell activation in the tumor microenvironment [20, 24, 58, 59]. Also, by transferring anti-tumor T cells into a lymphopenic host, T cell anergy may be reversed and cellular proliferation ensues and allows for effective tumor rejection *in vivo* [66].

The underlying cellular and molecular mechanisms which establish T cell anergy are currently being elucidated. T cells that are presented antigen in the context of suboptimal CD28 co-stimulation and/or high co-inhibition fall into anergic phenotypes as they have low IL-2 production and are arrested in the G1/S phase of their cell cycle [60, 61, 67]. Recent evidence points to T cell anergy being promoted by improper activation of mTOR and Ras/MAPK signaling pathways which lie directly downstream of TCR/CD28 engagement. It is thought that binding of TCR by MHC alone creates a Ca^{2+} imbalance in T cells and retention of active Rap-1 in the cytosol

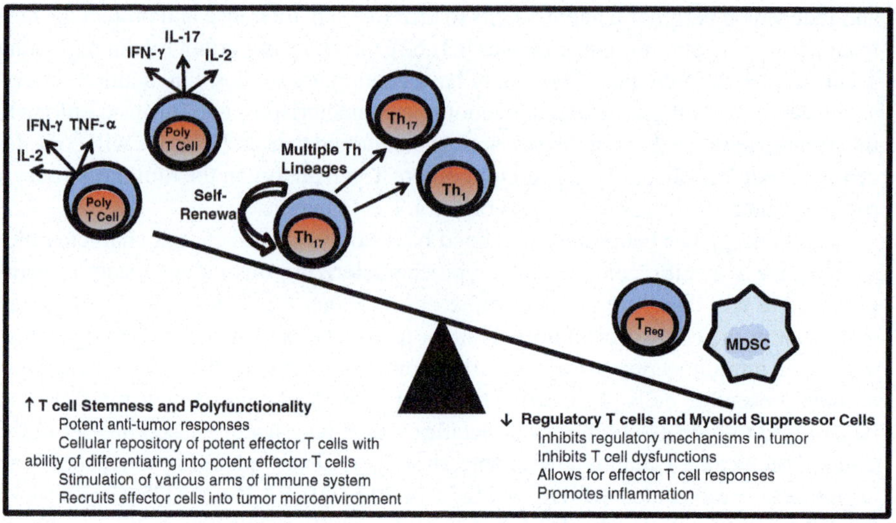

Fig. 3.3 Shifting immune imbalances in tumor. Immune suppression in the tumor microenvironment actively maintains T cell dysfunction in the tumor microenvironment. By targeting immune suppression mechanisms, anergic, exhausted, and senescent T cells may be rescued from their unresponsive state. This may be done by blockade of co-inhibitory molecules, as done for PD-1 and CTLA-4, and cellular depletion of immune suppressors like T_{Regs} and MDSC. Further, developing therapies and deeper understanding of stem-like and polyfunctional T cells will provide novel therapeutic value

which is normally shifted upon co-engagement of CD28 and activation of the Ras/MAPK pathways [61, 68, 69]. This imbalance reprograms these cells into an anergic state that is thought to be mediated by NFAT homodimer formation and transcription of anergy-inducing genes [70, 71]. Another anergy-inducing pathway may involve the E3 ubiquiting ligase family which affects PI3K, mTOR, and Ras/MAPK pathways and actively maintain anergy [70, 72, 73]. Current studies also have aimed at discerning the epigenetic factors which regulate anergy. IKAROS (through acetylation) and Sirt1 have been suggested to be involved in promoting T cell anergy [74–76]. Early growth response gene 2 (Egr2) has been suggested to be a crucial transcription factor which also regulates the T cell anergic state [77, 78]. Taken together, T cell anergy can be seen as the result of various factors synergistically promoting an active state of transcriptional silencing of effector genes due to improper TCR-coupled signal transductions in the cell [62] (Fig. 3.3).

To recapitulate, T cell anergy may be an operative dysfunction occurring in the tumor microenvironment that is initiated due to improper T cell activation. Given the lack of a specific surface mark which allows for the discreet study of this dysfunctional paradigm, it is yet incompletely understood how the anergic state is established. Furthermore, infiltrating MDSC may actively induce T cell anergy through expression of co-inhibitory molecules, affecting T cell metabolism, and increased ROS.

Exhausted T Cell

T cell exhaustion refers to effector T cells which have been strongly activated in the context of chronic inflammation such as cancer, autoimmune diseases, and chronic infection. Exhausted T cells are described as having decreased effector function as characterized by decreased cytokine expression and having developed resistance to reactivation [79]. It is thought that T cell exhaustion is actively enforced through the surface expression of various immune-suppressive enforcers which are currently being studied.

T cell exhaustion is thought to occur as a layered progressive process T cells fall due to repeated activation. Repeated activation of T cells promotes their acquisition of multiple inhibitory surface molecules as observed in persistent disease settings such as chronic infection and cancer [80–84]. This results in a highly repressed T cell. Initial mouse studies delineated PD-1 expression by $CD8^+$ T cells promotes functional exhaustion and as such was considered to be a marker for exhausted T cells [85]. In support of this, B7-H1 expression is observed by tumor cells, tumor-associated APC and MDSC which mediated immune suppression and its therapeutic blockade showed increased tumor rejection in both mouse models and recent clinical trials [20–22, 80, 86]. Furthermore, exhausted T cells have been identified in patients with melanoma, ovarian cancer and hepatocellular carcinoma (HCC) [20, 21, 87]. In studying $PD\text{-}1^+$ exhausted T cells, various immunosuppressive surface molecules have been found to associate with PD-1. Specifically, T cell immunoglobulin and mucin-domain containing molecule-3 (Tim-3), lymphocyte-activation gene (LAG-3), and the B and T-cell lymphocyte attenuator (BTLA, CD272), were co-expressed with PD-1 and highly correlated with immune dysfunction in patients with cancer [82, 83, 88–90]. In line with the idea of these cells being functionally exhausted, these T cells showed a significant decrease in IL-2, IFN-γ, and TNF-α expression and showed cell cycle arrest. Blockade of Tim-3 and PD-1 allowed these T cells to engage in cell cycle progression and rescue their effector cytokine expression and cytotoxicity, suggesting these surface molecules maintain T cell dysfunction [84, 90, 91]. Multiple other inhibitory receptors in T cell exhaustion may include PD-1, CTLA-4, CD160, 2B4 (CD244), BTLA, LAG-3, and Tim-3 [82, 89, 90, 92]. MDSC mechanisms in T cell exhaustion may rely in the engagement of co-inhibitory receptors by B7-H1 and B7-H4 which promotes a state of T cell dysfunction [20, 24, 58, 59]. MDSC may also promote T cell dysfunction through increased ROS production in the tumor and loss of T cell-derived cytokines. Nevertheless, it remains unanswered whether functional T cell exhaustion requires the co-expression of these suppressive surface molecules. For example, Tim-3 and PD-1 is minimally co-expressed by tumor infiltrating T cells in patients with HCC though HCC-associated $Tim\text{-}3^+$ T cells do show signs of early senescence as characterized by low CD28 expression [84]. Another question that remains is how whether the underlying molecular and genetic signature of these T cells is similar and whether they may be treated as such.

Although the molecular mechanism regulating T cell exhaustion is yet incompletely understood, it is thought that PD-1 ligation by its ligand recruits SH2-domain containing protein tyrosine phosphatases (SHP-1 and/or SHP-2) to the immunoreceptor tyrosine-based switch motif (ITSM) within the cytoplasmic tail of PD-1 which in turn inhibits T cell receptor signaling pathways PI3K/AKT and T cell activation [93]. It is also known that T cell activation results in the upregulation of the PD-1 receptor on the T cell surface with the resulting T cell remaining functional [94, 95]. It is therefore the result of high expression of inhibitory B7 family members by cancer cells and associated APC which is thought to promote and maintain T cell exhaustion in the tumor microenvironment.

Senescent T Cells

Cellular senescence is a naturally occurring physiological phenomenon due to cellular proliferation. T cell senescence has been characterized by telomere shortening, loss of CD28 expression on its surface, and the inability to enter the cell cycle [96, 97]. Telomere shortening is a known byproduct of cellular division affecting T cell function [98]. Cell cycle arrest is mainly mediated by the accumulation of cell cycle controlling proteins p16, p21, and p53 [99–101]. Senescent T cells also develop dysfunctional killing abilities and immune-suppressive abilities [102, 103].

Senescence is thought to naturally occur at a cellular level during the cell's natural life-span mainly due to exhaustion of the cell's proliferation capabilities. The fact that senescent T cells are observed in younger patients with either autoimmune diseases or chronic viral infections suggests chronic activation and proliferation of T cells may induce their dysfunctional senescence [104, 105]. Evidence to support this in the context of cancer has been observed as (1) *in vitro* co-incubation of T cells with tumor cells can induce T cell senescence [106], and (2) phenotypic recognition of $CD8^+$ $CD28^{-/dim}$ senescent T cells have been found in lung cancers and head and neck cancer [107, 108]. Senescence-like phenotype has also been observed on mouse thymic precursor lymphocytes which have experienced DNA damage [109]. Such damage was enough to promote cell cycle arrest however, whether this is an operative mechanism acting in T cells in the tumor microenvironment, has yet to be described.

Further phenotypic analyses have found Tim-3, CD57, killer cell lectin-like receptor subfamily G, member 1 (KLRG-1) highly associate with senescent dysfunction [84, 110–114]. $CD28^{lo}$T cells from HCC human patients showed high expression of Tim-3 and cyclin dependent kinase inhibitors and their inability to enter the cell cycle. These cells were further observed to interact with galectin-9^+ tumor associated myeloid APC [84]. These findings were recapitulated in melanoma and lymphoma patients [113, 114]. This suggests that $CD28^{lo}$Tim-3^+ T cells in cancer may contain senescent T cells unable to escape cell cycle arrest.

Though a specific genetic program has yet to be resolved in T cell senescence, recent studies link Tim-3 ligation to Galectin-9 may be involved. Specifically, Tim-3 ligation promotes dissociation of the human leukocyte antigen (HLA)-B-associated transcript 3 (Bat-3) and decreases effector cytokine secretion and proliferation [115].

This ligation also induces imbalances in calcium flux which is important for T cell effector functions, and may lead to cell death [116]. Though the implications of how this calcium imbalance and Bat3 dissociation directly affects gene expression, or whether Bat3 loss is due to senescence, is not yet known.

T Cell Stemness

Stem cells are a unique cell subset with self-renewal capabilities and multi-lineage differentiation potential [117]. T cell stemness is a new concept and observation that has been proposed for memory T cells which have increased self-renewal capabilities and are able to generate more differentiated memory T cells. These 'T memory stem cells' are rare in normal peripheral blood. In tumor, human and mouse studies suggest Th17 cells may present stemness characteristics as discussed in the next section. However, it is well illustrated that mouse central memory T cells have shown increased stemness characteristics as these seem to be arrested at a pre-differentiated stage and have a high production potential of effector T cells after secondary antigen challenge [118, 119]. Memory T cells with increased stem-cell properties allow for the continuous production of effector T cells throughout the human lifetime. This ensures the effector T cell pool is continuously replenished despite the finite lifespan of individual differentiated T cells in the periphery and decreased thymic output during adulthood [120]. Recent evidence in mouse models show some $CD44^{low}CD62L^{high}$ memory $CD8^{+}$ T cells have increased stem cell gene signature, specifically stem cell antigen-1 (Sca-1), Bcl-2 are also express the common IL-2 and IL-15 receptor β chain (CD122). Further, these cells showed increased self-renewal characteristics and the capacity to generate central memory and effector memory T cells [121]. In trying to understand how these cells are generated *in vivo*, co-engagement of CD3 and CD28 in the context of IL-7 and IL-15 produced T cells with increased stem cell gene expression patterns and self-renewal capacities [122]. Deeper study into these 'T memory stem cells' showed that blocking T cell differentiation allowed for Wnt-signaling expression and the generation of T cells with stem cell properties [123]. These studies shed light into better understanding the T memory stem cell population and the ability to promote their production to maintain long-lived, self-renewing, antigen experienced, memory T cells for the treatment of patients with cancer.

Th17 Cells

Furthermore, recent studies also report the presence of Th17 cells with increased stem cell markers in human tumors. In studying their biological function, mouse and human Th17 cells showed increased survival potential, persistence, and the ability of repopulating sublethally irradiated mice [124, 125]. More importantly, these cells showed greater anti-tumor responses when compared to central- and effector-memory T cells. In studying the underlying genetic networks, HIF1α/NOTCH/Bcl-2 were shown to mediate their stem-cell specific characteristics [124]. This also suggests the stem cell lineage may involve better immune responses in humans. In support of this, Th17 stem-like T cells were able to differentiate into distinct Th lineages, as measured through IFN-γ secretion and Foxp3^{+} cells, greater self-renewal potential and survival capabilities [124, 125].

Human Th17 cells in tumor were shown to have increased stem cell properties at the genetic, molecular, and functional levels. These cells were shown to be long-lived cells with increased self-renewal capacities and being able to differentiate into highly effective effector T cells and Th lineages. Manipulation of Th17 stemness may therefore be of interest at the therapeutic level in treating patients with cancer as well as patients with Th17-associated chronic illnesses such as multiple sclerosis and other autoimmune diseases.

Polyfunctional T Cells

Polyfunctional T cells are effector T cells capable of greater immune responses characterized by their transient co-expression of multiple cytokines [126]. T cell polyfunctionality has been mainly defined based on the specific cytokine expression patterns of CD4^{+} or CD8^{+} T cells, specifically IL-2, IFN-γ, and TNF-α or Granzyme-B (Gra-B), IFN-γ, and TNF-α, respectively. Such T cells are mainly studied in the context of viral infections such as HIV and vaccinia virus, and have been suggested to develop potent immune responses [127, 128].

The ability of polyfunctional T cells to simultaneously express several cytokines also allows for their simultaneous stimulation of several pathways in the immune response. CD4^{+} T cell expression of IL-2 promotes T cell survival in the ongoing immune response, IFN-γ actively promotes inflammation on its surroundings, while TNF-α effectively activates the surrounding epithelium and inhibits viral replication [126]. Similarly, polyfunctional CD8^{+} T cells express IFN-γ, TNF-α, and Gra-B which allows the direct cytolysis of target-infected cells. Such responses have been shown to correlate with more efficient pathogen clearance. It has also been suggested that induction of such cells during vaccinations improves subsequent infection against specific pathogens [127, 129]. Studies into T cell polyfunctionality in the context of cancer are limited. In one study, human ovarian cancer Th17 associated T cells were shown to synergistically produce IL-17 and IFN-γ which stimulated the production of CXCL9 and CXCL10 from surrounding tumor cells [130]. This chemokine gradient could then effectively recruit effector T cells from the periphery into the tumor microenvironment and promote anti-tumor responses. Furthermore in mice, polyfunctionality was observed to follow chemotherapeutic treatment against B-cell lymphoma [131]. Infiltration of tumors by T cells often results in the immune suppression of T cell responses. In light of this, blockade of immune suppressive molecules resulted in the increased response of polyfunctionality on tumor infiltrating T cells [132]. Specifically, blockade of CTLA-4 allowed for a co-expression pattern including IFN-γ, MIP-1β, and TNF-α in tissues from melanoma patients.

A phenotypical definition for the polyfunctional T cell subset remains to be found. It is currently understood that polyfunctionality is a transient stage activated T cells achieve upon TCR-driven activation [133]. Moreover, in defining these cellular responses, it was shown that polyfunctional T cells may develop from monofunctional, or single-cytokine expressing, T cells. It remains unknown whether polyfunctionality may be a pre-defined program T cells have or whether it is a byproduct of robust activation. Further, the underlying genetic signature regulating their specific effector functions remain to be determined.

Concluding Remarks

Understanding the signaling pathways associated with T cell dysfunctions, the molecular and cellular enforcers, is the focus of much recent research. Current literature has documented profound immune suppressive imbalances in the tumor microenvironment mainly in the form of co-inhibitory molecules, lack of co-stimulatory molecules, and increased cellular mediators of immune suppression such as MDSC and T_{Regs}. As such, many clinical trials and patient data aim to shift the microenvironment with lower co-inhibition through antibody blockade of specific co-inhibitory molecules or targeting immune suppressive cells.

There is a convincing amount of evidence arguing for the co-existence of T cellanergy, exhaustion, senescence, stemness, and polyfunctionality in the tumor microenvironment. As we move forward, new ways to identify specific functional characteristics of T cells will be explored. Current literature shows that PD-1$^+$ T cells represent functionally exhausted T cells, Tim-3$^+$ and KLRG-1$^+$ may be senescent T cells while Sca-1$^+$ T cells may be mouse stem-like T cells. However, as we have explored, these markers do not encompass their specific paradigm. It is probable that these T cell paradigms are functionally generated in their specific contexts and that there are specific genetic and functional patterns, not surface phenotypes, which regulate their biology and fate. One example resides in Th17 cells which have increased stem cell characteristics yet show surface markers of terminal differentiation. A second example is observed in that dysfunction does not follow phenotype. PD-1$^+$ cells may express Tim-3 and LAG-3. It is therefore probable that B7-H1/PD-1 and Tim-3/Galectin-9 signaling pathways may synergistically and/or additively promote T cell dysfunction and their blockade may improve T cell immunity. Both preclinical and clinical studies suggest T cell dysfunction may be functionally reversible. Finally, T cell dysfunction may be intertwined. It has been shown that dysfunctional T cells co-express varying levels of inhibitory molecules which include PD-1, Tim-3, LAG-3, 2B4, CD160, and KLRG-1. Similarly, Th17 cells show increased polyfunctional and stemness characteristics though a specific phenotype separating these two functions has yet to be defined.

Anti-tumor immunity relies on the effective stimulation of T cell responses. Th17 cells represent a highly immunogenic T cell subset with increased stem cell characteristics and polyfunctional capacities. Based on this, strategies which promote the acquisition of Th17 cells while simultaneously depleting regulatory T cells from the tumor microenvironment should allow for better treatment prognosis. Strategies which aim to shift the immune response from T_{Regs} to Th17 should therefore be actively explored.

In conclusion, tumors are able to hijack T cell tolerance mechanisms from the periphery which include regulatory T cells, MDSC, T cell anergy, exhaustion, and senescence, to survive. Deeper understanding of how these paradigms normally occur will have implications in the development of future therapeutics targeting human malignancies. One such strategy may be the relief of T cell dysfunction and promoting T cell stemness and/or polyfunctionality to treat patients with cancer.

References

1. Zou W (2005) Immunosuppressive networks in the tumour environment and their therapeutic relevance. Nat Rev Cancer 5(4):263–274. doi:10.1038/nrc1586, PubMed PMID: 15776005
2. Zou W, Chen L (2008) Inhibitory B7-family molecules in the tumour microenvironment. Nat Rev Immunol 8(6):467–477. doi:10.1038/nri2326, PubMed PMID: 18500231
3. Pardoll DM (2012) The blockade of immune checkpoints in cancer immunotherapy. Nat Rev Cancer 12(4):252–264. doi:10.1038/nrc3239, PubMed PMID: 22437870
4. Vanneman M, Dranoff G (2012) Combining immunotherapy and targeted therapies in cancer treatment. Nat Rev Cancer 12(4):237–251. doi:10.1038/nrc3237, PubMed PMID: 22437869
5. Crespo J, Sun H, Welling TH, Tian Z, Zou W (2013) T cell anergy, exhaustion, senescence, and stemness in the tumor microenvironment. Curr Opin Immunol 25(2):214–221. doi:10.1016/j.coi.2012.12.003, PubMed PMID: 23298609; PubMed Central PMCID: PMC3636159
6. Chen L, Flies DB (2013) Molecular mechanisms of T cell co-stimulation and co-inhibition. Nat Rev Immunol 13(4):227–242. doi:10.1038/nri3405, PubMed PMID: 23470321; PubMed Central PMCID: PMC3786574
7. June CH, Ledbetter JA, Gillespie MM, Lindsten T, Thompson CB (1987) T-cell proliferation involving the CD28 pathway is associated with cyclosporine-resistant interleukin 2 gene expression. Mol Cell Biol 7(12):4472–4481, PubMed PMID: 2830495; PubMed Central PMCID: PMC368131
8. Bretscher P, Cohn M (1970) A theory of self-nonself discrimination. Science 169(3950):1042–1049, PubMed PMID: 4194660
9. Mueller DL, Jenkins MK, Schwartz RH (1989) Clonal expansion versus functional clonal inactivation: a costimulatory signalling pathway determines the outcome of T cell antigen receptor occupancy. Annu Rev Immunol 7:445–480. doi:10.1146/annurev.iy.07.040189.002305, PubMed PMID: 2653373
10. Vallabhapurapu S, Karin M (2009) Regulation and function of NF-kappaB transcription factors in the immune system. Annu Rev Immunol 27:693–733. doi:10.1146/annurev.immunol.021908.132641, PubMed PMID: 19302050
11. Schulze-Luehrmann J, Ghosh S (2006) Antigen-receptor signaling to nuclear factor kappa B. Immunity 25(5):701–715. doi:10.1016/j.immuni.2006.10.010, PubMed PMID: 17098202
12. Feske S, Giltnane J, Dolmetsch R, Staudt LM, Rao A (2001) Gene regulation mediated by calcium signals in T lymphocytes. Nat Immunol 2(4):316–324. doi:10.1038/86318, PubMed PMID: 11276202

13. Boomer JS, Green JM (2010) An enigmatic tail of CD28 signaling. Cold Spring Harb Perspect Biol 2(8):a002436. doi:10.1101/cshperspect.a002436, PubMed PMID: 20534709; PubMed Central PMCID: PMC2908766
14. Yang K, Chi H (2012) mTOR and metabolic pathways in T cell quiescence and functional activation. Semin Immunol 24(6):421–428. doi:10.1016/j.smim.2012.12.004, PubMed PMID: 23375549; PubMed Central PMCID: PMC3855395
15. Hodi FS, O'Day SJ, McDermott DF, Weber RW, Sosman JA, Haanen JB et al (2010) Improved survival with ipilimumab in patients with metastatic melanoma. N Engl J Med 363(8):711–723. doi:10.1056/NEJMoa1003466, PubMed PMID: 20525992; PubMed Central PMCID: PMC3549297
16. Dong H, Zhu G, Tamada K, Chen L (1999) B7-H1, a third member of the B7 family, co-stimulates T-cell proliferation and interleukin-10 secretion. Nat Med 5(12):1365–1369. doi:10.1038/70932, PubMed PMID: 10581077
17. Azuma T, Yao S, Zhu G, Flies AS, Flies SJ, Chen L (2008) B7-H1 is a ubiquitous antiapoptotic receptor on cancer cells. Blood 111(7):3635–3643. doi:10.1182/blood-2007-11-123141, PubMed PMID: 18223165; PubMed Central PMCID: PMC2275025
18. Blank C, Brown I, Peterson AC, Spiotto M, Iwai Y, Honjo T et al (2004) PD-L1/B7H-1 inhibits the effector phase of tumor rejection by T cell receptor (TCR) transgenic CD8+ T cells. Cancer Res 64(3):1140–1145, PubMed PMID: 14871849
19. Blank C, Kuball J, Voelkl S, Wiendl H, Becker B, Walter B et al (2006) Blockade of PD-L1 (B7-H1) augments human tumor-specific T cell responses in vitro. Int J Cancer 119(2):317–327. doi:10.1002/ijc.21775, PubMed PMID: 16482562
20. Curiel TJ, Wei S, Dong H, Alvarez X, Cheng P, Mottram P et al (2003) Blockade of B7-H1 improves myeloid dendritic cell-mediated antitumor immunity. Nat Med 9(5):562–567. doi:10.1038/nm863, PubMed PMID: 12704383
21. Brahmer JR, Tykodi SS, Chow LQ, Hwu WJ, Topalian SL, Hwu P et al (2012) Safety and activity of anti-PD-L1 antibody in patients with advanced cancer. N Engl J Med 366(26):2455–2465. doi:10.1056/NEJMoa1200694, PubMed PMID: 22658128; PubMed Central PMCID: PMC3563263
22. Topalian SL, Hodi FS, Brahmer JR, Gettinger SN, Smith DC, McDermott DF et al (2012) Safety, activity, and immune correlates of anti-PD-1 antibody in cancer. N Engl J Med 366(26):2443–2454. doi:10.1056/NEJMoa1200690, PubMed PMID: 22658127; PubMed Central PMCID: PMC3544539
23. Dangaj D, Scholler N (2013) Blocking the B7-H4 pathway with novel recombinant antibodies enhances T cell-mediated antitumor responses. Oncoimmunology 2(8), e25913. doi:10.4161/onci.25913, PubMed PMID: 24083083; PubMed Central PMCID: PMC3782523
24. Kryczek I, Zou L, Rodriguez P, Zhu G, Wei S, Mottram P et al (2006) B7-H4 expression identifies a novel suppressive macrophage population in human ovarian carcinoma. J Exp Med 203(4):871–881. doi:10.1084/jem.20050930, PubMed PMID: 16606666; PubMed Central PMCID: PMC2118300
25. Tringler B, Zhuo S, Pilkington G, Torkko KC, Singh M, Lucia MS et al (2005) B7-h4 is highly expressed in ductal and lobular breast cancer. Clin Cancer Res 11(5):1842–1848. doi:10.1158/1078-0432.CCR-04-1658, PubMed PMID: 15756008
26. Mugler KC, Singh M, Tringler B, Torkko KC, Liu W, Papkoff J et al (2007) B7-h4 expression in a range of breast pathology: correlation with tumor T-cell infiltration. Appl Immunohistochem Mol Morphol 15(4):363–370. doi:10.1097/01.pai.0000213159.79557.71, PubMed PMID: 18091377
27. Simon I, Liu Y, Krall KL, Urban N, Wolfert RL, Kim NW et al (2007) Evaluation of the novel serum markers B7-H4, Spondin 2, and DcR3 for diagnosis and early detection of ovarian cancer. Gynecol Oncol 106(1):112–118. doi:10.1016/j.ygyno.2007.03.007, PubMed PMID: 17490732
28. Chen C, Qu QX, Shen Y, Mu CY, Zhu YB, Zhang XG et al (2012) Induced expression of B7-H4 on the surface of lung cancer cell by the tumor-associated macrophages: a potential

mechanism of immune escape. Cancer Lett 317(1):99–105. doi:10.1016/j.canlet.2011.11.017, PubMed PMID: 22108530

29. Sun Y, Wang Y, Zhao J, Gu M, Giscombe R, Lefvert AK et al (2006) B7-H3 and B7-H4 expression in non-small-cell lung cancer. Lung Cancer 53(2):143–151. doi:10.1016/j.lungcan.2006.05.012, PubMed PMID: 16782226
30. Kryczek I, Wei S, Zhu G, Myers L, Mottram P, Cheng P et al (2007) Relationship between B7-H4, regulatory T cells, and patient outcome in human ovarian carcinoma. Cancer Res 67(18):8900–8905. doi:10.1158/0008-5472.CAN-07-1866, PubMed PMID: 17875732
31. Zhu G, Augustine MM, Azuma T, Luo L, Yao S, Anand S et al (2009) B7-H4-deficient mice display augmented neutrophil-mediated innate immunity. Blood 113(8):1759–1767. doi:10.1182/blood-2008-01-133223, PubMed PMID: 19109567; PubMed Central PMCID: PMC2647680
32. Wei J, Loke P, Zang X, Allison JP (2011) Tissue-specific expression of B7x protects from CD4 T cell-mediated autoimmunity. J Exp Med 208(8):1683–1694. doi:10.1084/jem.20100639, PubMed PMID: 21727190; PubMed Central PMCID: PMC3149222
33. Zhang L, Wu H, Lu D, Li G, Sun C, Song H et al (2013) The costimulatory molecule B7-H4 promote tumor progression and cell proliferation through translocating into nucleus. Oncogene 32(46):5347–5358. doi:10.1038/onc.2012.600, PubMed PMID: 23318460; PubMed Central PMCID: PMC3898118
34. Gabrilovich DI, Nagaraj S (2009) Myeloid-derived suppressor cells as regulators of the immune system. Nat Rev Immunol 9(3):162–174. doi:10.1038/nri2506, PubMed PMID: 19197294; PubMed Central PMCID: PMC2828349
35. Youn JI, Gabrilovich DI (2010) The biology of myeloid-derived suppressor cells: the blessing and the curse of morphological and functional heterogeneity. Eur J Immunol 40(11):2969–2975. doi:10.1002/eji.201040895, PubMed PMID: 21061430; PubMed Central PMCID: PMC3277452
36. Sunderkotter C, Nikolic T, Dillon MJ, Van Rooijen N, Stehling M, Drevets DA et al (2004) Subpopulations of mouse blood monocytes differ in maturation stage and inflammatory response. J Immunol 172(7):4410–4417, PubMed PMID: 15034056
37. Almand B, Clark JI, Nikitina E, van Beynen J, English NR, Knight SC et al (2001) Increased production of immature myeloid cells in cancer patients: a mechanism of immunosuppression in cancer. J Immunol 166(1):678–689, PubMed PMID: 11123353
38. Goni O, Alcaide P, Fresno M (2002) Immunosuppression during acute Trypanosoma cruzi infection: involvement of Ly6G (Gr1(+))CD11b(+)immature myeloid suppressor cells. Int Immunol 14(10):1125–1134, PubMed PMID: 12356678
39. Giordanengo L, Guinazu N, Stempin C, Fretes R, Cerban F, Gea S (2002) Cruzipain, a major Trypanosoma cruzi antigen, conditions the host immune response in favor of parasite. Eur J Immunol 32(4):1003–1011. doi:10.1002/1521-4141(200204)32:4<1003::AID-IMMU1003>3.0.CO;2-P, PubMed PMID: 11920566
40. Yang L, DeBusk LM, Fukuda K, Fingleton B, Green-Jarvis B, Shyr Y et al (2004) Expansion of myeloid immune suppressor Gr+CD11b+cells in tumor-bearing host directly promotes tumor angiogenesis. Cancer Cell 6(4):409–421. doi:10.1016/j.ccr.2004.08.031, PubMed PMID: 15488763
41. Nakamura Y, Yasuoka H, Tsujimoto M, Yoshidome K, Nakahara M, Nakao K et al (2006) Nitric oxide in breast cancer: induction of vascular endothelial growth factor-C and correlation with metastasis and poor prognosis. Clin Cancer Res 12(4):1201–1207. doi:10.1158/1078-0432.CCR-05-1269, PubMed PMID: 16489074
42. Hoechst B, Ormandy LA, Ballmaier M, Lehner F, Kruger C, Manns MP et al (2008) A new population of myeloid-derived suppressor cells in hepatocellular carcinoma patients induces CD4(+)CD25(+)Foxp3(+) T cells. Gastroenterology 135(1):234–243. doi:10.1053/j.gastro.2008.03.020, PubMed PMID: 18485901
43. Filipazzi P, Valenti R, Huber V, Pilla L, Canese P, Iero M et al (2007) Identification of a new subset of myeloid suppressor cells in peripheral blood of melanoma patients with modulation by a granulocyte-macrophage colony-stimulation factor-based antitumor vaccine. J Clin Oncol 25(18):2546–2553. doi:10.1200/JCO.2006.08.5829, PubMed PMID: 17577033

44. Loercher AE, Nash MA, Kavanagh JJ, Platsoucas CD, Freedman RS (1999) Identification of an IL-10-producing HLA-DR-negative monocyte subset in the malignant ascites of patients with ovarian carcinoma that inhibits cytokine protein expression and proliferation of autologous T cells. J Immunol 163(11):6251–6260, PubMed PMID: 10570318
45. Vuk-Pavlovic S, Bulur PA, Lin Y, Qin R, Szumlanski CL, Zhao X et al (2010) Immunosuppressive CD14+HLA-DRlow/- monocytes in prostate cancer. Prostate 70(4):443–455. doi:10.1002/pros.21078, PubMed PMID: 19902470; PubMed Central PMCID: PMC2935631
46. Bronte V, Zanovello P (2005) Regulation of immune responses by L-arginine metabolism. Nat Rev Immunol 5(8):641–654. doi:10.1038/nri1668, PubMed PMID: 16056256
47. Rodriguez PC, Zea AH, Culotta KS, Zabaleta J, Ochoa JB, Ochoa AC (2002) Regulation of T cell receptor CD3zeta chain expression by L-arginine. J Biol Chem 277(24):21123–21129. doi:10.1074/jbc.M110675200, PubMed PMID: 11950832
48. Rodriguez PC, Quiceno DG, Ochoa AC (2007) L-arginine availability regulates T-lymphocyte cell-cycle progression. Blood 109(4):1568–1573. doi:10.1182/blood-2006-06-031856, PubMed PMID: 17023580; PubMed Central PMCID: PMC1794048
49. Serafini P, Mgebroff S, Noonan K, Borrello I (2008) Myeloid-derived suppressor cells promote cross-tolerance in B-cell lymphoma by expanding regulatory T cells. Cancer Res 68(13):5439–5449. doi:10.1158/0008-5472.CAN-07-6621, PubMed PMID: 18593947; PubMed Central PMCID: PMC2887390
50. Zoso A, Mazza EM, Bicciato S, Mandruzzato S, Bronte V, Serafini P et al (2014) Human fibrocytic myeloid-derived suppressor cells express IDO and promote tolerance via Treg-cell expansion. Eur J Immunol. doi:10.1002/eji.201444522, PubMed PMID: 25113564
51. Kusmartsev S, Nefedova Y, Yoder D, Gabrilovich DI (2004) Antigen-specific inhibition of CD8+ T cell response by immature myeloid cells in cancer is mediated by reactive oxygen species. J Immunol 172(2):989–999, PubMed PMID: 14707072
52. Mantovani G, Maccio A, Madeddu C, Mura L, Gramignano G, Lusso MR et al (2003) Antioxidant agents are effective in inducing lymphocyte progression through cell cycle in advanced cancer patients: assessment of the most important laboratory indexes of cachexia and oxidative stress. J Mol Med 81(10):664–673. doi:10.1007/s00109-003-0476-1, PubMed PMID: 12928788
53. Schmielau J, Finn OJ (2001) Activated granulocytes and granulocyte-derived hydrogen peroxide are the underlying mechanism of suppression of t-cell function in advanced cancer patients. Cancer Res 61(12):4756–4760, PubMed PMID: 11406548
54. Vickers SM, MacMillan-Crow LA, Green M, Ellis C, Thompson JA (1999) Association of increased immunostaining for inducible nitric oxide synthase and nitrotyrosine with fibroblast growth factor transformation in pancreatic cancer. Arch Surg 134(3):245–251, PubMed PMID: 10088562
55. Ekmekcioglu S, Ellerhorst J, Smid CM, Prieto VG, Munsell M, Buzaid AC et al (2000) Inducible nitric oxide synthase and nitrotyrosine in human metastatic melanoma tumors correlate with poor survival. Clin Cancer Res 6(12):4768–4775, PubMed PMID: 11156233
56. Bronte V, Kasic T, Gri G, Gallana K, Borsellino G, Marigo I et al (2005) Boosting antitumor responses of T lymphocytes infiltrating human prostate cancers. J Exp Med 201(8):1257–1268. doi:10.1084/jem.20042028, PubMed PMID: 15824085; PubMed Central PMCID: PMC2213151
57. Nagaraj S, Gupta K, Pisarev V, Kinarsky L, Sherman S, Kang L et al (2007) Altered recognition of antigen is a mechanism of CD8+ T cell tolerance in cancer. Nat Med 13(7):828–835. doi:10.1038/nm1609, PubMed PMID: 17603493; PubMed Central PMCID: PMC2135607
58. Sinha P, Clements VK, Bunt SK, Albelda SM, Ostrand-Rosenberg S (2007) Cross-talk between myeloid-derived suppressor cells and macrophages subverts tumor immunity toward a type 2 response. J Immunol 179(2):977–983, PubMed PMID: 17617589
59. Kuang DM, Zhao Q, Peng C, Xu J, Zhang JP, Wu C et al (2009) Activated monocytes in peritumoral stroma of hepatocellular carcinoma foster immune privilege and disease progression

through PD-L1. J Exp Med 206(6):1327–1337. doi:10.1084/jem.20082173, PubMed PMID: 19451266; PubMed Central PMCID: PMC2715058
60. Schwartz RH (2003) T cell anergy. Annu Rev Immunol 21:305–334. doi:10.1146/annurev.immunol.21.120601.141110, PubMed PMID: 12471050
61. Schwartz RH (1990) A cell culture model for T lymphocyte clonal anergy. Science 248(4961):1349–1356, PubMed PMID: 2113314
62. Wells AD (2009) New insights into the molecular basis of T cell anergy: anergy factors, avoidance sensors, and epigenetic imprinting. J Immunol 182(12):7331–7341. doi:10.4049/jimmunol.0803917, PubMed PMID: 19494254
63. Ozao-Choy J, Ma G, Kao J, Wang GX, Meseck M, Sung M et al (2009) The novel role of tyrosine kinase inhibitor in the reversal of immune suppression and modulation of tumor microenvironment for immune-based cancer therapies. Cancer Res 69(6):2514–2522. doi:10.1158/0008-5472.CAN-08-4709, PubMed PMID: 19276342
64. Chen L, Ashe S, Brady WA, Hellstrom I, Hellstrom KE, Ledbetter JA et al (1992) Costimulation of antitumor immunity by the B7 counterreceptor for the T lymphocyte molecules CD28 and CTLA-4. Cell 71(7):1093–1102, PubMed PMID: 1335364
65. Gajewski TF (1996) B7-1 but not B7-2 efficiently costimulates CD8+ T lymphocytes in the P815 tumor system in vitro. J Immunol 156(2):465–472, PubMed PMID: 8543795
66. Brown IE, Blank C, Kline J, Kacha AK, Gajewski TF (2006) Homeostatic proliferation as an isolated variable reverses CD8+ T cell anergy and promotes tumor rejection. J Immunol 177(7):4521–4529, PubMed PMID: 16982889
67. Greenwald RJ, Boussiotis VA, Lorsbach RB, Abbas AK, Sharpe AH (2001) CTLA-4 regulates induction of anergy in vivo. Immunity 14(2):145–155, PubMed PMID: 11239447
68. Boussiotis VA, Freeman GJ, Berezovskaya A, Barber DL, Nadler LM (1997) Maintenance of human T cell anergy: blocking of IL-2 gene transcription by activated Rap1. Science 278(5335):124–128, PubMed PMID: 9311917
69. Dolmetsch RE, Xu K, Lewis RS (1998) Calcium oscillations increase the efficiency and specificity of gene expression. Nature 392(6679):933–936. doi:10.1038/31960, PubMed PMID: 9582075
70. Anandasabapathy N, Ford GS, Bloom D, Holness C, Paragas V, Seroogy C et al (2003) GRAIL: an E3 ubiquitin ligase that inhibits cytokine gene transcription is expressed in anergic CD4+ T cells. Immunity 18(4):535–547, PubMed PMID: 12705856
71. Soto-Nieves N, Puga I, Abe BT, Bandyopadhyay S, Baine I, Rao A et al (2009) Transcriptional complexes formed by NFAT dimers regulate the induction of T cell tolerance. J Exp Med 206(4):867–876. doi:10.1084/jem.20082731, PubMed PMID: 19307325; PubMed Central PMCID: PMC2715123
72. Jeon MS, Atfield A, Venuprasad K, Krawczyk C, Sarao R, Elly C et al (2004) Essential role of the E3 ubiquitin ligase Cbl-b in T cell anergy induction. Immunity 21(2):167–177. doi:10.1016/j.immuni.2004.07.013, PubMed PMID: 15308098
73. Mueller DL (2004) E3 ubiquitin ligases as T cell anergy factors. Nat Immunol 5(9):883–890. doi:10.1038/ni1106, PubMed PMID: 15334084
74. Thomas RM, Chunder N, Chen C, Umetsu SE, Winandy S, Wells AD (2007) Ikaros enforces the costimulatory requirement for IL2 gene expression and is required for anergy induction in CD4+ T lymphocytes. J Immunol 179(11):7305–7315, PubMed PMID: 18025173
75. Gao B, Kong Q, Kemp K, Zhao YS, Fang D (2012) Analysis of sirtuin 1 expression reveals a molecular explanation of IL-2-mediated reversal of T-cell tolerance. Proc Natl Acad Sci U S A 109(3):899–904. doi:10.1073/pnas.1118462109, PubMed PMID: 22219356; PubMed Central PMCID: PMC3271862
76. Bandyopadhyay S, Dure M, Paroder M, Soto-Nieves N, Puga I, Macian F (2007) Interleukin 2 gene transcription is regulated by Ikaros-induced changes in histone acetylation in anergic T cells. Blood 109(7):2878–2886. doi:10.1182/blood-2006-07-037754, PubMed PMID: 17148585; PubMed Central PMCID: PMC1852212
77. Zheng Y, Zha Y, Driessens G, Locke F, Gajewski TF (2012) Transcriptional regulator early growth response gene 2 (Egr2) is required for T cell anergy in vitro and in vivo. J Exp Med

209(12):2157–2163. doi:10.1084/jem.20120342, PubMed PMID: 23129747; PubMed Central PMCID: PMC3501351
78. Zheng Y, Zha Y, Spaapen RM, Mathew R, Barr K, Bendelac A et al (2013) Egr2-dependent gene expression profiling and ChIP-Seq reveal novel biologic targets in T cell anergy. Mol Immunol 55(3–4):283–291. doi:10.1016/j.molimm.2013.03.006, PubMed PMID: 23548837; PubMed Central PMCID: PMC3646929
79. Wherry EJ (2011) T cell exhaustion. Nat Immunol 12(6):492–499, PubMed PMID: 21739672
80. Wu K, Kryczek I, Chen L, Zou W, Welling TH (2009) Kupffer cell suppression of CD8+ T cells in human hepatocellular carcinoma is mediated by B7-H1/programmed death-1 interactions. Cancer Res 69(20):8067–8075. doi:10.1158/0008-5472.CAN-09-0901, PubMed PMID: 19826049
81. Wherry EJ, Ha SJ, Kaech SM, Haining WN, Sarkar S, Kalia V et al (2007) Molecular signature of CD8+ T cell exhaustion during chronic viral infection. Immunity 27(4):670–684. doi:10.1016/j.immuni.2007.09.006, PubMed PMID: 17950003
82. Blackburn SD, Shin H, Haining WN, Zou T, Workman CJ, Polley A et al (2009) Coregulation of CD8+ T cell exhaustion by multiple inhibitory receptors during chronic viral infection. Nat Immunol 10(1):29–37. doi:10.1038/ni.1679, PubMed PMID: 19043418; PubMed Central PMCID: PMC2605166
83. Sakuishi K, Apetoh L, Sullivan JM, Blazar BR, Kuchroo VK, Anderson AC (2010) Targeting Tim-3 and PD-1 pathways to reverse T cell exhaustion and restore anti-tumor immunity. J Exp Med 207(10):2187–2194. doi:10.1084/jem.20100643, PubMed PMID: 20819927; PubMed Central PMCID: PMC2947065
84. Li H, Wu K, Tao K, Chen L, Zheng Q, Lu X et al (2012) Tim-3/galectin-9 signaling pathway mediates T-cell dysfunction and predicts poor prognosis in patients with hepatitis B virus-associated hepatocellular carcinoma. Hepatology 56(4):1342–1351. doi:10.1002/hep.25777, PubMed PMID: 22505239
85. Barber DL, Wherry EJ, Masopust D, Zhu B, Allison JP, Sharpe AH et al (2006) Restoring function in exhausted CD8 T cells during chronic viral infection. Nature 439(7077):682–687. doi:10.1038/nature04444, PubMed PMID: 16382236
86. Duraiswamy J, Freeman GJ, Coukos G (2013) Therapeutic PD-1 pathway blockade augments with other modalities of immunotherapy T-cell function to prevent immune decline in ovarian cancer. Cancer Res 73(23):6900–6912. doi:10.1158/0008-5472.CAN-13-1550, PubMed PMID: 23975756; PubMed Central PMCID: PMC3851914
87. Baitsch L, Baumgaertner P, Devevre E, Raghav SK, Legat A, Barba L et al (2011) Exhaustion of tumor-specific CD8(+) T cells in metastases from melanoma patients. J Clin Invest 121(6):2350–2360. doi:10.1172/JCI46102, PubMed PMID: 21555851; PubMed Central PMCID: PMC3104769
88. Zhou Q, Munger ME, Veenstra RG, Weigel BJ, Hirashima M, Munn DH et al (2011) Coexpression of Tim-3 and PD-1 identifies a CD8+ T-cell exhaustion phenotype in mice with disseminated acute myelogenous leukemia. Blood 117(17):4501–4510. doi:10.1182/blood-2010-10-310425, PubMed PMID: 21385853; PubMed Central PMCID: PMC3099570
89. Fourcade J, Sun Z, Pagliano O, Guillaume P, Luescher IF, Sander C et al (2012) CD8(+) T cells specific for tumor antigens can be rendered dysfunctional by the tumor microenvironment through upregulation of the inhibitory receptors BTLA and PD-1. Cancer Res 72(4):887–896. doi:10.1158/0008-5472.CAN-11-2637, PubMed PMID: 22205715; PubMed Central PMCID: PMC3288235
90. Derre L, Rivals JP, Jandus C, Pastor S, Rimoldi D, Romero P et al (2010) BTLA mediates inhibition of human tumor-specific CD8+ T cells that can be partially reversed by vaccination. J Clin Invest 120(1):157–167. doi:10.1172/JCI40070, PubMed PMID: 20038811; PubMed Central PMCID: PMC2799219
91. Jin HT, Anderson AC, Tan WG, West EE, Ha SJ, Araki K et al (2010) Cooperation of Tim-3 and PD-1 in CD8 T-cell exhaustion during chronic viral infection. Proc Natl Acad Sci U S A 107(33):14733–14738. doi:10.1073/pnas.1009731107, PubMed PMID: 20679213; PubMed Central PMCID: PMC2930455

92. Woo SR, Turnis ME, Goldberg MV, Bankoti J, Selby M, Nirschl CJ et al (2012) Immune inhibitory molecules LAG-3 and PD-1 synergistically regulate T-cell function to promote tumoral immune escape. Cancer Res 72(4):917–927. doi:10.1158/0008-5472.CAN-11-1620, PubMed PMID: 22186141; PubMed Central PMCID: PMC3288154
93. Riley JL (2009) PD-1 signaling in primary T cells. Immunol Rev 229(1):114–125. doi:10.1111/j.1600-065X.2009.00767.x, PubMed PMID: 19426218; PubMed Central PMCID: PMC3424066
94. Haymaker C, Wu R, Bernatchez C, Radvanyi L (2012) PD-1 and BTLA and CD8(+) T-cell "exhaustion" in cancer: "exercising" an alternative viewpoint. Oncoimmunology 1(5):735–738. doi:10.4161/onci.20823, PubMed PMID: 22934265; PubMed Central PMCID: PMC3429577
95. Goldberg MV, Maris CH, Hipkiss EL, Flies AS, Zhen L, Tuder RM et al (2007) Role of PD-1 and its ligand, B7-H1, in early fate decisions of CD8 T cells. Blood 110(1):186–192. doi:10.1182/blood-2006-12-062422, PubMed PMID: 17392506; PubMed Central PMCID: PMC1896112
96. Adibzadeh M, Pohla H, Rehbein A, Pawelec G (1995) Long-term culture of monoclonal human T lymphocytes: models for immunosenescence? Mech Ageing Dev 83(3):171–183, PubMed PMID: 8583835
97. Effros RB (1998) Replicative senescence in the immune system: impact of the Hayflick limit on T-cell function in the elderly. Am J Hum Genet 62(5):1003–1007. doi:10.1086/301845, PubMed PMID: 9545415; PubMed Central PMCID: PMC1377102
98. Hayflick L, Moorhead PS (1961) The serial cultivation of human diploid cell strains. Exp Cell Res 25:585–621, PubMed PMID: 13905658
99. Passos JF, Nelson G, Wang C, Richter T, Simillion C, Proctor CJ et al (2010) Feedback between p21 and reactive oxygen production is necessary for cell senescence. Mol Syst Biol 6:347. doi:10.1038/msb.2010.5, PubMed PMID: 20160708; PubMed Central PMCID: PMC2835567
100. Rodier F, Coppe JP, Patil CK, Hoeijmakers WA, Munoz DP, Raza SR et al (2009) Persistent DNA damage signalling triggers senescence-associated inflammatory cytokine secretion. Nat Cell Biol 11(8):973–979. doi:10.1038/ncb1909, PubMed PMID: 19597488; PubMed Central PMCID: PMC2743561
101. Beausejour CM, Krtolica A, Galimi F, Narita M, Lowe SW, Yaswen P et al (2003) Reversal of human cellular senescence: roles of the p53 and p16 pathways. EMBO J 22(16):4212–4222. doi:10.1093/emboj/cdg417, PubMed PMID: 12912919; PubMed Central PMCID: PMC175806
102. Appay V, Nixon DF, Donahoe SM, Gillespie GM, Dong T, King A et al (2000) HIV-specific CD8(+) T cells produce antiviral cytokines but are impaired in cytolytic function. J Exp Med 192(1):63–75, PubMed PMID: 10880527; PubMed Central PMCID: PMC1887711
103. Akbar AN, Henson SM (2011) Are senescence and exhaustion intertwined or unrelated processes that compromise immunity? Nat Rev Immunol 11(4):289–295. doi:10.1038/nri2959, PubMed PMID: 21436838
104. Pawelec G, Solana R (2001) Immunoageing – the cause or effect of morbidity. Trends Immunol 22(7):348–349, PubMed PMID: 11505929
105. Vallejo AN, Weyand CM, Goronzy JJ (2004) T-cell senescence: a culprit of immune abnormalities in chronic inflammation and persistent infection. Trends Mol Med 10(3):119–124. doi:10.1016/j.molmed.2004.01.002, PubMed PMID: 15102354
106. Montes CL, Chapoval AI, Nelson J, Orhue V, Zhang X, Schulze DH et al (2008) Tumor-induced senescent T cells with suppressor function: a potential form of tumor immune evasion. Cancer Res 68(3):870–879. doi:10.1158/0008-5472.CAN-07-2282, PubMed PMID: 18245489
107. Meloni F, Morosini M, Solari N, Passadore I, Nascimbene C, Novo M et al (2006) Foxp3 expressing CD4+ CD25+ and CD8+CD28- T regulatory cells in the peripheral blood of patients with lung cancer and pleural mesothelioma. Hum Immunol 67(1–2):1–12. doi:10.1016/j.humimm.2005.11.005, PubMed PMID: 16698419
108. Tsukishiro T, Donnenberg AD, Whiteside TL (2003) Rapid turnover of the CD8(+)CD28(−) T-cell subset of effector cells in the circulation of patients with head and neck cancer. Cancer

Immunol Immunother 52(10):599–607. doi:10.1007/s00262-003-0395-6, PubMed PMID: 12827303
109. Van Nguyen T, Puebla-Osorio N, Pang H, Dujka ME, Zhu C (2007) DNA damage-induced cellular senescence is sufficient to suppress tumorigenesis: a mouse model. J Exp Med 204(6):1453–1461. doi:10.1084/jem.20062453, PubMed PMID: 17535972; PubMed Central PMCID: PMC2118600
110. Brenchley JM, Karandikar NJ, Betts MR, Ambrozak DR, Hill BJ, Crotty LE et al (2003) Expression of CD57 defines replicative senescence and antigen-induced apoptotic death of CD8+ T cells. Blood 101(7):2711–2720. doi:10.1182/blood-2002-07-2103, PubMed PMID: 12433688
111. Heffner M, Fearon DT (2007) Loss of T cell receptor-induced Bmi-1 in the KLRG1(+) senescent CD8(+) T lymphocyte. Proc Natl Acad Sci U S A 104(33):13414–13419. doi:10.1073/pnas.0706040104, PubMed PMID: 17686974; PubMed Central PMCID: PMC1941641
112. Voehringer D, Blaser C, Brawand P, Raulet DH, Hanke T, Pircher H (2001) Viral infections induce abundant numbers of senescent CD8 T cells. J Immunol 167(9):4838–4843, PubMed PMID: 11673487
113. Fourcade J, Sun Z, Benallaoua M, Guillaume P, Luescher IF, Sander C et al (2010) Upregulation of Tim-3 and PD-1 expression is associated with tumor antigen-specific CD8+ T cell dysfunction in melanoma patients. J Exp Med 207(10):2175–2186. doi:10.1084/jem.20100637, PubMed PMID: 20819923; PubMed Central PMCID: PMC2947081
114. Huang X, Bai X, Cao Y, Wu J, Huang M, Tang D et al (2010) Lymphoma endothelium preferentially expresses Tim-3 and facilitates the progression of lymphoma by mediating immune evasion. J Exp Med 207(3):505–520. doi:10.1084/jem.20090397, PubMed PMID: 20176801; PubMed Central PMCID: PMC2839144
115. Rangachari M, Zhu C, Sakuishi K, Xiao S, Karman J, Chen A et al (2012) Bat3 promotes T cell responses and autoimmunity by repressing Tim-3-mediated cell death and exhaustion. Nat Med 18(9):1394–1400. doi:10.1038/nm.2871, PubMed PMID: 22863785; PubMed Central PMCID: PMC3491118
116. Zhu C, Anderson AC, Schubart A, Xiong H, Imitola J, Khoury SJ et al (2005) The Tim-3 ligand galectin-9 negatively regulates T helper type 1 immunity. Nat Immunol 6(12):1245–1252. doi:10.1038/ni1271, PubMed PMID: 16286920
117. Weissman IL (2000) Stem cells: units of development, units of regeneration, and units in evolution. Cell 100(1):157–168, PubMed PMID: 10647940
118. Fearon DT, Manders P, Wagner SD (2001) Arrested differentiation, the self-renewing memory lymphocyte, and vaccination. Science 293(5528):248–250. doi:10.1126/science.1062589, PubMed PMID: 11452114
119. Gattinoni L, Restifo NP (2013) Moving T memory stem cells to the clinic. Blood 121(4):567–568. doi:10.1182/blood-2012-11-468660, PubMed PMID: 23349370
120. Lanzavecchia A, Sallusto F (2002) Progressive differentiation and selection of the fittest in the immune response. Nat Rev Immunol 2(12):982–987. doi:10.1038/nri959, PubMed PMID: 12461571
121. Zhang Y, Joe G, Hexner E, Zhu J, Emerson SG (2005) Host-reactive CD8+ memory stem cells in graft-versus-host disease. Nat Med 11(12):1299–1305. doi:10.1038/nm1326, PubMed PMID: 16288282
122. Cieri N, Camisa B, Cocchiarella F, Forcato M, Oliveira G, Provasi E et al (2013) IL-7 and IL-15 instruct the generation of human memory stem T cells from naive precursors. Blood 121(4):573–584. doi:10.1182/blood-2012-05-431718, PubMed PMID: 23160470
123. Gattinoni L, Zhong XS, Palmer DC, Ji Y, Hinrichs CS, Yu Z et al (2009) Wnt signaling arrests effector T cell differentiation and generates CD8+ memory stem cells. Nat Med 15(7):808–813. doi:10.1038/nm.1982, PubMed PMID: 19525962; PubMed Central PMCID: PMC2707501
124. Kryczek I, Zhao E, Liu Y, Wang Y, Vatan L, Szeliga W et al (2011) Human TH17 cells are long-lived effector memory cells. Sci Transl Med 3(104):104ra0. doi:10.1126/scitranslmed.3002949. PubMed PMID: 21998407; PubMed Central PMCID: PMC3345568

125. Muranski P, Borman ZA, Kerkar SP, Klebanoff CA, Ji Y, Sanchez-Perez L et al (2011) Th17 cells are long lived and retain a stem cell-like molecular signature. Immunity 35(6):972–985. doi:10.1016/j.immuni.2011.09.019, PubMed PMID: 22177921; PubMed Central PMCID: PMC3246082
126. Harari A, Dutoit V, Cellerai C, Bart PA, Du Pasquier RA, Pantaleo G (2006) Functional signatures of protective antiviral T-cell immunity in human virus infections. Immunol Rev 211:236–254. doi:10.1111/j.0105-2896.2006.00395.x, PubMed PMID: 16824132
127. Precopio ML, Betts MR, Parrino J, Price DA, Gostick E, Ambrozak DR et al (2007) Immunization with vaccinia virus induces polyfunctional and phenotypically distinctive CD8(+) T cell responses. J Exp Med 204(6):1405–1416. doi:10.1084/jem.20062363, PubMed PMID: 17535971; PubMed Central PMCID: PMC2118607
128. Duvall MG, Precopio ML, Ambrozak DA, Jaye A, McMichael AJ, Whittle HC et al (2008) Polyfunctional T cell responses are a hallmark of HIV-2 infection. Eur J Immunol 38(2): 350–363. doi:10.1002/eji.200737768, PubMed PMID: 18200635; PubMed Central PMCID: PMC2362391
129. Makedonas G, Betts MR (2006) Polyfunctional analysis of human T cell responses: importance in vaccine immunogenicity and natural infection. Springer Semin Immunopathol 28(3):209–219. doi:10.1007/s00281-006-0025-4, PubMed PMID: 16932955
130. Kryczek I, Banerjee M, Cheng P, Vatan L, Szeliga W, Wei S et al (2009) Phenotype, distribution, generation, and functional and clinical relevance of Th17 cells in the human tumor environments. Blood 114(6):1141–1149. doi:10.1182/blood-2009-03-208249, PubMed PMID: 19470694; PubMed Central PMCID: PMC2723011
131. Ding ZC, Huang L, Blazar BR, Yagita H, Mellor AL, Munn DH et al (2012) Polyfunctional CD4(+) T cells are essential for eradicating advanced B-cell lymphoma after chemotherapy. Blood 120(11):2229–2239. doi:10.1182/blood-2011-12-398321, PubMed PMID: 22859605; PubMed Central PMCID: PMC3447781
132. Yuan J, Gnjatic S, Li H, Powel S, Gallardo HF, Ritter E et al (2008) CTLA-4 blockade enhances polyfunctional NY-ESO-1 specific T cell responses in metastatic melanoma patients with clinical benefit. Proc Natl Acad Sci U S A 105(51):20410–20415. doi:10.1073/pnas.0810114105, PubMed PMID: 19074257; PubMed Central PMCID: PMC2629307
133. Han Q, Bagheri N, Bradshaw EM, Hafler DA, Lauffenburger DA, Love JC (2012) Polyfunctional responses by human T cells result from sequential release of cytokines. Proc Natl Acad Sci U S A 109(5):1607–1612. doi:10.1073/pnas.1117194109, PubMed PMID: 22160692; PubMed Central PMCID: PMC3277116

Chapter 4
Influence of Antigen Receptor Avidity, Affinity, and Specificity on Genetically Engineered T Cells

Kaoru Nagato, Timothy T. Spear, and Michael I. Nishimura

Abstract Affinity of a T cell receptor (TCR) refers to the strength of binding between a single TCR and a peptide/MHC (pMHC) complex while avidity of a TCR refers to the overall strength of binding between multiple cell-bound receptors and their ligands. Affinity of a TCR plays a role in thymic T cell selection and the generation of the TCR repertoire. In the absence of sensitivity to strong or weak antigen receptor signals, the homeostasis of the immune system is compromised and the risk of autoimmunity and/or infection ensues. Over the past few decades, T cells which have been genetically modified to target tumor antigens have been used to treat cancer patients. Antibody-based chimeric antigen receptors (CAR) were the first molecules used to redirect the specificity of normal T cells. CAR gene modified T cells can direct tumor rejection in mice and humans. Another class of receptors used to redirect the specificity of T cells is the T cell receptor (TCR). TCR gene modified T cells can also direct tumor rejection in mice and humans. CAR and TCR engineered T cells reactive against tumors have emerged as a promising advance in tumor immunotherapy. The rationale of this chapter is to study how CAR and TCR gene modified T cells modulate tumor immunity.

Keywords T cell • T cell receptor • TCR affinity • T cellavidity • Gene modified T cells • Adoptive cell therapy • Tumor immunity

K. Nagato
Department of Surgery, Cardinal Bernardin Cancer Center, Loyola University Medical Center, Maywood, IL 60153, USA

T.T. Spear • M.I. Nishimura (✉)
Department of Surgery, Cardinal Bernardin Cancer Center, Loyola University Medical Center, Maywood, IL 60153, USA
e-mail: mnishimura@luc.edu

P.A. Ascierto et al. (eds.), *Developments in T Cell Based Cancer Immunotherapies*, Cancer Drug Discovery and Development,
DOI 10.1007/978-3-319-21167-1_4

Adoptive T Cell Transfer for Cancer

Over the past decade, adoptive cell therapy (ACT) using ex vivo activated and expanded tumor infiltrating lymphocytes (TIL) has emerged as a promising advance in tumor immunotherapy [1–3]. While ACT has been effective in patients with metastatic melanoma [4, 5], when combined with nonmyeloablative chemotherapy preconditioning rendering the patient lymphopenic, the adoptive transfer of ex vivo tumor-infiltrating lymphocytes can mediate objective clinical responses in 50–70 % of patients with advanced melanoma [4–6]. More recently clinical responses have been reported in cervical cancer patients treated with TIL [7]. Despite these encouraging results, it is difficult to isolate tumor reactive T cells from all melanoma patients and expand their TIL to therapeutic numbers ex vivo. TIL therapy has been even more difficult for non-melanoma patients. Therefore, other sources of tumor reactive T cells are needed for ACT to become a therapy for most cancer patients.

Strategies to Genetically Engineer T Cells

One strategy which can overcome our inability to isolate and expand tumor reactive T cells for adoptive T cell transfer is to genetically modify a patients T cells with receptors capable of redirecting the specificity and function of the engineered T cells. The three main approaches developed include antibody based chimeric antigen receptors (CAR's), T cell antigen receptors (TCR's), and NKG2D-based CAR's. It has been shown that introducing these receptors into T cells can redirect the reactivity of any individual's T cells to recognize the target antigens of choice [8–10].

Antibody-Based Chimeric Antigen Receptors

Antibody-based CAR's were the first antigen recognition molecules used to redirect the specificity of normal T cells [8]. Antibody-based CAR's consist of a single chain monoclonal antibody extracellular domain fused to an intracellular signaling domain capable of activating the T cell. Because the antigen recognition domain is derived from a monoclonal antibody, CAR's directly recognize cell surface antigens without the need for antigen processing and presentation by MHC molecules [11, 12]. This limits the mechanisms of immune escape to loss of target antigen expression.

The first antibody-based CAR's were designed using the signaling domain from the Fcγ receptor [8]. More recent antibody-based CAR's use the CD3ζ signaling domain which have three immunoreceptor tyrosine-based activation motifs (ITAM's) [13] which more efficiently activate T cells than the Fcγ signaling domain which contains a single ITAM [14–16]. Antibody-based CAR's have been further extensively modified to include co-stimulatory domains and other features designed to extend the survival and function of CAR engineered T cells in vivo [17, 18]. Studies have shown that antibody-based CAR gene modified T cells can facilitate tumor rejection in mouse and humans [17, 19–25]. There are both positive and negative aspects of using antibody-based CAR engineered T cells for adoptive immunotherapy. It is clear that the affinity and specificity of the antigen binding properties of an antibody-based CAR as well as the function of antibody-based CAR engineered T cells impact on the effectiveness and the adverse events associated with antibody-based CAR-based T cell therapy.

NK Receptor-Based Chimeric Antigen Receptors

A new variant on the CAR design uses the extracellular portion of the NK activating receptor NKG2D [10]. NK cells have broad reactivity against tumor and virus-infected cells but not normal cells because the expression of the ligands for most NK activating receptors is associated with cellular stress [26, 27]. Therefore, it has been proposed that NKG2D CAR gene modified T cells would have broad anti-tumor reactivity with little or no killing of normal tissues [28]. Preclinical studies have found that NKG2D CAR gene modified T cells efficiently recognize mouse and human lymphoma [29], myeloma [30], and ovarian cancer cells [31–33] in vitro. Adoptive transfer of NKG2D CAR gene modified T cells inhibited tumor growth in vivo with no evidence of autoimmunity [10, 31]. Despite these encouraging results with NKG2D CAR, there is room for improvement. Similar modifications that were made to antibody-based CAR to improve their persistence and function in vivo have been made to NKG2D CAR's by including cassettes from other signaling/costimulatory molecules such as the DAP10 adapter protein [34]. While this new CAR approach has shown promise in animal models, its safety and effectiveness still needs to be evaluated in humans.

T Cell Receptors

Another important class of receptors used to redirect the specificity of T cells and the focus of this chapter is the TCR. The TCR is a cell surface heterodimer expressed on the cell surface of T lymphocytes that mediates the target cell recognition and specificity of a T cell [35]. There are two types of T cells which can be distinguished by the TCR's expressed on their surface. αβ T cells express a highly variable αβ TCR whereas γδ T cells express a highly conserved γδ TCR [36, 37]. TCR's mediate recognition of foreign or self-peptides bound to MHC molecule displayed by antigen presenting cells [38–43]. The interaction between the TCR and its peptide/MHC ligand plays a crucial role in determining the reactivity and specificity of an individual T cell. In this chapter, we will focus on TCR-gene modified T cells and the properties that are important when considering using them for laboratory and/or clinical studies.

TCR Gene Modified T Cells

Using a retroviral vector encoding a T cell receptor isolated from an HLA-A2 restricted, MART-1 reactive T cell clone [44], we were the first to demonstrate that the specificity of normal PBL-derived T cells could be redirected using a TCR directed against the melanoma antigen MART-1 [9, 45]. Since then, a large number of TCR's have been cloned which recognize numerous viral and tumor antigens [45–76]. Subsequently, TCR gene modified T cells have been shown to promote tumor rejection in both mouse and human studies [69, 74, 77–84]. Koya et al. adoptive transferred mouse T cells transduced to express an HLA-A2 restricted, tyrosinase reactive TCR into HLA-A2 transgenic mice bearing established B16 A2 melanoma tumors [79]. Using PET imaging, it was shown that these TCR transduced T cells trafficked specifically to melanoma tumors and mediated their rejection. More importantly, no signs of autoimmunity were found despite the strong anti-tumor reactivity against a melanoma/melanocyte differentiation antigen. The first use of TCR gene modified T cells in humans used an HLA-A2 restricted, MART-1 reactive TCR to establish a melanoma reactive allogeneic T cell line [77]. When injected directly into tumor lesions, this allogeneic line mediated complete regression many of the treated lesions. Of greater significance was the induction of systemic anti-tumor immunity leading to the regression of untreated tumors in two patients and progressive vitiligo in another two patients. In all subsequent clinical trials using TCR gene modified T cells, the cells were delivered systemically. As predicted, TCR gene modified T cells could mediate objective clinical responses in melanoma [64, 78, 80, 84, 85] as well as colorectal carcinoma [69], synovial cell sarcoma [80] and esophageal cancer [86] patients. Based on these animal and human studies, TCR gene transfer is rapidly becoming an exciting new therapeutic option for treating patients with advanced malignancies.

Cell surface expression of the TCR requires its association with the CD3 complex [87–92], Upon pMHC engagement by the TCR, a cascade begins with the initiation of a series of signaling events starting with the phosphorylation of ITAM's on the CD3 ζ chain leading to full T cell activation and function [93], The CD4 and CD8 co-receptors play a critical role in T cell activation and function by enhancing the binding of the TCR to the pMHC [94–97] and promoting the signaling by localizing *lck* to the TCR/CD3 complex [97–99]. There are many important factors involved in antigen recognition and T cell function which should be considered and thoroughly evaluated when considering how to generate the most effective anti-tumor T cells by introducing a foreign TCR.

Selecting a TCR for Use in TCR Gene Transfer Studies

The early question we faced by investigators in the field was how do we select the best TCR for our studies. For practical reasons, we and others cloned TCR's because we already had a T cell clones reactive against an antigen known to be expressed by tumor cells [46–49, 51–54, 56, 57, 59–73, 75, 76, 80] or T cell clones which recognized unknown antigens but the parent clone was known to be tumor reactive [45, 100]. In our case, the first TCR cloned was isolated from a T cell clone (TIL 5) known to be melanoma reactive but the target antigen was unknown [44]. Subsequent analysis of T cells expressing the TIL 5 TCR found that the target antigen was the melanoma/melanocyte differentiation antigen MART-1 [45] and the T cells transduced with the TIL 5 TCR required expression of the CD8 coreceptor for tumor recognition indicating it was a low affinity TCR [9]. These and other results from the analysis of TCR transduced T cells led most of the field to select a T clones for TCR cloning based on predicted TCR affinity as well as the target antigen they recognize (a more detailed discussion of TCR cloning strategies will follow in later sections) [47, 50, 57, 66, 69]. It is quite clear that many factors must be considered when embarking on a TCR gene transfer project, especially if the eventual goal is use is patients.

TCR Diversity

All higher vertebrates maintain a large and diverse repertoire of T cells capable of recognizing most of the pathogens we will ever encounter. This tremendous diversity is due to the extreme variability in the TCR expressed by each T cell. Each TCR chain consists of a variable (V) gene segment, a joining (J) region, a diversity (D) region in the β chain only, and a constant (C) region [101]. In the thymus, the different TCR α and β chains are generated randomly by germline rearrangements which bring together one of many Vα genes with one of many Jα regions for the TCRα chain or one of many Vβ genes, one of the two Dβ regions, and one of the thirteen Jβ regions for the TCRβ chain [102, 103]. The numerous combinations of elements in each TCR chain together with the combinations of α and β chain pairing contribute to some of the TCR diversity observed in nature [43]. However, the majority of the diversity in TCR's expressed by mature T cell results from the addition and deletion of bases at the Vα-Jα, Vβ-Dβ, and, Dβ-Jβ junctions which occurs during TCR gene rearrangement [43]. This hypervariable region of the TCR α chain or TCR β chain is the third complementarity determining region (CDR3) region. The TCR α and β chain CDR3 regions are considered to be the most important regions of the TCR for antigen recognition.

The random TCR rearrangements that lead to such a large potential TCR repertoire means we should have T cell immunity against most pathogens presented by most MHC molecules. Because antigen recognition by T cells is restricted by self-MHC molecules, the random generation of TCR chain results in a high percentage of the TCR's that are restricted by other, nonself-MHC molecules. Positive selection enables developing thymocytes expressing TCR's capable of binding antigen/self-MHC molecules in the thymus with the "correct" affinity to be protected from programmed cell death. These positively selected T cells then complete T cell development resulting in a pool of mature T cells restricted only by self-MHC. In contrast, thymocytes expressing TCR's whose affinity for the host antigen/MHC is too low (death by neglect) or too high (negative selection) are not protected from programmed cell death and do not complete T cell development eliminating them from the pool of mature T cells [104]. The net effect of thymic selection is to save "immunologic space" for T cells that are beneficial to the host by eliminating T cells that are not self-MHC restricted and those that have the potential for autoimmunity. These are important factors to consider for TCR gene transfer studies given that most human tumor reactive T cells recognize normal, non-mutated self-antigens [105–107].

Given the extreme TCR diversity, two critical questions initially raised by the field are do we have a choice in TCR's that target a single antigen and if we have a choice, does it matter which TCR we select to use for a given target. We and others initially focused on the diversity of the TCR repertoire against a single target antigen by testing the concept of restricted TCR V gene usage [44, 108–112]. Because of clonal selection theory, it makes sense that our immune system would select for one or a limited number of TCR rearrangements that "best" recognize a given antigen [113]. Using the reactivity of the first TCR we cloned as a starting point, we found that the TCR diversity among MART-1 reactive T cell clones to be very high [109, 111]. More importantly, using a panel peptides with homology to the MART-1 peptide, we found individual MART-1 reactive T cells clones recognized different subsets of the peptide homologs [114]. Furthermore, peptides recognized commonly by the different MART-1 reactive T cell clones were recognized by each clone with different efficiencies [114]. Diverse TCR V gene usage is not a property of MART-1 reactive T cells since we also found high TCR diversity among gp100 and HCV reactive T cell clones (unpublished). These results indicate that each TCR's can recognize a single antigen differently and the how a T cell recognizes its target might make a difference in the effectiveness and the crossreactivity (safety) of the resulting TCR gene modified T cells in vivo.

Given that melanoma TIL cultures are known to be effective in treating patients with advanced disease, we and others investigated the TCR repertoire in melanoma TIL with the goal of identifying therapeutic T cells which would be great candidates for TCR cloning. Expanded TIL cultures used to treat patients had at least 10–12 distinct T cell clonotypes [110]. Included in TIL cultures were a mixture of T cells which recognized shared melanoma antigens as well as those that recognize only their autologous tumor [44, 112]. As TIL cultures expand, we found that the relative frequency of each individual clonotype can vary dramatically [115]. Therefore, it is difficult to predict which T cell clone from a TIL culture might be therapeutic making the decision of which TCR to clone from a TIL culture difficult.

Next generation DNA sequencing techniques have enable the field to perform a more detailed analysis of the TCR repertoire of TIL. Several studies have found very large TCR diversity among TIL in tumor lesions [116–118]. It is not surprising that fresh TIL samples were composed of T cells which recognized shared tumor antigens as well as those that recognized only their autologous tumor, presumably due to mutations in the target antigens [116–118]. Despite new technologies which can sort for T cells based on their antigen reactivity, it would seem that this large and diverse TCR repertoire poses a significant challenge for selecting the "right" TCR for gene transfer studies.

Isolating High Affinity TCR's

During the course of our studies, we identified three CD8-independent TCR's which have high affinity for their antigen [50, 57, 66]. The rationale for cloning high affinity TCR's came from the analysis of T cells transduced with our CD8-dependent low affinity TIL 5 TCR. We found it was possible to engineer both $CD4^+$ and $CD8^+$ T cells to recognize target cells loaded with the MART-1:27–35 peptide [9]. However, only $CD8^+$ T cells could recognize MART-1^+ tumor cells indicating the affinity of the introduced TCR was inadequate to recognize the levels of antigen naturally presented by tumor cells [9]. Since tumor recognition by TIL 5 TCR transduced T cells was dependent on the expression of the CD8 co-receptor, we speculated that if we could identify CD8 independent or high affinity TCR, the TCR gene transfer approach could be used to make better $CD8^+$ T cells as well as MHC class I restricted $CD4^+$ T cells for adoptive T cell transfer.

The first CD8 independent TCR we identified was isolated from TIL 1383I, an MHC class I restricted $CD4^+$T cell [119]. We correctly hypothesized that the TIL 1383I TCR could transfer MHC class I restricted anti-tumor reactivity to both $CD8^+$ and $CD4^+$ T cells [50, 55]. Others have used the strategy of CD8 independence to identify and clone high affinity TCR's [53, 120]. It is known that mouse CD8 does not bind to the α3 domain of human MHC class I molecules [121–123]. Thus, HLA-A2 restricted mouse T cell clones that recognize human tumor cell lines would be CD8 independent and would likely express high affinity TCR's. The approach of vaccinating HLA-A2 transgenic mice was used successfully to identify T cell clones expressing high affinity TCR's targeting p53 [53] and CEA [120]. While CD8 independent tumor cell recognition by a T cell clone should be a hallmark of a high affinity TCR, we have found it not always to be true. We identified a gp100 reactive T cell clone designated T4H2 that was a $CD4^-$/$CD8^-$ T cell that efficiently recognized HLA-$A2^+$ $gp100^+$ human melanoma cells in vitro [65, 124]. However, when the T4H2 TCR was cloned and expressed in human T cells, we were surprised to find it required CD8 expression for tumor cell recognition [124]. Therefore, each TCR must be evaluated individually to ensure that a CD8-independent/high affinity TCR has been cloned.

More recently, TCR's have been genetically modified to improve their physical properties [60, 61, 74, 80, 125–128]. Using phage display, TCR's have undergone "evolution" to select for high affinity binding to pMHC [126, 129]. Although it might be predicted that changes to the CDR3 region (which is responsible for the majority of the TCR diversity) would lead to the highest affinity TCR's, mutations in all three CDR regions can lead to high affinity TCR's [125]. Collectively, these results indicate that TCR affinity can make a difference in target recognition by TCR transduced T cells and each TCR must be carefully evaluated for how its properties impact T cells.

Consequences of TCR Chain Mispairing

The ideal function of a TCR gene-modified T cell depends on adequate expression of the introduced TCR, so that correct pairing of the introduced TCR α and β chains and proper association with the CD3 complex occurs. Since a mature T cell expresses its own TCR, the introduced TCR α and β chains compete with the endogenous TCR α and β chains for expression and assembly on cell surface [130, 131]. Another disadvantage of introducing another set of TCR α and β chains into T cells is the potential for mispairing of introduced and endogenous TCR chains. Mispairing of TCR chains reduces the level of expression of the introduced TCR [45, 132] and can lead to the creation the unfavorable T cells with self-reactive TCR's, potentially inducing autoimmunity [133, 134]. Several strategies have been developed to ensure proper pairing of the introduced TCR. It has been shown that small interfering RNA (siRNA) constructs reduces expression of endogenous TCR chains thus facilitating the pairing of this introduced TCR α and β chains [135]. Codon optimizing the TCR α and β chains genes increases translation of the TCR transgenes leading to increased expression and a competitive advantage in the assembly of the introduced TCR on the T cell surface [79, 136]. Another set of strategies directly promotes pairing of the introduced TCR α and β chains. These strategies include adding cysteine residues to the constant regions in order to promote inter-chain disulfide bridge [137], modifying the glycosylation of the TCR [138], and including leucine zippers [139] to improve pairing of the introduced TCR α and β chains and enhance the expression and function of the introduced TCR.

TCR expression is limited by the expression of elements of CD3 complex [140]. Since T cells have endogenous TCR's, the introduced TCR has to compete with the endogenous TCR for association with the CD3 complex on the T cell surface. This issue has been addressed in several ways. It was shown that human TCR's will not express on mouse T cells unless the human constant regions have been replaced by mouse constant regions [141, 142]. Using this approach, we and others found that human TCR genes can achieve high expression and function when the introduced TCR has mouse constant regions instead of human constant regions [50, 123, 132, 143]. It was subsequently shown that elements of the human TCR constant regions do not efficiently pair with the mouse CD3 complex [144]. A definitive approach to solve the TCR pairing issue was a recombinant set of TCR chains consisting of a single chain Vα-Vβ-Cβ and a Cα chain [144]. These two TCR chains can only pair with each other and not the endogenous TCR α and β chains. These different strategies are being incorporated into TCR's individually or in various combinations with the goal of obtaining higher expression, pairing, and function of the introduced TCR.

Relationship Between TCR Affinity and T Cell Function

TCR affinity plays a significant role in determining the sensitivity of a T cell to antigen recognition. As discussed above, TCR's have different affinities for the pMHC with the range of affinities of TCR's expressed by mature T cells being tightly regulated by thymic selection. Generally, the affinity of TCR for pMHC is lower than Ab/Ag interactions [43, 104, 145]. Our ability to engineer T cells with high affinity TCR's allows us to circumvent the limitations thymic selection places on the affinity of TCR's that target most tumor antigens (self antigens) by generating T cells which would not normally be found in the periphery. While it is clear that we can make more efficient tumor reactive $CD8^+$ T cells [55, 61, 74, 80, 128] and the novel population of MHC class I restricted $CD4^+$ T cells [50, 66, 120] by engineering them with high affinity TCR's, there could be negative consequences as well. T cells expressing TCR's with extremely high affinity for pMHC can undergo activation-induced cell death (AICD) upon encounter with antigen. AICD would lead to the destruction of the very effector cells which are intended to destroy the patient's cancer [146, 147]. T cells expressing high affinity TCR's can also lead to autoimmunity as observed in one transgenic mouse model [148] and two clinical trials [64, 69]. Given these observations, the question of the need to use a high affinity TCR for effective anti-tumor immunity remains unresolved.

There are several important facts to consider regarding the relationship between TCR affinity and T cell function if we are to understand how TCR affinity relates to T cell function. Contrary to the predictions of thymic selection, T cells reactive with self-antigens are not always deleted in the thymus since many found in the periphery. In both animal and human systems, self-reactive T cells can be induced to mediate tumor rejection and in some cases, autoimmunity. These T cells are generally CD8 dependent meaning they express TCR's with relatively low affinity for pMHC. Their anti-self-reactivity is limited until their physiology or the host tumor environment is altered. Another key observation is that T cells derived from a single T cell clone can have varied T cell function. Under certain circumstances, T cells can be very antigen reactive whereas in other circumstances they are weakly antigen reactive [149]. Therefore, it is likely that the biology of T cells dictates T cell function and their TCR affinity plays a less role than previously thought.

There are several mechanisms which can explain how the function of a T cell is influenced by its environment. It is well known that T cells become refractory to immune function resulting from the level of immune suppression in the tumor bearing host [150–156]. Furthermore, reactive oxygen and nitrogen species in the tumor bearing host can promote T cell death [73, 157–161]. While none of these suppressive mechanisms require changes in TCR affinity to reduce immune function of a T cell, there are examples of immune suppression mechanisms that do seem to impact on TCR affinity. We know that one key role of CD8 is to stabilize the TCR/pMHC complex. It has been observed that the spatial relationship of the TCR and CD8 on the surface of a T cell can vary leading to differences in the relative stability of the TCR/pMHC complex [162]. Similarly, the ratio of the high affinity form of CD8 (CD8αβ) vs the low affinity form of CD8 (CD8αα) can vary on each T cells expressing the same TCR [163]. The result is CD8αβ have higher functional avidity than CD8αα T cells which translates to higher anti-tumor activity. And finally, MDSC have been shown to modify the TCR proteins by nitration of tyrosine residues leading to weaker binding to pMHC [164]. In all these suppressive mechanisms, the function of a T cell expressing a high affinity TCR could be as easily overcome as the function of a T cell expressing a low affinity TCR. Therefore, the real need for using a high affinity TCR in TCR gene transfer studies remains unclear.

While you might be left with the impression that selecting a high affinity TCR may not be the best choice for TCR gene transfer studies, there are clear advantages to using a high affinity TCR to engineer T cells for adoptive transfer studies. The ability to generate tumor reactive MHC class I restricted $CD4^+$ T cells is one benefit to a high affinity TCR [55]. Their ability to produce helper cytokines upon antigen stimulation offers the real opportunity to provide MHC class I restricted T cell help at the tumor site [55, 165, 166]. Therefore, TCR transduced $CD4^+$ T cells might facilitate with cross priming or epitope spreading leading to broad systemic anti-tumorimmunity. Another application for a high affinity TCR would be to engineer $CD4^+$ Treg cells [167]. Like T_h cells, MHC class I restricted Treg cells could directly inhibit an immune reaction at sites of autoimmunity thus reducing the severity of the disease. Despite the potential negatives for using a high affinity TCR, these significant benefits which should be considered for any TCR gene transfer study.

Tumor Rejection by TCR Transduced T Cells

We have addressed how the specificity of normal T cells can be redirected using TCR genes. We have also discussed the factors that influence the selection of a TCR for TCR gene transfer studies and how TCR affinity impacts T cell function. The next major topic is the actual impact of TCR gene modified T cells on the growth of tumors in vivo and to consider the adverse events that have been observed.

Mouse Tumor Treatment Studies Using TCR Gene Modified T Cells

Adoptive transfer of TCR gene modified T cells has been used by several investigators to treat mice with established tumors. Kessels et al. [168] described the first study using TCR gene transfer to treat tumors in mice. The F5 TCR used in this study recognized the immunodominant H-2D^b restricted epitope from the influenza NP. Mouse splenic T cells, transduced to express the F5 TCR persisted in vivo and mediated the elimination of virus in influenza infected mice and mediated the rejection of NP expressing EL-4 cells. No side effects due to autoimmunity or other factors were found in treated mice. These results suggested that the use of TCR gene modified T cells could be safe and effective in vivo.

Mouse T cells transduced to express the "murinized" version of the human TIL 1383I TCR which specifically recognize the MHC class I-restricted tyrosinase 368–376 peptide presented by HLA-A2 were evaluated for safety, specificity, and anti-tumor efficacy in an HLA-A2 transgenic mouse tumor model [79]. Encoded in the vector was a modified thymidine kinase gene which has a unique F18-labeled substrate that allowed the tracking of the engineered T cells using PET imaging. The TIL 1383I TCR transduced T cells were found to persist in vivo. Furthermore, PET imaging demonstrated the TIL 1383I TCR transduced T cells specifically trafficked to B16 A2 melanoma tumors (HLA-A2$^+$, tyrosinase$^+$) but not EL4 A2 thymoma tumors (HLA-A2$^+$, tyrosinase$^-$). The TIL 1383I TCR transduced T cells also exhibited robust anti-tumor activity and the treated mice had improved survival. Despite the specific homing of the TIL 1383I TCR transduced T cells to sites of antigen and their strong anti-tumor activity, no vitiligo was observed meaning that T cells expressing the high TIL 1383I TCR did not lead to melanocyte destruction or autoimmunity. Several other groups have also reported that TCR gene modified T cells can effectively eliminate established tumors in mouse models [71, 74, 169–174]. While these results further support the safety and efficacy of TCR gene modified T cells in vivo, there have been a few reports of harmful self-reactivity which has been attributed to TCR mispairing [133, 134]. In summary, while TCR transduced mouse T cells can be effective in treating established tumors, the issue of autoimmunity needs more careful study. However, given that the animal models do not often adequately mimic the human disease situation, ultimately the true test of safety and effectiveness of the approach must be evaluated in patients.

Clinical Trials Using TCR Gene Modified T Cells

While still early in development, TCR gene modified T cells have been evaluated in humans. The first two clinical trials to treat cancer patients with TCR transduced T cells were reported in 2006. In the first trial, an allogeneic T cell line was transduced to express the TIL 5 TCR to recognize the melanoma/melanocyte differentiation antigen MART-1 [77]. This allogeneic line could effectively kill any HLA-A2$^+$ MART-1$^+$ tumor so the trial was designed to deliver the T cells via intratumoral injection. The goal of injecting TIL 5 TCR gene modified T cells into individual tumor lesions was the destruction of the treated tumor and to induce local and systemic anti-tumor immunity through cross priming leading to regression of untreated lesions. In this trial, one of 15 patients obtained a partial response (PR) and another patient had regression of uninjected lesions. The only treatment related, adverse events noted were two patients that developed vitiligo. These observations supported the premise that local tumor destruction can lead to systemic anti-tumor immunity and in one case, tumor regression. The second clinical trial reported in 2006 used T cells engineered with a different MART-1 reactive TCR. The TCR transduced T cells were delivered by systemic infusion into stage IV melanoma patients pretreated with nonmyeloablative chemotherapy [78]. Seventeen patients were treated with systemic infusion of autologous TCR gene modified T cells. Two of the 17 patients had an objective PR, and there were no adverse events observed. It was noted that the two patients that responded had prolonged survival of the infused TCR gene modified T cells [78]. These first clinical trials indicated that the use of TCR gene modified T cells was feasible, could lead to objective clinical responses, and was generally safe.

In 2009, a third TCR gene transfer study was reported targeting the melanoma differentiation antigens MART-1 and gp100 using two distinct TCR's [64]. The anti-MART-1 TCR had higher affinity (CD8 independent) than the TCR's used in the previous trials but the anti-gp100 TCR was a low affinity CD8 dependent TCR.

In this clinical trial, 6 of the 20 patients treated (30 %) had objective clinical responses. The anti-gp100 TCR reported in this study was of mouse origin and was selected for CD8 co-receptor independent tumor cell recognition. In patients treated with the anti-gp100 TCR, 3 of the 16 (19 %) had objective clinical responses. It should also be noted that the use of T cells engineered with these high affinity TCR's was implicated in the observed toxicity in the eye and inner ear which was associated with the destruction of normal melanocytes. However, almost all patients resolved these toxicities naturally or with the use of topical steroids. This study concluded that using a high affinity TCR was better than a low affinity TCR for engineering T cells. These clinical trials further supported the assertion that TCR gene modified T cells are generally safe and well tolerated by the patients. More important, they have the potential to be an effective therapeutic for cancer patients. While T cells engineered with high-affinity TCR's did lead to improved clinical outcomes for melanoma patients, the adverse events need to be closely monitored.

Since these early clinical trials, additional clinical trials have been reported using other TCR's. One trial use a high affinity mouse TCR which recognizes an HLA-A2 restricted epitope from CEA which is expressed by colon cancer cells. When infused into patients with metastatic colon cancer, CEA TCR transduced T cells mediated tumor regression one of the three patients treated [69]. However, a severe transient inflammatory colitis in normal colon was induced in all three patients leading to the clinical trial being discontinued.

A subsequent clinical trial reported using a modified TCR targeting NY-ESO-1 to engineer T cells for treating patients with metastatic melanoma or metastatic synovial cell sarcoma patients [80]. Tumor regression was seen in four of the six patients with synovial cell sarcoma and in five of 11 patients with melanoma. It was important to note that none of these patients had any on-target toxicities. It is important to note that this study was the first clinical trial to demonstrate that TCR gene modified T cells could lead to clinical responses in patients with non-melanoma tumors.

Adverse Events

The adverse events in clinical trials using TCR gene modified T cells warranted further investigation into targets considered to be tumor specific. Based on the previous clinical trials, it speculated that targeting cancer testis antigens such as NY-ESO-1, MAGE-A3, MAGE-A4 with TCR gene modified T cells would be generally safe and well tolerated by the patients [83, 175–177]. However, a clinical trial using an affinity-enhanced MAGE-A3 TCR induced severe neurologic toxicity [84]. The cause of this neurotoxicity is considered to be crossreactivity with another member of the MAGE-A family (MAGE-A12). The severe adverse events were due to unexpected MAGE-A12 on cells of the central nerve system. More recently, it was observed that the trial using high affinity MAGE-A3 TCR also induced severe cardiac toxicity. This toxicity was not due to off-tumor antigen expression or recognition of epitopes from related cancer-testis antigens but recognition of the unrelated human protein titin in cardiac tissue [178]. It is clear that while the affinity-enhanced TCR modified T cells is an attractive treatment for advanced malignancies [179, 180], it is necessary to identify appropriate epitopes to ensure on-target specificity that will lead to improve clinical outcomes.

Conclusion

While the field of TCR gene transfer is still quite new, the use of TCR transduced T represents a promising new approach cells for treating cancer patients. It clearly circumvents the hurdles problems of obtaining tumor reactive T cells that are associated with TIL therapy and other forms of adoptive T cell transfer. The use of viral vector to engineer T cells with TCR genes enables us to generate populations of autologous antigen-reactive T cells for any patient regardless of their ability to naturally mount an immune response against the target antigen. It is clear that the TCR gene transfer approach is feasible for treating patients and the TCR transduced T cells can be delivered safely. However, one of the most important aspects of the approach is that the objective clinical responses have been obtained in all trials indicates these genetically engineered T cells can be effective. As we learn more about how the TCR influences T cell function, we will be able to design better strategies for using TCR gene modified T cells for treating patients.

References

1. Rosenberg SA, Restifo NP, Yang JC, Morgan RA, Dudley ME (2008) Adoptive cell transfer: a clinical path to effective cancer immunotherapy. Nat Rev Cancer 8(4):299–308
2. June CH, Blazar BR, Riley JL (2009) Engineering lymphocyte subsets: tools, trials and tribulations. Nat Rev Immunol 9(10):704–716
3. Weber J, Atkins M, Hwu P, Radvanyi L, Sznol M, Yee C (2011) White paper on adoptive cell therapy for cancer with tumor-infiltrating lymphocytes: a report of the CTEP subcommittee on adoptive cell therapy. Clin Cancer Res 17(7):1664–1673
4. Dudley ME, Wunderlich JR, Robbins PF, Yang JC, Hwu P, Schwartzentruber DJ, Topalian SL, Sherry R, Restifo NP, Hubicki AM, Robinson MR, Raffeld M, Duray P, Seipp CA, Rogers-Freezer L, Morton KE, Mavroukakis SA, White DE, Rosenberg SA (2002) Cancer regression and autoimmunity in patients after clonal repopulation with antitumor lymphocytes. Science 298(5594):850–854
5. Dudley ME, Wunderlich JR, Yang JC, Sherry RM, Topalian SL, Restifo NP, Royal RE, Kammula U, White DE, Mavroukakis SA, Rogers LJ, Gracia GJ, Jones SA, Mangiameli DP, Pelletier MM, Gea-Banacloche J, Robinson MR, Berman DM, Filie AC, Abati A, Rosenberg SA (2005) Adoptive cell transfer therapy following non-myeloablative but lymphodepleting chemotherapy for the treatment of patients with refractory metastatic melanoma. J Clin Oncol 23(10):2346–2357
6. Dudley ME, Yang JC, Sherry R, Hughes MS, Royal R, Kammula U, Robbins PF, Huang J, Citrin DE, Leitman SF, Wunderlich J, Restifo NP, Thomasian A, Downey SG, Smith FO, Klapper J, Morton K, Laurencot C, White DE, Rosenberg SA (2008) Adoptive cell therapy

for patients with metastatic melanoma: evaluation of intensive myeloablative chemoradiation preparative regimens. J Clin Oncol 26(32):5233–5239
7. Stevanovic S, Draper LM, Langhan MM, Campbell TE, Kwong ML, Wunderlich JR, Dudley ME, Yang JC, Sherry RM, Kammula US, Restifo NP, Rosenberg SA, Hinrichs CS (2015) Complete regression of metastatic cervical cancer after treatment with human papillomavirus-targeted tumor-infiltrating T cells. J Clin Oncol 33(14):1543–1550
8. Gross G, Waks T, Eshhar Z (1989) Expression of immunoglobulin-T-cell receptor chimeric molecules as functional receptors with antibody-type specificity. Proc Natl Acad Sci U S A 86(24):10024–10028
9. Clay TM, Custer MC, Sachs J, Hwu P, Rosenberg SA, Nishimura MI (1999) Efficient transfer of a tumor antigen-reactive TCR to human peripheral blood lymphocytes confers anti-tumor reactivity. J Immunol 163(1):507–513
10. Zhang T, Lemoi BA, Sentman CL (2005) Chimeric NK-receptor-bearing T cells mediate antitumor immunotherapy. Blood 106(5):1544–1551
11. Curran KJ, Pegram HJ, Brentjens RJ (2012) Chimeric antigen receptors for T cell immunotherapy: current understanding and future directions. J Gene Med 14(6):405–415
12. Jensen MC, Riddell SR (2015) Designing chimeric antigen receptors to effectively and safely target tumors. Curr Opin Immunol 33:9–15
13. Cambier JC (1995) Antigen and Fc receptor signaling. The awesome power of the immunoreceptor tyrosine-based activation motif (ITAM). J Immunol 155(7):3281–3285
14. Romeo C, Seed B (1991) Cellular immunity to HIV activated by CD4 fused to T cell or Fc receptor polypeptides. Cell 64(5):1037–1046
15. Romeo C, Kolanus W, Amiot M, Seed B (1992) Activation of immune system effector function by T-cell or Fc receptor intracellular domains. Cold Spring Harb Symp Quant Biol 57:117–125
16. Stancovski I, Schindler DG, Waks T, Yarden Y, Sela M, Eshhar Z (1993) Targeting of T lymphocytes to Neu/HER2-expressing cells using chimeric single chain Fv receptors. J Immunol 151(11):6577–6582
17. Eshhar Z (2010) Adoptive cancer immunotherapy using genetically engineered designer T-cells: first steps into the clinic. Curr Opin Mol Ther 12(1):55–63
18. Pegram HJ, Park JH, Brentjens RJ (2014) CD28z CARs and armored CARs. Cancer J 20(2):127–133
19. Haynes NM, Trapani JA, Teng MW, Jackson JT, Cerruti L, Jane SM, Kershaw MH, Smyth MJ, Darcy PK (2002) Single-chain antigen recognition receptors that costimulate potent rejection of established experimental tumors. Blood 100(9):3155–3163
20. Moeller M, Haynes NM, Trapani JA, Teng MW, Jackson JT, Tanner JE, Cerutti L, Jane SM, Kershaw MH, Smyth MJ, Darcy PK (2004) A functional role for CD28 costimulation in tumor recognition by single-chain receptor-modified T cells. Cancer Gene Ther 11(5):371–379
21. Moeller M, Haynes NM, Kershaw MH, Jackson JT, Teng MW, Street SE, Cerutti L, Jane SM, Trapani JA, Smyth MJ, Darcy PK (2005) Adoptive transfer of gene-engineered CD4+ helper T cells induces potent primary and secondary tumor rejection. Blood 106(9):2995–3003
22. Dotti G, Gottschalk S, Savoldo B, Brenner MK (2014) Design and development of therapies using chimeric antigen receptor-expressing T cells. Immunol Rev 257(1):107–126
23. Jena B, Moyes JS, Huls H, Cooper LJ (2014) Driving CAR-based T-cell therapy to success. Curr Hematol Malig Rep 9(1):50–56
24. Jensen MC, Riddell SR (2014) Design and implementation of adoptive therapy with chimeric antigen receptor-modified T cells. Immunol Rev 257(1):127–144
25. Zhao A, Tohidkia MR, Siegel DL, Coukos G, Omidi Y (2014) Phage antibody display libraries: a powerful antibody discovery platform for immunotherapy. Crit Rev Biotechnol 1–14
26. Iannello A, Raulet DH (2013) Immune surveillance of unhealthy cells by natural killer cells. Cold Spring Harb Symp Quant Biol 78:249–257
27. Paschen A, Baingo J, Schadendorf D (2014) Expression of stress ligands of the immunoreceptor NKG2D in melanoma: regulation and clinical significance. Eur J Cell Biol 93(1–2):49–54

28. Sentman CL, Meehan KR (2014) NKG2D CARs as cell therapy for cancer. Cancer J 20(2):156–159
29. Zhang T, Barber A, Sentman CL (2007) Chimeric NKG2D modified T cells inhibit systemic T-cell lymphoma growth in a manner involving multiple cytokines and cytotoxic pathways. Cancer Res 67(22):11029–11036
30. Barber A, Meehan KR, Sentman CL (2011) Treatment of multiple myeloma with adoptively transferred chimeric NKG2D receptor-expressing T cells. Gene Ther 18(5):509–516
31. Barber A, Zhang T, Sentman CL (2008) Immunotherapy with chimeric NKG2D receptors leads to long-term tumor-free survival and development of host antitumor immunity in murine ovarian cancer. J Immunol 180(1):72–78
32. Barber A, Rynda A, Sentman CL (2009) Chimeric NKG2D expressing T cells eliminate immunosuppression and activate immunity within the ovarian tumor microenvironment. J Immunol 183(11):6939–6947
33. Song DG, Ye Q, Santoro S, Fang C, Best A, Powell DJ Jr (2013) Chimeric NKG2D CAR-expressing T cell-mediated attack of human ovarian cancer is enhanced by histone deacetylase inhibition. Hum Gene Ther 24(3):295–305
34. Chang YH, Connolly J, Shimasaki N, Mimura K, Kono K, Campana D (2013) A chimeric receptor with NKG2D specificity enhances natural killer cell activation and killing of tumor cells. Cancer Res 73(6):1777–1786
35. McIntyre BW, Allison JP (1983) The mouse T cell receptor: structural heterogeneity of molecules of normal T cells defined by xenoantiserum. Cell 34(3):739–746
36. Sangster RN, Minowada J, Suciu-Foca N, Minden M, Mak TW (1986) Rearrangement and expression of the alpha, beta, and gamma chain T cell receptor genes in human thymic leukemia cells and functional T cells. J Exp Med 163(6):1491–1508
37. Toyonaga B, Mak TW (1987) Genes of the T-cell antigen receptor in normal and malignant T cells. Annu Rev Immunol 5:585–620
38. Bjorkman PJ, Saper MA, Samraoui B, Bennett WS, Strominger JL, Wiley DC (1987) The foreign antigen binding site and T cell recognition regions of class I histocompatibility antigens. Nature 329(6139):512–518
39. Davis MM, Bjorkman PJ (1989) A model for T cell receptor and MHC/peptide interaction. Adv Exp Med Biol 254:13–16
40. Davis MM, Chien YH, Bjorkman PJ, Elliott JF, Iwashima M, Rock EP, Patten PA (1989) A possible basis for major histocompatibility complex-restricted T-cell recognition. Philos Trans R Soc Lond B Biol Sci 323(1217):521–524
41. Brown JH, Jardetzky TS, Gorga JC, Stern LJ, Urban RG, Strominger JL, Wiley DC (1993) Three-dimensional structure of the human class II histocompatibility antigen HLA-DR1. Nature 364(6432):33–39
42. Stern LJ, Brown JH, Jardetzky TS, Gorga JC, Urban RG, Strominger JL, Wiley DC (1994) Crystal structure of the human class II MHC protein HLA-DR1 complexed with an influenza virus peptide. Nature 368(6468):215–221
43. Davis MM, Boniface JJ, Reich Z, Lyons D, Hampl J, Arden B, Chien Y (1998) Ligand recognition by alpha beta T cell receptors. Annu Rev Immunol 16:523–544
44. Shilyansky J, Nishimura MI, Yannelli JR, Kawakami Y, Jacknin LS, Charmley P, Rosenberg SA (1994) T-cell receptor usage by melanoma-specific clonal and highly oligoclonal tumor-infiltrating lymphocyte lines. Proc Natl Acad Sci U S A 91(7):2829–2833
45. Cole DJ, Weil DP, Shilyansky J, Custer M, Kawakami Y, Rosenberg SA, Nishimura MI (1995) Characterization of the functional specificity of a cloned T-cell receptor heterodimer recognizing the MART-1 melanoma antigen. Cancer Res 55(4):748–752
46. Orentas RJ, Roskopf SJ, Nolan GP, Nishimura MI (2001) Retroviral transduction of a T cell receptor specific for an Epstein-Barr virus-encoded peptide. Clin Immunol 98(2):220–228
47. Stanislawski T, Voss RH, Lotz C, Sadovnikova E, Willemsen RA, Kuball J, Ruppert T, Bolhuis RL, Melief CJ, Huber C, Stauss HJ, Theobald M (2001) Circumventing tolerance to a human MDM2-derived tumor antigen by TCR gene transfer. Nat Immunol 2(10):962–970

48. Heemskerk MH, Hoogeboom M, de Paus RA, Kester MG, van der Hoorn MA, Goulmy E, Willemze R, Falkenburg JH (2003) Redirection of antileukemic reactivity of peripheral T lymphocytes using gene transfer of minor histocompatibility antigen HA-2-specific T-cell receptor complexes expressing a conserved alpha joining region. Blood 102(10):3530–3540
49. Morgan RA, Dudley ME, Yu YY, Zheng Z, Robbins PF, Theoret MR, Wunderlich JR, Hughes MS, Restifo NP, Rosenberg SA (2003) High efficiency TCR gene transfer into primary human lymphocytes affords avid recognition of melanoma tumor antigen glycoprotein 100 and does not alter the recognition of autologous melanoma antigens. J Immunol 171(6):3287–3295
50. Roszkowski JJ, Yu DC, Rubinstein MP, McKee MD, Cole DJ, Nishimura MI (2003) CD8-independent tumor cell recognition is a property of the T cell receptor and not the T cell. J Immunol 170(5):2582–2589
51. Heemskerk MH, Hoogeboom M, Hagedoorn R, Kester MG, Willemze R, Falkenburg JH (2004) Reprogramming of virus-specific T cells into leukemia-reactive T cells using T cell receptor gene transfer. J Exp Med 199(7):885–894
52. Hughes MS, Yu YY, Dudley ME, Zheng Z, Robbins PF, Li Y, Wunderlich J, Hawley RG, Moayeri M, Rosenberg SA, Morgan RA (2005) Transfer of a TCR gene derived from a patient with a marked antitumor response conveys highly active T-cell effector functions. Hum Gene Ther 16(4):457–472
53. Kuball J, Schmitz FW, Voss RH, Ferreira EA, Engel R, Guillaume P, Strand S, Romero P, Huber C, Sherman LA, Theobald M (2005) Cooperation of human tumor-reactive CD4+ and CD8+ T cells after redirection of their specificity by a high-affinity p53A2.1-specific TCR. Immunity 22(1):117–129
54. Mommaas B, van Halteren AG, Pool J, van der Veken L, Wieles B, Heemskerk MH, Goulmy E (2005) Adult and cord blood T cells can acquire HA-1 specificity through HA-1 T-cell receptor gene transfer. Haematologica 90(10):1415–1421
55. Roszkowski JJ, Lyons GE, Kast WM, Yee C, Van Besien K, Nishimura MI (2005) Simultaneous generation of CD8+ and CD4+ melanoma-reactive T cells by retroviral-mediated transfer of a single T-cell receptor. Cancer Res 65(4):1570–1576
56. Zhao Y, Zheng Z, Robbins PF, Khong HT, Rosenberg SA, Morgan RA (2005) Primary human lymphocytes transduced with NY-ESO-1 antigen-specific TCR genes recognize and kill diverse human tumor cell lines. J Immunol 174(7):4415–4423
57. Callender GG, Rosen HR, Roszkowski JJ, Lyons GE, Li M, Moore T, Brasic N, McKee MD, Nishimura MI (2006) Identification of a hepatitis C virus-reactive T cell receptor that does not require CD8 for target cell recognition. Hepatology 43(5):973–981
58. Lyons GE, Roszkowski JJ, Man S, Yee C, Kast WM, Nishimura MI (2006) T-cell receptor tetramer binding or the lack there of does not necessitate antigen reactivity in T-cell receptor transduced T cells. Cancer Immunol Immunother 55(9):1142–1150
59. Zhao Y, Zheng Z, Khong HT, Rosenberg SA, Morgan RA (2006) Transduction of an HLA-DP4-restricted NY-ESO-1-specific TCR into primary human CD4+ lymphocytes. J Immunother 29(4):398–406
60. Gostick E, Cole DK, Hutchinson SL, Wooldridge L, Tafuro S, Laugel B, Lissina A, Oxenius A, Boulter JM, Price DA, Sewell AK (2007) Functional and biophysical characterization of an HLA-A*6801-restricted HIV-specific T cell receptor. Eur J Immunol 37(2):479–486
61. Purbhoo MA, Li Y, Sutton DH, Brewer JE, Gostick E, Bossi G, Laugel B, Moysey R, Baston E, Liddy N, Cameron B, Bennett AD, Ashfield R, Milicic A, Price DA, Classon BJ, Sewell AK, Jakobsen BK (2007) The HLA A*0201-restricted hTERT(540–548) peptide is not detected on tumor cells by a CTL clone or a high-affinity T-cell receptor. Mol Cancer Ther 6(7):2081–2091
62. Zhao Y, Bennett AD, Zheng Z, Wang QJ, Robbins PF, Yu LY, Li Y, Molloy PE, Dunn SM, Jakobsen BK, Rosenberg SA, Morgan RA (2007) High-affinity TCRs generated by phage display provide CD4+ T cells with the ability to recognize and kill tumor cell lines. J Immunol 179(9):5845–5854

63. Hiasa A, Hirayama M, Nishikawa H, Kitano S, Nukaya I, Yu SS, Mineno J, Kato I, Shiku H (2008) Long-term phenotypic, functional and genetic stability of cancer-specific T-cell receptor (TCR) alphabeta genes transduced to CD8+ T cells. Gene Ther 15(9):695–699
64. Johnson LA, Morgan RA, Dudley ME, Cassard L, Yang JC, Hughes MS, Kammula US, Royal RE, Sherry RM, Wunderlich JR, Lee CC, Restifo NP, Schwarz SL, Cogdill AP, Bishop RJ, Kim H, Brewer CC, Rudy SF, VanWaes C, Davis JL, Mathur A, Ripley RT, Nathan DA, Laurencot CM, Rosenberg SA (2009) Gene therapy with human and mouse T-cell receptors mediates cancer regression and targets normal tissues expressing cognate antigen. Blood 114(3):535–546
65. Voelkl S, Moore TV, Rehli M, Nishimura MI, Mackensen A, Fischer K (2009) Characterization of MHC class-I restricted TCRalphabeta+CD4–CD8- double negative T cells recognizing the gp100 antigen from a melanoma patient after gp100 vaccination. Cancer Immunol Immunother 58(5):709–718
66. Zhang Y, Liu Y, Moxley KM, Golden-Mason L, Hughes MG, Liu T, Heemskerk MH, Rosen HR, Nishimura MI (2010) Transduction of human T cells with a novel T-cell receptor confers anti-HCV reactivity. PLoS Pathog 6(7), e1001018
67. Chinnasamy N, Wargo JA, Yu Z, Rao M, Frankel TL, Riley JP, Hong JJ, Parkhurst MR, Feldman SA, Schrump DS, Restifo NP, Robbins PF, Rosenberg SA, Morgan RA (2011) A TCR targeting the HLA-A*0201-restricted epitope of MAGE-A3 recognizes multiple epitopes of the MAGE-A antigen superfamily in several types of cancer. J Immunol 186(2):685–696
68. Ochi T, Fujiwara H, Okamoto S, An J, Nagai K, Shirakata T, Mineno J, Kuzushima K, Shiku H, Yasukawa M (2011) Novel adoptive T-cell immunotherapy using a WT1-specific TCR vector encoding silencers for endogenous TCRs shows marked antileukemia reactivity and safety. Blood 118(6):1495–1503
69. Parkhurst MR, Yang JC, Langan RC, Dudley ME, Nathan DA, Feldman SA, Davis JL, Morgan RA, Merino MJ, Sherry RM, Hughes MS, Kammula US, Phan GQ, Lim RM, Wank SA, Restifo NP, Robbins PF, Laurencot CM, Rosenberg SA (2011) T cells targeting carcinoembryonic antigen can mediate regression of metastatic colorectal cancer but induce severe transient colitis. Mol Ther 19(3):620–626
70. Scholten KB, Turksma AW, Ruizendaal JJ, van den Hende M, van der Burg SH, Heemskerk MH, Meijer CJ, Hooijberg E (2011) Generating HPV specific T helper cells for the treatment of HPV induced malignancies using TCR gene transfer. J Transl Med 9:147
71. Spranger S, Jeremias I, Wilde S, Leisegang M, Starck L, Mosetter B, Uckert W, Heemskerk MH, Schendel DJ, Frankenberger B (2012) TCR-transgenic lymphocytes specific for HMMR/Rhamm limit tumor outgrowth in vivo. Blood 119(15):3440–3449
72. Linnemann C, Heemskerk B, Kvistborg P, Kluin RJ, Bolotin DA, Chen X, Bresser K, Nieuwland M, Schotte R, Michels S, Gomez-Eerland R, Jahn L, Hombrink P, Legrand N, Shu CJ, Mamedov IZ, Velds A, Blank CU, Haanen JB, Turchaninova MA, Kerkhoven RM, Spits H, Hadrup SR, Heemskerk MH, Blankenstein T, Chudakov DM, Bendle GM, Schumacher TN (2013) High-throughput identification of antigen-specific TCRs by TCR gene capture. Nat Med 19(11):1534–1541
73. Miyazaki Y, Fujiwara H, Asai H, Ochi F, Ochi T, Azuma T, Ishida T, Okamoto S, Mineno J, Kuzushima K, Shiku H, Yasukawa M (2013) Development of a novel redirected T-cell-based adoptive immunotherapy targeting human telomerase reverse transcriptase for adult T-cell leukemia. Blood 121(24):4894–4901
74. Schmitt TM, Aggen DH, Stromnes IM, Dossett ML, Richman SA, Kranz DM, Greenberg PD (2013) Enhanced-affinity murine T-cell receptors for tumor/self-antigens can be safe in gene therapy despite surpassing the threshold for thymic selection. Blood 122(3):348–356
75. Abate-Daga D, Speiser DE, Chinnasamy N, Zheng Z, Xu H, Feldman SA, Rosenberg SA, Morgan RA (2014) Development of a T cell receptor targeting an HLA-A*0201 restricted epitope from the cancer-testis antigen SSX2 for adoptive immunotherapy of cancer. PLoS One 9(3), e93321

76. Rosati SF, Parkhurst MR, Hong Y, Zheng Z, Feldman SA, Rao M, Abate-Daga D, Beard RE, Xu H, Black MA, Robbins PF, Schrump DA, Rosenberg SA, Morgan RA (2014) A novel murine T-cell receptor targeting NY-ESO-1. J Immunother 37(3):135–146
77. Duval L, Schmidt H, Kaltoft K, Fode K, Jensen JJ, Sorensen SM, Nishimura MI, von der Maase H (2006) Adoptive transfer of allogeneic cytotoxic T lymphocytes equipped with a HLA-A2 restricted MART-1 T-cell receptor: a phase I trial in metastatic melanoma. Clin Cancer Res 12(4):1229–1236
78. Morgan RA, Dudley ME, Wunderlich JR, Hughes MS, Yang JC, Sherry RM, Royal RE, Topalian SL, Kammula US, Restifo NP, Zheng Z, Nahvi A, de Vries CR, Rogers-Freezer LJ, Mavroukakis SA, Rosenberg SA (2006) Cancer regression in patients after transfer of genetically engineered lymphocytes. Science 314(5796):126–129
79. Koya RC, Mok S, Comin-Anduix B, Chodon T, Radu CG, Nishimura MI, Witte ON, Ribas A (2010) Kinetic phases of distribution and tumor targeting by T cell receptor engineered lymphocytes inducing robust antitumor responses. Proc Natl Acad Sci U S A 107(32):14286–14291
80. Robbins PF, Morgan RA, Feldman SA, Yang JC, Sherry RM, Dudley ME, Wunderlich JR, Nahvi AV, Helman LJ, Mackall CL, Kammula US, Hughes MS, Restifo NP, Raffeld M, Lee CC, Levy CL, Li YF, El-Gamil M, Schwarz SL, Laurencot C, Rosenberg SA (2011) Tumor regression in patients with metastatic synovial cell sarcoma and melanoma using genetically engineered lymphocytes reactive with NY-ESO-1. J Clin Oncol 29(7):917–924
81. Chinnasamy D, Yu Z, Kerkar SP, Zhang L, Morgan RA, Restifo NP, Rosenberg SA (2012) Local delivery of interleukin-12 using T cells targeting VEGF receptor-2 eradicates multiple vascularized tumors in mice. Clin Cancer Res 18(6):1672–1683
82. Shirakura Y, Mizuno Y, Wang L, Imai N, Amaike C, Sato E, Ito M, Nukaya I, Mineno J, Takesako K, Ikeda H, Shiku H (2012) T-cell receptor gene therapy targeting melanoma-associated antigen-A4 inhibits human tumor growth in non-obese diabetic/SCID/gammac-null mice. Cancer Sci 103(1):17–25
83. Chinnasamy D, Tran E, Yu Z, Morgan RA, Restifo NP, Rosenberg SA (2013) Simultaneous targeting of tumor antigens and the tumor vasculature using T lymphocyte transfer synergize to induce regression of established tumors in mice. Cancer Res 73(11):3371–3380
84. Morgan RA, Chinnasamy N, Abate-Daga D, Gros A, Robbins PF, Zheng Z, Dudley ME, Feldman SA, Yang JC, Sherry RM, Phan GQ, Hughes MS, Kammula US, Miller AD, Hessman CJ, Stewart AA, Restifo NP, Quezado MM, Alimchandani M, Rosenberg AZ, Nath A, Wang T, Bielekova B, Wuest SC, Akula N, McMahon FJ, Wilde S, Mosetter B, Schendel DJ, Laurencot CM, Rosenberg SA (2013) Cancer regression and neurological toxicity following anti-MAGE-A3 TCR gene therapy. J Immunother 36(2):133–151
85. Chodon T, Comin-Anduix B, Chmielowski B, Koya RC, Wu Z, Auerbach M, Ng C, Avramis E, Seja E, Villanueva A, McCannel TA, Ishiyama A, Czernin J, Radu CG, Wang X, Gjertson DW, Cochran AJ, Cornetta K, Wong DJ, Kaplan-Lefko P, Hamid O, Samlowski W, Cohen PA, Daniels GA, Mukherji B, Yang L, Zack JA, Kohn DB, Heath JR, Glaspy JA, Witte ON, Baltimore D, Economou JS, Ribas A (2014) Adoptive transfer of MART-1 T-cell receptor transgenic lymphocytes and dendritic cell vaccination in patients with metastatic melanoma. Clin Cancer Res 20(9):2457–2465
86. Kageyama S, Ikeda H, Miyahara Y, Imai N, Ishihara M, Saito K, Sugino S, Ueda S, Ishikawa T, Kokura S, Naota H, Ohishi K, Shiraishi T, Inoue N, Tanabe M, Kidokoro T, Yoshioka H, Tomura D, Nukaya I, Mineno J, Takesako K, Katayama N, Shiku H (2015) Adoptive transfer of MAGE-A4 T-cell receptor gene-transduced lymphocytes in patients with recurrent esophageal cancer. Clin Cancer Res 21(10):2268–2277
87. Borst J, Alexander S, Elder J, Terhorst C (1983) The T3 complex on human T lymphocytes involves four structurally distinct glycoproteins. J Biol Chem 258(8):5135–5141
88. Oettgen HC, Kappler J, Tax WJ, Terhorst C (1984) Characterization of the two heavy chains of the T3 complex on the surface of human T lymphocytes. J Biol Chem 259(19):12039–12048
89. Samelson LE, Harford JB, Klausner RD (1985) Identification of the components of the murine T cell antigen receptor complex. Cell 43(1):223–231

90. Leontsini E, Brown T, Biddison WE (1986) Physical and functional association of the T cell receptor and the T3 molecular complex on cytotoxic T cell clones that are differentially inhibitable by anti-T3 antibodies. Cell Immunol 102(1):21–32
91. Oettgen HC, Pettey CL, Maloy WL, Terhorst C (1986) A T3-like protein complex associated with the antigen receptor on murine T cells. Nature 320(6059):272–275
92. Lanier LL, Federspiel NA, Ruitenberg JJ, Phillips JH, Allison JP, Littman D, Weiss A (1987) The T cell antigen receptor complex expressed on normal peripheral blood CD4-, CD8- T lymphocytes. A CD3-associated disulfide-linked gamma chain heterodimer. J Exp Med 165(4):1076–1094
93. Weiss A, Littman DR (1994) Signal transduction by lymphocyte antigen receptors. Cell 76(2):263–274
94. Dialynas DP, Wilde DB, Marrack P, Pierres A, Wall KA, Havran W, Otten G, Loken MR, Pierres M, Kappler J et al (1983) Characterization of the murine antigenic determinant, designated L3T4a, recognized by monoclonal antibody GK1.5: expression of L3T4a by functional T cell clones appears to correlate primarily with class II MHC antigen-reactivity. Immunol Rev 74:29–56
95. Swain SL (1983) T cell subsets and the recognition of MHC class. Immunol Rev 74:129–142
96. Emmrich F, Strittmatter U, Eichmann K (1986) Synergism in the activation of human CD8 T cells by cross-linking the T-cell receptor complex with the CD8 differentiation antigen. Proc Natl Acad Sci U S A 83(21):8298–8302
97. de Vries JE, Yssel H, Spits H (1989) Interplay between the TCR/CD3 complex and CD4 or CD8 in the activation of cytotoxic T lymphocytes. Immunol Rev 109:119–141
98. Veillette A, Bookman MA, Horak EM, Bolen JB (1988) The CD4 and CD8 T cell surface antigens are associated with the internal membrane tyrosine-protein kinase p56lck. Cell 55(2):301–308
99. Barber EK, Dasgupta JD, Schlossman SF, Trevillyan JM, Rudd CE (1989) The CD4 and CD8 antigens are coupled to a protein-tyrosine kinase (p56lck) that phosphorylates the CD3 complex. Proc Natl Acad Sci U S A 86(9):3277–3281
100. Taniguchi RT, Anderson MS (2011) The role of Aire in clonal selection. Immunol Cell Biol 89(1):40–44
101. Davis MM, Bjorkman PJ (1988) T-cell antigen receptor genes and T-cell recognition. Nature 334(6181):395–402
102. Toyonaga B, Yoshikai Y, Vadasz V, Chin B, Mak TW (1985) Organization and sequences of the diversity, joining, and constant region genes of the human T-cell receptor beta chain. Proc Natl Acad Sci U S A 82(24):8624–8628
103. Koop BF, Rowen L, Wang K, Kuo CL, Seto D, Lenstra JA, Howard S, Shan W, Deshpande P, Hood L (1994) The human T-cell receptor TCRAC/TCRDC (C alpha/C delta) region: organization, sequence, and evolution of 97.6 kb of DNA. Genomics 19(3):478–493
104. Eisen HN, Sykulev Y, Tsomides TJ (1996) Antigen-specific T-cell receptors and their reactions with complexes formed by peptides with major histocompatibility complex proteins. Adv Protein Chem 49:1–56
105. Rosenberg SA (2000) Identification of cancer antigens: impact on development of cancer immunotherapies. Cancer J 6(Suppl 3):S200–S207
106. Van Der Bruggen P, Zhang Y, Chaux P, Stroobant V, Panichelli C, Schultz ES, Chapiro J, Van Den Eynde BJ, Brasseur F, Boon T (2002) Tumor-specific shared antigenic peptides recognized by human T cells. Immunol Rev 188:51–64
107. Coulie PG, Van den Eynde BJ, van der Bruggen P, Boon T (2014) Tumour antigens recognized by T lymphocytes: at the core of cancer immunotherapy. Nat Rev Cancer 14(2):135–146
108. Kawakami Y, Nishimura MI, Restifo NP, Topalian SL, O'Neil BH, Shilyansky J, Yannelli JR, Rosenberg SA (1993) T-cell recognition of human melanoma antigens. J Immunother Emphasis Tumor Immunol 14(2):88–93
109. Cole DJ, Weil DP, Shamamian P, Rivoltini L, Kawakami Y, Topalian S, Jennings C, Eliyahu S, Rosenberg SA, Nishimura MI (1994) Identification of MART-1-specific T-cell receptors:

T cells utilizing distinct T-cell receptor variable and joining regions recognize the same tumor epitope. Cancer Res 54(20):5265–5268
110. Nishimura MI, Kawakami Y, Charmley P, O'Neil B, Shilyansky J, Yannelli JR, Rosenberg SA, Hood L (1994) T-cell receptor repertoire in tumor-infiltrating lymphocytes. Analysis of melanoma-specific long-term lines. J Immunother Emphasis Tumor Immunol 16(2):85–94
111. Cole DJ, Wilson MC, Rivoltini L, Custer M, Nishimura MI (1997) T-cell receptor repertoire in matched MART-1 peptide-stimulated peripheral blood lymphocytes and tumor-infiltrating lymphocytes. Cancer Res 57(23):5320–5327
112. Nishimura MI, Custer MC, Schwarz SL, Parker LL, Mixon A, Clay TM, Yannelli JR, Rosenberg SA (1998) T cell-receptor V gene use by CD4+ melanoma-reactive clonal and oligoclonal T-cell lines. J Immunother 21(5):352–362
113. Sensi M, Castelli C, Anichini A, Grossberger D, Mazzocchi A, Mortarini R, Parmiani G (1991) Two autologous melanoma-specific and MHC-restricted human T cell clones with identical intra-tumour reactivity do not share the same TCR V alpha and V beta gene families. Melanoma Res 1(4):261–271
114. Loftus DJ, Castelli C, Clay TM, Squarcina P, Marincola FM, Nishimura MI, Parmiani G, Appella E, Rivoltini L (1996) Identification of epitope mimics recognized by CTL reactive to the melanoma/melanocyte-derived peptide MART-1(27–35). J Exp Med 184(2):647–657
115. McKee MD, Clay TM, Diamond RA, Rosenberg SA, Nishimura MI (2000) Quantitation of T-cell receptor frequencies by competitive polymerase chain reaction: dynamics of T-cell clonotype frequencies in an expanding tumor-infiltrating lymphocyte culture. J Immunother 23(4):419–429
116. Gros A, Robbins PF, Yao X, Li YF, Turcotte S, Tran E, Wunderlich JR, Mixon A, Farid S, Dudley ME, Hanada K, Almeida JR, Darko S, Douek DC, Yang JC, Rosenberg SA (2014) PD-1 identifies the patient-specific CD8(+) tumor-reactive repertoire infiltrating human tumors. J Clin Invest 124(5):2246–2259
117. Linnemann C, Mezzadra R, Schumacher TN (2014) TCR repertoires of intratumoral T-cell subsets. Immunol Rev 257(1):72–82
118. Linnemann C, van Buuren MM, Bies L, Verdegaal EM, Schotte R, Calis JJ, Behjati S, Velds A, Hilkmann H, Atmioui DE, Visser M, Stratton MR, Haanen JB, Spits H, van der Burg SH, Schumacher TN (2015) High-throughput epitope discovery reveals frequent recognition of neo-antigens by CD4+ T cells in human melanoma. Nat Med 21(1):81–85
119. Nishimura MI, Avichezer D, Custer MC, Lee CS, Chen C, Parkhurst MR, Diamond RA, Robbins PF, Schwartzentruber DJ, Rosenberg SA (1999) MHC class I-restricted recognition of a melanoma antigen by a human CD4+ tumor infiltrating lymphocyte. Cancer Res 59(24):6230–6238
120. Parkhurst MR, Joo J, Riley JP, Yu Z, Li Y, Robbins PF, Rosenberg SA (2009) Characterization of genetically modified T-cell receptors that recognize the CEA:691–699 peptide in the context of HLA-A2.1 on human colorectal cancer cells. Clin Cancer Res 15(1):169–180
121. Vitiello A, Marchesini D, Furze J, Sherman LA, Chesnut RW (1991) Analysis of the HLA-restricted influenza-specific cytotoxic T lymphocyte response in transgenic mice carrying a chimeric human-mouse class I major histocompatibility complex. J Exp Med 173(4):1007–1015
122. Sherman LA, Hesse SV, Irwin MJ, La Face D, Peterson P (1992) Selecting T cell receptors with high affinity for self-MHC by decreasing the contribution of CD8. Science 258(5083):815–818
123. Cohen CJ, Zhao Y, Zheng Z, Rosenberg SA, Morgan RA (2006) Enhanced antitumor activity of murine-human hybrid T-cell receptor (TCR) in human lymphocytes is associated with improved pairing and TCR/CD3 stability. Cancer Res 66(17):8878–8886
124. Moore TV, Lyons GE, Brasic N, Roszkowski JJ, Voelkl S, Mackensen A, Kast WM, Le Poole IC, Nishimura MI (2009) Relationship between CD8-dependent antigen recognition, T cell functional avidity, and tumor cell recognition. Cancer Immunol Immunother 58(5):719–728
125. Chlewicki LK, Holler PD, Monti BC, Clutter MR, Kranz DM (2005) High-affinity, peptide-specific T cell receptors can be generated by mutations in CDR1, CDR2 or CDR3. J Mol Biol 346(1):223–239

126. Jones LL, Brophy SE, Bankovich AJ, Colf LA, Hanick NA, Garcia KC, Kranz DM (2006) Engineering and characterization of a stabilized alpha1/alpha2 module of the class I major histocompatibility complex product Ld. J Biol Chem 281(35):25734–25744
127. Malecek K, Zhong S, McGary K, Yu C, Huang K, Johnson LA, Rosenberg SA, Krogsgaard M (2013) Engineering improved T cell receptors using an alanine-scan guided T cell display selection system. J Immunol Methods 392(1–2):1–11
128. Malecek K, Grigoryan A, Zhong S, Gu WJ, Johnson LA, Rosenberg SA, Cardozo T, Krogsgaard M (2014) Specific increase in potency via structure-based design of a TCR. J Immunol 193(5):2587–2599
129. Dunn SM, Rizkallah PJ, Baston E, Mahon T, Cameron B, Moysey R, Gao F, Sami M, Boulter J, Li Y, Jakobsen BK (2006) Directed evolution of human T cell receptor CDR2 residues by phage display dramatically enhances affinity for cognate peptide-MHC without increasing apparent cross-reactivity. Protein Sci 15(4):710–721
130. Langerman A, Callender GG, Nishimura MI (2004) Retroviral transduction of peptide stimulated t cells can generate dual t cell receptor-expressing (bifunctional) t cells reactive with two defined antigens. J Transl Med 2(1):42
131. Heemskerk MH, Hagedoorn RS, van der Hoorn MA, van der Veken LT, Hoogeboom M, Kester MG, Willemze R, Falkenburg JH (2007) Efficiency of T-cell receptor expression in dual-specific T cells is controlled by the intrinsic qualities of the TCR chains within the TCR-CD3 complex. Blood 109(1):235–243
132. Hart DP, Xue SA, Thomas S, Cesco-Gaspere M, Tranter A, Willcox B, Lee SP, Steven N, Morris EC, Stauss HJ (2008) Retroviral transfer of a dominant TCR prevents surface expression of a large proportion of the endogenous TCR repertoire in human T cells. Gene Ther 15(8):625–631
133. Bendle GM, Linnemann C, Hooijkaas AI, Bies L, de Witte MA, Jorritsma A, Kaiser AD, Pouw N, Debets R, Kieback E, Uckert W, Song JY, Haanen JB, Schumacher TN (2010) Lethal graft-versus-host disease in mouse models of T cell receptor gene therapy. Nat Med 16(5): 565–570, 561 p following 570
134. van Loenen MM, de Boer R, Amir AL, Hagedoorn RS, Volbeda GL, Willemze R, van Rood JJ, Falkenburg JH, Heemskerk MH (2010) Mixed T cell receptor dimers harbor potentially harmful neoreactivity. Proc Natl Acad Sci U S A 107(24):10972–10977
135. Okamoto S, Mineno J, Ikeda H, Fujiwara H, Yasukawa M, Shiku H, Kato I (2009) Improved expression and reactivity of transduced tumor-specific TCRs in human lymphocytes by specific silencing of endogenous TCR. Cancer Res 69(23):9003–9011
136. Meyerhuber P, Conrad H, Starck L, Leisegang M, Busch DH, Uckert W, Bernhard H (2010) Targeting the epidermal growth factor receptor (HER) family by T cell receptor gene-modified T lymphocytes. J Mol Med (Berl) 88(11):1113–1121
137. Kuball J, Dossett ML, Wolfl M, Ho WY, Voss RH, Fowler C, Greenberg PD (2007) Facilitating matched pairing and expression of TCR chains introduced into human T cells. Blood 109(6):2331–2338
138. Kuball J, Hauptrock B, Malina V, Antunes E, Voss RH, Wolfl M, Strong R, Theobald M, Greenberg PD (2009) Increasing functional avidity of TCR-redirected T cells by removing defined N-glycosylation sites in the TCR constant domain. J Exp Med 206(2): 463–475
139. Chang HC, Bao Z, Yao Y, Tse AG, Goyarts EC, Madsen M, Kawasaki E, Brauer PP, Sacchettini JC, Nathenson SG et al (1994) A general method for facilitating heterodimeric pairing between two proteins: application to expression of alpha and beta T-cell receptor extracellular segments. Proc Natl Acad Sci U S A 91(24):11408–11412
140. Rutledge T, Cosson P, Manolios N, Bonifacino JS, Klausner RD (1992) Transmembrane helical interactions: zeta chain dimerization and functional association with the T cell antigen receptor. EMBO J 11(9):3245–3254
141. Blank U, Boitel B, Mege D, Ermonval M, Acuto O (1993) Analysis of tetanus toxin peptide/DR recognition by human T cell receptors reconstituted into a murine T cell hybridoma. Eur J Immunol 23(12):3057–3065

142. Hastings AE, Hurley CK, Robinson ED, Salerno K, Hernandez E, Richert JR (1996) Molecular interactions between transfected human TCR, immunodominant myelin basic protein peptide 152–165, and HLA-DR13. J Immunol 157(8):3460–3471
143. Sommermeyer D, Uckert W (2010) Minimal amino acid exchange in human TCR constant regions fosters improved function of TCR gene-modified T cells. J Immunol 184(11):6223–6231
144. Voss RH, Thomas S, Pfirschke C, Hauptrock B, Klobuch S, Kuball J, Grabowski M, Engel R, Guillaume P, Romero P, Huber C, Beckhove P, Theobald M (2010) Coexpression of the T-cell receptor constant alpha domain triggers tumor reactivity of single-chain TCR-transduced human T cells. Blood 115(25):5154–5163
145. Foote J, Eisen HN (2000) Breaking the affinity ceiling for antibodies and T cell receptors. Proc Natl Acad Sci U S A 97(20):10679–10681
146. Lenardo MJ, Boehme S, Chen L, Combadiere B, Fisher G, Freedman M, McFarland H, Pelfrey C, Zheng L (1995) Autocrine feedback death and the regulation of mature T lymphocyte antigen responses. Int Rev Immunol 13(2):115–134
147. Combadiere B, Reis e Sousa C, Trageser C, Zheng LX, Kim CR, Lenardo MJ (1998) Differential TCR signaling regulates apoptosis and immunopathology during antigen responses in vivo. Immunity 9(3):305–313
148. Mehrotra S, Al-Khami AA, Klarquist J, Husain S, Naga O, Eby JM, Murali AK, Lyons GE, Li M, Spivey ND, Norell H, Martins da Palma T, Onicescu G, Diaz-Montero CM, Garrett-Mayer E, Cole DJ, Le Poole IC, Nishimura MI (2012) A coreceptor-independent transgenic human TCR mediates anti-tumor and anti-self immunity in mice. J Immunol 189(4):1627–1638
149. Kroger CJ, Alexander-Miller MA (2007) Cutting edge: CD8+ T cell clones possess the potential to differentiate into both high- and low-avidity effector cells. J Immunol 179(2):748–751
150. Marigo I, Dolcetti L, Serafini P, Zanovello P, Bronte V (2008) Tumor-induced tolerance and immune suppression by myeloid derived suppressor cells. Immunol Rev 222:162–179
151. Riley JL, June CH, Blazar BR (2009) Human T regulatory cell therapy: take a billion or so and call me in the morning. Immunity 30(5):656–665
152. Ostrand-Rosenberg S (2010) Myeloid-derived suppressor cells: more mechanisms for inhibiting antitumor immunity. Cancer Immunol Immunother 59(10):1593–1600
153. Leguern C (2011) Regulatory T cells for tolerance therapy: revisiting the concept. Crit Rev Immunol 31(3):189–207
154. Josefowicz SZ, Lu LF, Rudensky AY (2012) Regulatory T cells: mechanisms of differentiation and function. Annu Rev Immunol 30:531–564
155. Raber P, Ochoa AC, Rodriguez PC (2012) Metabolism of L-arginine by myeloid-derived suppressor cells in cancer: mechanisms of T cell suppression and therapeutic perspectives. Immunol Invest 41(6–7):614–634
156. Kumar V, Gabrilovich DI (2014) Hypoxia-inducible factors in regulation of immune responses in tumour microenvironment. Immunology 143(4):512–519
157. Kono K, Salazar-Onfray F, Petersson M, Hansson J, Masucci G, Wasserman K, Nakazawa T, Anderson P, Kiessling R (1996) Hydrogen peroxide secreted by tumor-derived macrophages down-modulates signal-transducing zeta molecules and inhibits tumor-specific T cell-and natural killer cell-mediated cytotoxicity. Eur J Immunol 26(6):1308–1313
158. Kusmartsev S, Nefedova Y, Yoder D, Gabrilovich DI (2004) Antigen-specific inhibition of CD8+ T cell response by immature myeloid cells in cancer is mediated by reactive oxygen species. J Immunol 172(2):989–999
159. Lu T, Gabrilovich DI (2012) Molecular pathways: tumor-infiltrating myeloid cells and reactive oxygen species in regulation of tumor microenvironment. Clin Cancer Res 18(18):4877–4882
160. De Sanctis F, Sandri S, Ferrarini G, Pagliarello I, Sartoris S, Ugel S, Marigo I, Molon B, Bronte V (2014) The emerging immunological role of post-translational modifications by reactive nitrogen species in cancer microenvironment. Front Immunol 5:69
161. Mao Y, Poschke I, Kiessling R (2014) Tumour-induced immune suppression: role of inflammatory mediators released by myelomonocytic cells. J Intern Med 276(2):154–170

162. Pecht I, Gakamsky DM (2005) Spatial coordination of CD8 and TCR molecules controls antigen recognition by CD8+ T-cells. FEBS Lett 579(15):3336–3341
163. Cawthon AG, Alexander-Miller MA (2002) Optimal colocalization of TCR and CD8 as a novel mechanism for the control of functional avidity. J Immunol 169(7):3492–3498
164. Nagaraj S, Gupta K, Pisarev V, Kinarsky L, Sherman S, Kang L, Herber DL, Schneck J, Gabrilovich DI (2007) Altered recognition of antigen is a mechanism of CD8+ T cell tolerance in cancer. Nat Med 13(7):828–835
165. Chhabra A, Yang L, Wang P, Comin-Anduix B, Das R, Chakraborty NG, Ray S, Mehrotra S, Yang H, Hardee CL, Hollis R, Dorsky DI, Koya R, Kohn DB, Ribas A, Economou JS, Baltimore D, Mukherji B (2008) CD4+CD25- T cells transduced to express MHC class I-restricted epitope-specific TCR synthesize Th1 cytokines and exhibit MHC class I-restricted cytolytic effector function in a human melanoma model. J Immunol 181(2):1063–1070
166. Ray S, Chhabra A, Chakraborty NG, Hegde U, Dorsky DI, Chodon T, von Euw E, Comin-Anduix B, Koya RC, Ribas A, Economou JS, Rosenberg SA, Mukherji B (2010) MHC-I-restricted melanoma antigen specific TCR-engineered human CD4+ T cells exhibit multifunctional effector and helper responses, in vitro. Clin Immunol 136(3):338–347
167. Brusko TM, Koya RC, Zhu S, Lee MR, Putnam AL, McClymont SA, Nishimura MI, Han S, Chang LJ, Atkinson MA, Ribas A, Bluestone JA (2010) Human antigen-specific regulatory T cells generated by T cell receptor gene transfer. PLoS One 5(7), e11726
168. Kessels HW, Wolkers MC, van den Boom MD, van der Valk MA, Schumacher TN (2001) Immunotherapy through TCR gene transfer. Nat Immunol 2(10):957–961
169. Abad JD, Wrzensinski C, Overwijk W, De Witte MA, Jorritsma A, Hsu C, Gattinoni L, Cohen CJ, Paulos CM, Palmer DC, Haanen JB, Schumacher TN, Rosenberg SA, Restifo NP, Morgan RA (2008) T-cell receptor gene therapy of established tumors in a murine melanoma model. J Immunother 31(1):1–6
170. de Witte MA, Bendle GM, van den Boom MD, Coccoris M, Schell TD, Tevethia SS, van Tinteren H, Mesman EM, Song JY, Schumacher TN (2008) TCR gene therapy of spontaneous prostate carcinoma requires in vivo T cell activation. J Immunol 181(4):2563–2571
171. Stauss HJ, Thomas S, Cesco-Gaspere M, Hart DP, Xue SA, Holler A, King J, Wright G, Perro M, Pospori C, Morris E (2008) WT1-specific T cell receptor gene therapy: improving TCR function in transduced T cells. Blood Cells Mol Dis 40(1):113–116
172. Frankel TL, Burns WR, Peng PD, Yu Z, Chinnasamy D, Wargo JA, Zheng Z, Restifo NP, Rosenberg SA, Morgan RA (2010) Both CD4 and CD8 T cells mediate equally effective in vivo tumor treatment when engineered with a highly avid TCR targeting tyrosinase. J Immunol 184(11):5988–5998
173. Kerkar SP, Sanchez-Perez L, Yang S, Borman ZA, Muranski P, Ji Y, Chinnasamy D, Kaiser AD, Hinrichs CS, Klebanoff CA, Scott CD, Gattinoni L, Morgan RA, Rosenberg SA, Restifo NP (2011) Genetic engineering of murine CD8+ and CD4+ T cells for preclinical adoptive immunotherapy studies. J Immunother 34(4):343–352
174. Xue SA, Gao L, Ahmadi M, Ghorashian S, Barros RD, Pospori C, Holler A, Wright G, Thomas S, Topp M, Morris EC, Stauss HJ (2013) Human MHC Class I-restricted high avidity CD4 T cells generated by co-transfer of TCR and CD8 mediate efficient tumor rejection in vivo. Oncoimmunology 2(1), e22590
175. Calogero A, Hospers GA, Kruse KM, Schrier PI, Mulder NH, Hooijberg E, de Leij LF (2000) Retargeting of a T cell line by anti MAGE-3/HLA-A2 alpha beta TCR gene transfer. Anticancer Res 20(3A):1793–1799
176. Straetemans T, van Brakel M, van Steenbergen S, Broertjes M, Drexhage J, Hegmans J, Lambrecht BN, Lamers C, van Der Bruggen P, Coulie PG, Debets R (2012) TCR gene transfer: MAGE-C2/HLA-A2 and MAGE-A3/HLA-DP4 epitopes as melanoma-specific immune targets. Clin Dev Immunol 2012:586314
177. Sommermeyer D, Conrad H, Kronig H, Gelfort H, Bernhard H, Uckert W (2013) NY-ESO-1 antigen-reactive T cell receptors exhibit diverse therapeutic capability. Int J Cancer 132(6):1360–1367

178. Linette GP, Stadtmauer EA, Maus MV, Rapoport AP, Levine BL, Emery L, Litzky L, Bagg A, Carreno BM, Cimino PJ, Binder-Scholl GK, Smethurst DP, Gerry AB, Pumphrey NJ, Bennett AD, Brewer JE, Dukes J, Harper J, Tayton-Martin HK, Jakobsen BK, Hassan NJ, Kalos M, June CH (2013) Cardiovascular toxicity and titin cross-reactivity of affinity-enhanced T cells in myeloma and melanoma. Blood 122(6):863–871
179. Li Y, Moysey R, Molloy PE, Vuidepot AL, Mahon T, Baston E, Dunn S, Liddy N, Jacob J, Jakobsen BK, Boulter JM (2005) Directed evolution of human T-cell receptors with picomolar affinities by phage display. Nat Biotechnol 23(3):349–354
180. Restifo NP, Dudley ME, Rosenberg SA (2012) Adoptive immunotherapy for cancer: harnessing the T cell response. Nat Rev Immunol 12(4):269–281

Chapter 5
Toward the Identification of Genetic Determinants of Responsiveness to Cancer Immunotherapy

Davide Bedognetti, Sara Tomei, Wouter Hendrickx, Francesco M. Marincola, and Ena Wang

Abstract For a long time, the lack of a human-centered translational approach to tumor immunology has led to the accumulation of a multitude of disjointed clinical and basic researches, followed by a disharmonic growth of scientific knowledge. The huge amount of conjectural hypotheses has competitively interfered with the few evidence-based concepts generated by studies conducted in humans, resulting in several clinical failures and some 'mysterious' successes. During the last few years, refined immunotherapies have been shown to induce clinical response and/or improve survival in a significant proportion of cancer patients. *Pari passu,* high-throughput approaches applied to the analysis of tumor specimens have unveiled unexpected and paradoxical relations between cancer and the immune system. Such studies have described a cancer immune phenotype typified by better prognosis and increased responsiveness to immunotherapeutic approaches. Whether the favorable cancer immune phenotype is dictated by the intrinsic genetics of the tumor or by the genetic makeup of the individual bearing the disease is presently unclear. Here, we focus on molecular biomarkers derived from genomic and genetic studies to summarize the recent advances in our understanding of the mechanisms associated with distinct outcomes in the context of cancer immunotherapy.

D. Bedognetti (✉) • W. Hendrickx
Tumor Biology, Immunology and Therapy Section, Division of Translational Medicine, Research Branch, Sidra Medical and Research Center, PO Box 26999, Doha, Qatar
e-mail: dbedognetti@sidra.org; whendrickx@sidra.org

S. Tomei
Omics Core and Biorepository, Division of Translational Medicine, Research Branch, Sidra Medical and Research Center, Doha, Qatar
e-mail: stomei@sidra.org

F.M. Marincola
Research Branch, Sidra Medical and Research Center, Doha, Qatar
e-mail: fmarincola@sidra.org

E. Wang
Division of Translational Medicine, Research Branch, Sidra Medical and Research Center, Doha, Qatar
e-mail: ewang@sidra.org

P.A. Ascierto et al. (eds.), *Developments in T Cell Based Cancer Immunotherapies*, Cancer Drug Discovery and Development,
DOI 10.1007/978-3-319-21167-1_5

Introduction

The first attempts to prove that manipulations of host immune system can be used to induce tumor rejection date back to the end of nineteenth century. In the 1890s William Coley observed that the injection of bacterial products could induce considerable tumor regression, at least in some patients. In the 1980s, almost one century after Coley's experiments, clinical trials investigating pro–inflammatory cytokines (interleukin-(IL)-2 and interferon-(IF)-α) demonstrated that the elicitation of the immune system can lead, in some cases, to an efficient tumor clearance [1–5]. The molecular era of tumor immunology began a decade after, with the sequencing and cloning of tumor antigens recognized by autologous T-cells (tumor antigens: TAs). These studies conclusively proved that CD8+ T cells can recognize and eventually kill cancer cells [6, 7]. The subsequent natural step was the development of anti-cancer vaccines [8–12] directed against those antigens. The induction of a clonal expansion of CD8+ T cells able to recognize cancer cells enabled researchers to study the dynamics of anti-tumor response. A paradigm emerged from these studies. In fact, results of clinical trials testing TA-based vaccines showed that this strategy is extremely effective in inducing antigen-specific cellular responses but is rarely followed by a clinically detectable tumor rejection, which occurs in about 4 % of patients [13, 14]. It was evident that the generation of TA-specific immune responses may represent a necessary but not sufficient condition for the induction of cancer rejection. It is now clear that other factors of tumor-host interactions dictate the final clinical outcome.

Since the last decade, a number of studies have assessed the dynamisms of tumor-host interaction by profiling tumor biopsies from patients enrolled in immunotherapy trials. An active tumor microenvironment conducive to immune recognition was shown to be predictive of response to immunotherapy [15–19]. Furthermore, the evaluation of adaptive immune cell infiltration was even more accurate than traditional histopathological score in prognosticating the clinical outcome [20, 21]. Overall, a flurry of studies has indicated that the presence of an inflammatory status in established tumors is strongly associated with a more favorable prognosis and/or treatment responsiveness [22–24].

It appeared evident that the effectiveness of a given treatment strongly depends upon its ability to enhance intratumoral inflammation. Rather than using pro-inflammatory cytokines or TA vaccines, immunologists have explored the possibility of enhancing tumor inflammation by targeting immune-regulatory pathways, now broadly known as immune checkpoints. The unexpected results of such approaches have revolutionized the field of tumor immunology. In metastatic melanoma patients, the blockade of T-cell inhibitory receptor CTLA-4 was shown to induce prolonged responses and increased overall survival in phase III trials [25, 26]. More recently, the inhibition of another T-cell immune-checkpoint (PD-1 pathway) was found to mediate durable tumor regression in several tumor types, including melanoma, renal cell carcinoma, ovarian, lung, head and neck, and gastrointestinal tumors [27–31]. The remarkable successes of other refined immunotherapies, such as adoptive transfer therapy with autologous tumor infiltrating lymphocytes (TILs) or with engineered T cells have amplified the enthusiasm in the field [32–35]. Still, in solid tumors, objective responses occur in only a minority of patients (less than 40 %) and complete remission is still, and unfortunately, a rare phenomenon [29, 30]. The precise definition of molecular features associated with treatment response is a critical step toward the development of a more efficient, personalized, immunotherapy. In this chapter we summarize the recent advances in our understanding of molecular mechanisms associated with differential immune responsiveness in the context of cancer immunotherapy.

Signatures Associated with Responsiveness to Cancer Immunotherapy

Signatures of Immune-Mediated Tissue Destruction: The Immunologic Constant of Rejection (ICR) hypothesis

One of the main contributions of high-throughput gene expression profile studies lies in the definition of generalizable mechanisms conducive to immune-mediated tumor rejection [36–39]. Several speculations derived from experimental models could be conceived to explain how cancer cells antagonize host immune reaction. Nevertheless, the most compelling mechanistic insights come from correlative studies conducted in humans through a *Bedside to Bench and Back approach*. In the *Bedside to Bench and Back* approach, hypothesis-generating investigations aim at defining theories able to explain the target observation generated at the patient's bedside. The hypothesis formulated by such studies can be tested in vitro or in-vivo in animals models (to *Bench*) but need to be conclusively validated in humans (and *Back* (to the bedside)) [40–43]. Ex vivo gene expression profile studies of tumor biopsies have shown that the activation of specific inflammatory pathways is necessary to mediate tumor rejection while resistance to treatment is characterized by the lack of their coordinate activation. Lesions undergoing regression following immune manipulations exhibit, early after treatment, a powerful acute inflammatory process symbolized by the activation of specific molecular pathways [15, 18, 19, 36, 44–46]. Qualitatively overlapping molecular pathways involved in the activation of this acute inflammation in post-treatment rejecting lesions are found to be partially activated in pre-treatment lesions that are more likely to respond to treatment [15–17, 19]. Therefore, it is likely that tumors intrinsically unable to display such a polarized inflammatory status in steady state condition (i.e., before treatment) are also quite refractory to *switch on* those pathways following treatment administration.

Observational studies in humans have corroborated the experimental hypothesis that chronic inflammation promotes tumor development [47]. For years, the presence of immune cells in tumor lesions has been seen in a reductionist way as proof of the pro-tumorigenic effect of tumor-inflammation [39, 48]. However, when a temporal vector is added to this one-dimensional observation and patients are prospectively followed, intriguing hypotheses on the role of tumor inflammation in counteracting tumor spreading emerge [20, 22, 40, 41, 49]. Correlative investigations analyzing thousands of patients in different tumor settings have repeatedly shown that the presence of a T-cell infiltrate is a favorable prognostic factor in several cancer types including melanoma [50], breast [51, 52] ovarian [53] and colorectal cancer [20, 48, 54]. Furthermore, in breast cancers, the density of T cell infiltrates, strongly correlate with response to chemotherapy [55]. Analysis of tumor transcriptional program has added molecular precision to these observations. Tumor transcriptome has been extensively studied in (1) excised primary tumor, (2) baseline biopsies of primary or metastatic tumors from patients receiving chemotherapy or immunotherapy, and (3) post-treatment tumor biopsies from patients treated with immunotherapy. Overall, these three approaches led to the definition of, respectively (1) prognostic signatures (aimed at assessing patient outcome such as recurrence of death independent of therapy) (2) predictive signatures (aimed at defining the likelihood of treatment effectiveness) (3) mechanistic signatures (aimed at discriminating the effect of a given treatment by assessing the behavior of target lesions following treatment) [24, 40, 41].

Such studies have revealed that prognostic, predictive and mechanistic immune signatures (defined in the immunotherapeutic setting) qualitatively overlap with each other [24]. In turn, molecular pathways captured by such signatures share strong analogies with those responsible for the development of other forms of immune-mediated tissue destructions such as allograft rejection, graft versus host disease (GVHD), and flares of autoimmunity [37, 39, 56, 57]. These pathways reflect a process typified by the coordinated modular activation of interferon stimulated genes (ISGs), the recruitment of cytotoxic cells through the production of specific chemokine ligands (i.e. the CCR5 and CXCR3 ligands CCL3-5, and CXCL9-11) with a consecutive T helper 1 (Th1) polarization, and the activation of immune effector function (IEF) genes (granzyme, granulysis and perforin; genes expressed by NK cells and $CD8^+$ and Th1 $CD4^+$ cells upon activation). Regulation of these genes is orchestrated by the activation of the transcription factors IRF1 and STAT1. We refer to these modules as the Immunologic Constant of Rejection (ICR) [17, 24, 38, 39]. A number of studies have described prognostic immune signatures recapitulating the ICR pathways in breast [58–61], ovarian [62–64], melanoma [65–67], lung [68], hepatocellular [69], and colorectal cancer [20, 54, 70–72], as reviewed elsewhere [24]. In breast cancer, similar signatures can predict the outcome to neoadjuvant chemotherapy [55, 73–75], neo-adjuvant molecular therapy (anti-Her2/neu) (Gianni et al. SABC, 2012) and adjuvant therapy [76].

Importantly, all the immunotherapeutic approaches assessed so far seem to act through the induction of the ICR modules, and the degree of activation positively correlates with treatment activity. The lack of a certain extent of activation configures a non-inflammatory cancer phenotype that is not only more resistant to immunotherapeutic treatments but that is also characterized by a poor prognosis. Below we provide an overview of signatures associated with immunotherapy responsiveness.

Signatures of Responsiveness to IL2-Based Therapy and Vaccination

The first gene expression profile studies in the context of cancer immunotherapy were performed a decade ago in metastatic melanoma patients treated with IL-2 and vaccination [18]. We observed that lesions more likely to respond to therapy carried an inflammatory status as reflected by the pre-activation of cytotoxic mechanisms and the upregulation of IFN signaling. However, expression of lymphoid cell markers (e.g., CD3D and CD8A) was similar in pre-treatment responding versus non responding lesions suggesting that intrinsic features of tumor lesions are critical determinants of reactivity to IL-2 administration. We prospectively validated these concepts in a subsequent study using a comprehensive genome-wide expression profile approach allowing a more refined networking analysis [19]. Biopsies of IL-2 responding lesions displayed a signature of immune activation centered on IFN-γ signaling [19]. Additionally, Sullivan et al. have described an immune-gene signature predictive of favorable outcome in melanoma patients treated with IL-2 (Sullivan et al. ASCO 2009). In a slightly different therapeutic setting, Gajewski et al. showed that the overexpression of CXCR3 and CCR5 ligands (i.e., CXCL9, CXCL10 and CCL4, and CCL5, respectively) by pre-treatment melanoma lesion was associated with the presence of $CD8^+$ cells expressing CCR5 and CXCR3, and in turn, correlate with response to IL-12-based vaccination (Gajewski et al. ASCO 2007) [77]. However, a similar inflammatory profile consisting of specific chemokines, T-cell markers and interferon-related genes was associated with clinical benefit in metastatic melanoma patients treated with a dendritic-cell based vaccine (Gajewski et al., ASCO 2009) [78]. Recently, global transcriptome analysis has been employed to identify pre-treatment markers of responsiveness in cancer patients treated with MAGE-A3 vaccination within two randomized phase II trials [16, 17]. A classifier consisting of 84 genes was able to discriminate responding vs non responding metastatic melanoma patients and was also correlated with prolonged survival. IRF1 and STAT1 were identified as the master regulators of the differentially expressed genes, which recapitulate the ICR pathways. These genes reflect an active microenviroment, polarized toward a Th1 response, defined by the expression of chemokines (CXCL9, CXCL10, and CCL5), cytotoxic granules (GZMK), HLA class I and II molecules, T cell-surface markers (CD3D, CD8A, and IL2RG), T cell-activation markers (ICOS and CD86), NK cell-associated genes (KLRD1 and KLRB1), and other classical IFN-stimulated genes (STAT1, IRF1, JAK2, PSMB9, GBP1, GBP5, and FAM26F) [16, 17]. The predictive role of the gene panel, validated through real time PCR (61 genes), was further confirmed in a parallel study assessing the efficacy of adjuvant MAGE-A3 administration in patients affected by early stage non small lung carcinoma [16]. Unfortunately, the final validation is lacking due to the premature interruption of the subsequent MAGRIT phase III trial [79].

While these data have been generated from clinical trials testing therapeutics with direct pro-inflammatory properties, it is extremely intriguing to notice that overlapping signatures have been described in the context of immune-checkpoint blockade studies. Ji et al. profiled melanoma of patients receiving anti-CTLA4 mAb ipilimumab [15]. By comparing responding and non responding patients, strong analogies with the ICR modules were observed not only in term of pathways but also in term of single genes. The top 26 genes upregulated in samples from patients responding to ipilimumab include CD8A, HLADQA1, CCL4, and CCL5 (CCR5 ligands); CXCL9, CXCL10, and CXCL11 (CXCR3 ligands); and NKG7, GZMB, and PRF1 (immune effector genes). Furthermore, Hodi and colleagues, by analyzing pre-treatment biopsies of metastatic patients treated with anti-PD-L1 MPDL3280A showed that biopsies from responding patients have a higher expression of IFN-γ, GZMA, CD8A, and CXCL9 [80]. In line with these findings, a coordinated upregulation of the Th1 chemoattractant chemokines CCL5, CXCL9, CXCL10, and CXCL11 in pretreatment lesions segregated melanoma patients who responded to adoptive-transfer therapy and IL-2 from those that did not [40, 41]. These studies highlight the permissive role of a Th1-polarizing microenvironment in the immunotherapeutic setting, independent of the type of treatment.

IL-2 studies in melanoma have shown that sensitive tumors (i.e. those that will subsequently undergo regression) promptly react to immune stimulation by shifting toward an acute Th1 inflammatory status through the activation of the key transcription factor IRF1 [18, 19]. Similarly, following ipilimumab administration, the increment of effector function and chemokine genes is stronger in responding vs non-responding patients [15]. Accordingly, in the aforementioned anti-PD-L1 trial in multiple cancer types, on-treatment responding tumors showed increasing PD-L1 expression and a Th1-dominant immune infiltrate, while non-responding tumors had a lack of tumor CD8+ T-cell infiltration and T-cell activation markers (i.e. GZMA and perforin, CXCL9, CXCL10, ICOS). Interestingly, FOXP3 neither increased nor decreased in regressing lesions [80].

Signatures of Responsiveness to Checkpoint Inhibitors

Molecular analyses of tumor samples and clinical data from immune-checkpoint inhibitors have prompted us to revisit the assumption that the expression of classical inhibitory molecules such as IDO and/or FOXP3 denotes a cancer more resistant to immune manipulation.

In fact, tumors bearing the inflammatory phenotype also display activation of pathways associated with immune-suppressive mechanisms, suggesting an ongoing immune response in the tumor microenvironment concomitant with tumor escape mechanisms [78, 81]. Hamid et al. showed that the pre-treatment FOXP3 (a bona fide markers of T regulatory cells) and IDO expression (evaluated by immunohistochemistry) by tumor infiltrating immune-cells is positively associated with clinical outcome in melanoma patients treated with ipilimumab [82]. Pre-treatment gene expression levels of IDO1 also correlated to response to ipilimumab and anti-PDL1 [15, 80]. In order to interpret these findings, it should be taken in account that IDO1 is an IFN-γ inducible gene [83], and its over-expression could be due to the excess of IFN-γ consequent to the T helper 1 infiltration. Accordingly, Gajewsky et al. recently demonstrated that the expression of (IDO), PD-L1/B7-H1 (both IFN-γ inducible) and the presence of FOXP3 regulatory T cells in melanoma are the consequence of a counter-regulatory mechanism that follow, rather than precede, CD8 T cell infiltration. However, gene-expression profile studies have shown that the immune-favorable phenotype is also characterized by the presence of B cell signature such as immunoglobulin genes (e.g., IGKC and IGL@), and surface markers (i.e., CD19). Those signatures have been recently associated with good prognosis in breast [59, 60, 84], colon [85], and non-small lung cancer [86]. The presence of B cell signatures also correlates with responsiveness to IL-2 [19] and ipilimumab [15] in melanoma, and to neo-adjuvant chemotherapy or molecular therapy (anti-Her2/neu) in breast cancer (Bianchini JCO 2011; [87]). A direct relationship between the expression of immune-suppressive genes (IDO1, PD-L1, PD-1, CTLA4, and FOXP3), pro-inflammatory/Th1 genes (e.g., CXCL10, CCL5, STAT1, IRF1, IFNG, etc.), and B cell-related genes (CD19, IGKC) is shown in Fig. 5.1.

Besides the negative-feedback mechanisms, it is worth mentioning that IDO exerts pleiotropic functions [83]. In fact, IDO can mediate pro-inflammatory mechanisms, in particular those driven by auto reactive B-cells, as shown by several mechanistic studies performed by Prendergast and coworkers [83, 89, 90].

Regarding FOXP3, a number of researches recently described a paradoxical association between the infiltration of colon cancer by FOXP3 T-cell and favorable outcome after primary tumor excision or after chemo- or chemo- immunotherapy in advanced stages [72, 91–93], in contrast with that observed in other type of cancers, as reviewed by Ladoire et al. [94]. Although the presence of T regulatory cells could be interpreted as a counter-regulatory response after a powerful immune-reaction, the evidence that CD4+ T cells can transiently express FOXP3 without acquisition of suppressive functions [95, 96] and that CD8+/FOXP-3+ T cells with effector function were detected in the context of an effective anti-tumor response [97] suggest caution in the interpretation of the aforementioned results in the absence of functional cell-specific analyses.

In the setting of anti-PD-1/PD-L1 therapy, several studies across multiple cancers have observed a positive correlation between the expression of the PD-L1 and the likelihood to response to treatment [31, 80, 98] (Cho et al. ASCO 2013; Grosso et al. ASCO 201; Daud et al. AACR 2014; Seiwert et al. ASCO 2014; Gandh et al. AACR 2014; Soria et al. ECC 2013; Powles et al. ASCO 2013; Segal et al. ASCO 2014). PD-L1 binds to PD-1 expressed by activated T cells and decrease their effector functions by triggering inhibitory signaling downstream of the T cell receptor (TCR). The expression of PD-L1 by tumor cells correlate with the density of T cells infiltration and also with the expression of PD-1 by tumor infiltrating lymphocytes (TILs) [98]. Interestingly, Tumeh et al. recently showed that a high clonality of the TCR, which was not directly correlated with the density of tumor infiltrating lymphocytes, was associated with responsiveness to PD-1 blockade treatment in melanoma patients [99]. However, all patients with low density of TILs and TCR clonality did not respond to treatment [99]. The expression of the inhibitory molecule CTLA4 is also associated with responsiveness to anti-PDL1 therapy [80].

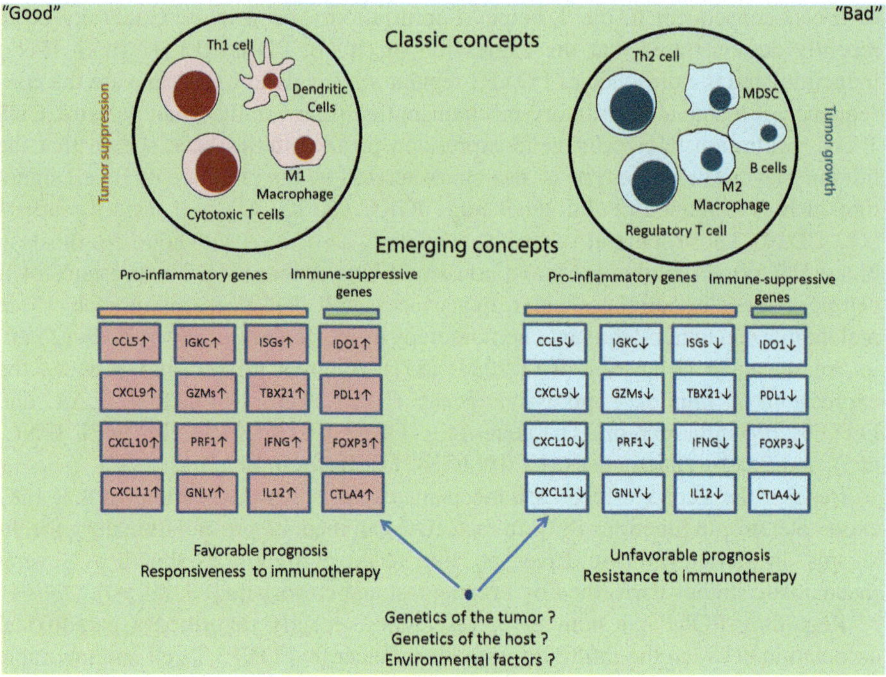

Fig. 5.1 "Bad" and "Good" tumors: classic vs emerging concepts. The classic view of tumor microenvironment postulated the existence of two opposite microenvironments; one sustaining tumor growth and exemplified by the presence of myeloid-derived suppressor cells (MDSC), M2 macrophages Th2, and regulatory T cells and the other one promoting tumor suppression and characterized by the presence of Cytotoxic T cells, Th1 cells, Type-1 dendritic cells, and M1 macrophages [88]. The current view, derived from gene-expression studies across multiple human cancers, dichotomizes tumors in two different phenotypes with distinct prognostic and predictive connotations. The first phenotype is typified by the presence of an inflammatory status driven by the expression of IFNG stimulated genes, T helper 1 chemokine genes (i.e. CXCL9, CXCL10, CCL5) and by the activation of the immune effector function genes (i.e., PRF1, GNZMs). This phenotype is also characterized by the counter-activation of suppressive mechanisms (e.g., IDO, CTLA4, PDL1, and FoxP3), and by the co-expression of B cell molecules (e.g., IGKC, CD19). This inflammatory phenotype is distinguished by a more favorable prognosis and responsiveness to immunotherapy. The lack of the activation of these genes is associated with unfavorable prognosis and resistance to immune manipulations

Altogether, studies in the context of checkpoint inhibitors have conclusively demonstrated that the more responsive tumors bear an inflammatory status accompanied by the concomitant counter activation of immune-suppressive mechanisms. Tumors lacking these two characteristics are insensitive to therapeutic immune manipulations. A schematic representation of these emerging concepts is shown in Fig. 5.2. Tumoral features associated with responsiveness to immunotherapy are listed in Table 5.1.

Even though a link between an active immune microenvironment and clinical outcome has been clearly established, it is presently unclear whether the genetics of the individual bearing the disease, the genetics of the tumor, or other influencing factors,

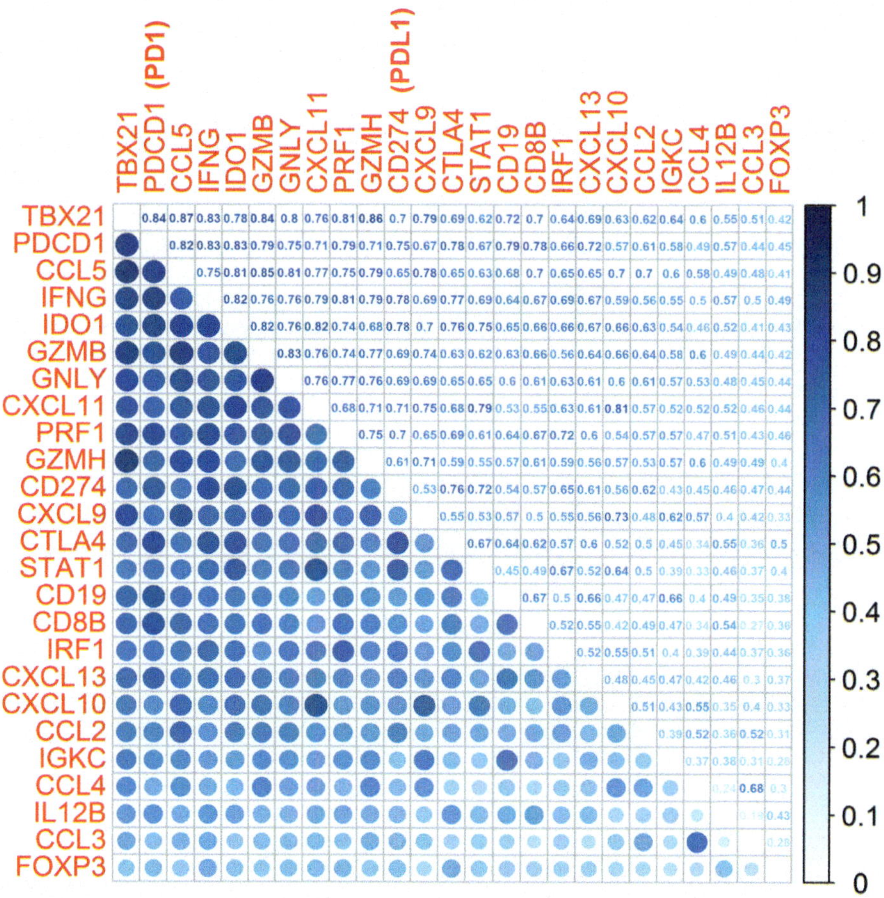

Fig. 5.2 Correlation between Th1 pro-inflammatory, immune-suppressive, and B-cell genes in 531 breast cancer samples. Gene-gene correlation matrix based on Pearson's correlation coefficient of genes related to immune response. (TCGA breast cancer micro array dataset retrieved from TCGA using TCGA Assembler and plotted using *R* corrplot_0.73, and first principal component [FCP] ordering; N=531). These genes include the immunologic-constant of rejection genes (TBX21, IL12B, IFNG, STAT1, IRF1, CCL3-5, CXCL9-11, GZMB, GZMH, GNLY), immune-suppressive genes (PDL1/CD274, PD1/PDCD1, CTLA4, FOXP3, IDO1), and B-cell related molecules (CD19, IGKC). A person R>than 0.3 is equal to a nominal P value<than 0.00001

plays a dominant/driving role in defining these two opposite cancer immunophenotypes [38, 101]. These variables are closely related considering that, for example, the genetics of cancer depends also upon the genetics of the individual and that host environmental factors (e.g., smoking, viral infection, dysbiosis) can influence the carcinogenetic process and the repertoire of somatic alterations. Recently, data from animal models have shown that gut microbiota can influence the response to immunotherapy or chemotherapy [102, 103] and this variable is expected to be further evaluated in humans [104]. Below we summarize findings from recent reports assessing the contribution of tumor host genetics in shaping anti-tumor immune response.

Table 5.1 Pre-treatment intratumoral features associated with responsiveness to immunotherapy

	Immunologic constant of rejection molecules (gene expression)				Immune-suppressive molecules			T cells		Tumor genetics	
	STAT1/ IRF1/ IFNG-SG pathway	CXCR3/ CXCL 9–11 pathway	CCR5/ CCL3-5 pathway	Granzyme/ Perforin/ Granulysin /TIA1 pathway	IDO (gene expression or IHC)	FOXP3 (IHC)	PD-L1 (IHC)	T-cell receptor clonality (sequencing)	CD8 T cell or T cell density (IHC)	Mutational and neoantigen load (sequencing)	References
IL-2-based treatment/vaccination (metastatic melanoma)	+			+							Wang et al. [18]
MAGE-A3 vaccination (metastatic melanoma and NSCLC, adjuvant)	+	+	+	+							Ulloa-Montoya et al. [16]
IL-2 (metastatic melanoma)	+										Weiss et al. [19]
Adoptive therapy and IL-2 (metastatic melanoma)		+									Bedognetti et al. [40, 41]
IL-12-based vaccination (melanoma)		+									Gajewski ASCO (2007)
Anti-CTLA4 (metastatic melanoma)	+	+	+	+	+						Ji et al. [15]
Anti-PDL1 (metastatic tumors)	+	+		+	+		+				Herbst et al. [80]

Anti-CTLA4 (metastatic melanoma)					+	+					Hamid et al. [82]
Anti-CTLA4 (metastatic melanoma)										+	Snyder et al. [100]
Anti-PD1 (metastatic melanoma)							+	+	+		Tumeh et al. [99]
Anti-PD1a (metastatic tumors)							+				Taube et al. [98]; Topalian et al. [31]; Weber et al. 2013; Daud et al. AACR 2014; Seiwert et al. ASCO 2014; Gandh et al. AACR 2014; Grosso et al. ASCO 2014
Anti-PDL1 (metastatic tumors)					+		+				Soria et al. ECC 2013; Powles et al. ASCO 2013; Segal et al ASCO 2014; Cho et al. ASCO 2013

NSCLC non-small-cell lung cancer, *IHC* Immunohistochemistry
[a]Overall response (complete response + partial response)

Genetic Drivers of Tumor Immune Responsiveness

Genetics of the Tumor

The transcriptional profile of the whole tumor tissue cannot clarify the source of the immune genes that identify the immune-favorable cancer phenotype. The assumption that the activation of immune-related genes detected in these studies only depends upon the activation or presence of immune-cells likely represents an oversimplification of a more complex phenomenon. Cancer cells, stromal cells, and different subsets of immune cells interact with each other and collectively contribute to the development of the desirable immune phenotype [105, 106]. Zeimet et al. noticed that the density of T cells in ovarian cancer correlates with the expression of IRF1. However only cancer cells stained positively for IRF1, which was expressed by a limited proportion of lymphocytes and not expressed at all by stromal cells [107]. Similarly, Callahan et al., by microdissecting ovarian cancer epithelium, observed that cancer cells expresses HLA class II molecules IRF1 and, which in turn correlate with the abundance of CD8 T cells [108]. In addition, cultured tumor cells can produce a large amount of chemokines that are able to recruit T cells [77]. If these studies imply that tumor cells contribute to the genesis of the immune signatures, they cannot clarify whether the activation of interferon-stimulated genes and the production of other immune stimulant by tumor cells drive the T cell response or rather represent the consequence of the release of IFN-γ and other pro-inflammatory molecules by activated T cells. Interestingly, pancreatic cell lines can segregate in different categories according to the intrinsic activation of interferon-stimulated-genes including interferon-regulatory-factors [109]. Concordantly, constitutive activation of the JAK1-STAT1 pathway can be found in a considerable proportion of melanoma cells cultured in unconditioned medium (De Giorgi et al. SITC 2012). The correlation between the signatures detected in vivo and in vitro is however not linear. When we correlated the expression of CXCL9, CXCL10, CXCL11 (CXCR3 ligands) and CCL3, CCL4, and CCL5 (CCR5 ligands) between melanoma tumor cell lines and the parental tumor tissue we did not find any significant correlation between the in vivo and in vitro data (Bedognetti et al. ASCO 2012). Altogether these findings provide the notion that the origin of the immune signature in some tumors is a complex, non-linear, multi-factorial and dynamic in vivo phenomenon but also support the hypothesis that it is at least in part dictated by the intrinsic biology of cancer cells. We identified a list of 968 genes (genomic delegates) that display correlation between copy number and gene expression in melanoma tumor cell lines and that strongly correlate with the gene expression of the parental tumor metastases [110]. When we re-classified tumor metastases according to the genomic delegates, we observed that they segregate in two opposite categories. One, named TARA A (transcriptional adjustments related to amplification/deletion class A), with prevalent expression of cancer testis antigens, enhanced cyclin activity, WNT signaling, and a Th17 immune phenotype (e.g., IL17A, IL17B,

WNT5A). This phenotype expressed, therefore, transcripts previously associated to more aggressive cancer [72, 111, 112]. The second class (TARA B) prevalently expresses genes associated with melanoma signaling (i.e., MITF), and with Th1 immune response (e.g., STAT1, GBP1, CXCL9, −10, −11), which are classically associated with better prognosis and likelihood to respond to Immunotherapy [24, 106, 110]. Again, this observation proposes the existence of a stable trait of melanoma genetics that can in turn modulate immune reaction in vivo with consequent lack of linearity between in vivo and ex vivo transcripts of the immune-related genes. Interestingly, Curtis et al. showed that a particular breast cancer in vivo phenotype characterized by a flat copy number landscape is associated with an enrichment of ICR genes, and in turn, with a favorable prognosis [60]. Although copy number aberrations could represent the effect of a dilution of cancer aberrations consequent to an enrichment of immune-cell germinal DNA, this observation is intriguing and could support the driving role of cancer-cell genetics in determining the in vivo immune-phenotype. When we compared signatures of melanoma metastases according to the presence or absence of BRAF and NRAS mutations, genes differentially expressed by tumors carrying BRAF or NRAS mutations were mostly represented by constituents or regulators of MAPK and related pathways [113]. However, when testing gene lists distinctive of BRAF, NRAS and MAPK alterations, we found that 112 BRAF-specific transcripts were able to discriminate the two immune-related phenotypes previously described in melanoma (i.e. TARA B/Th1 and TARA A/Th17), with the poor phenotype (TARA A) associated mostly with BRAF mutation. Such association was stronger in samples displaying low BRAF mRNA expression [113]. This observation indicates that pathways related with driver oncogenes can antagonize the development of the immune-favorable cancer phenotype. Recently, we observed that overexpression of NOS, driven by genomic amplification of NOS1 locus within segment 12q22-24, impairs IFN-α responsiveness of peripheral blood mononuclear cells. Interestingly, baseline expression of NOS1 in tumor metastases was negatively associated with response to adoptive therapy in a cohort of 113 metastatic melanoma patients, therefore linking genetic of tumors with specific immune-dysfunctions [114]. Other studies have substantiated the perception that the immune-favorable cancer phenotype is driven by tumor genetics. By applying a modular approach to the analysis TCGA melanoma datasets, Linsley et al. defined sets of co-regulated immune genes associated with prolonged survival [115]. These modules consist of sub-networks of type I in Interferon-stimulated-genes (ISG), and T/T regulatory and T/NK effector-function genes. Interestingly, tumors with reduced ISG displayed a significant copy number loss of the interferon gene cluster located at chromosome 9p21.3, suggesting that the ISG derived from the tumor itself. This study also postulated a causal connection between genomic instability of the tumor cells and expression of immune-related genes. By mining exome-sequencing and microarray data Spranger et al. observed that melanoma lacking a T cell signature show dysregulation of the Wnt/β-catenin signaling pathway, consistent with our earlier observations (Spranger et al. SITC 2014) [110]. However, investigators convincingly demonstrated that T cell priming was lacking in experimental melanoma models activating β-catenin signaling, therefore highlighting a strong relationship between oncogenic pathways and the development of an effective T cell response against tumor (Spranger et al. SITC 2014). Galon's group recently inspected the role of genomic instability of chemokine, chemokine receptor and interleukin genes in colorectal cancer. In Bindea at al. the authors found that CCL24 and CXCL9 have higher expression in tumors that had a gain, while tumors with deletion of CCL1, CCL26, CCR6, CXCL2, CXCL12, CXCL13, and CXCL16 displayed a significantly lower expression of the corresponding genes [111]. In addition, CXCL13 was correlated with B, T follicular helper, T helper 1, and cytotoxic T cells, and patients with CXCL13 deletions experienced a shorter disease-free survival [111]. A similar nexus between CXCL13, T follicular helper cells and prognosis has also been reported in breast cancer [73, 116]. In a separate report, Galon and colleagues showed that more than 10 % of patients affected by colorectal cancer have a gain of TNF, IFN, IL, and TGF family genes, while TNF was the only family deleted in more than 10 % of patients [117]. The highest level of gain and loss was observed for IL29 and IL15. Deletion of IL15, IL21, and IL2 was associated with higher risk of relapse. IL15 was the only cytokine in which the gene loss was associated with lower gene expression. Interestingly, IL-15 levels were correlated with density of cytotoxic T, activated T/NK, T helper 1, and memory T cells and with the expression of immune-effector function genes (i.e., GZMs, and PRFs) [117]. Furthermore, IL-15 seemed to modulate the local proliferation of infiltrating lymphocytes as patients with high level of IL15 showed a significantly higher density of proliferating T and B cells [117]. In a massive analysis of the TCGA genomic and proteomic database including 12 cancer types, Hoadley et al., introduced a pan-cancer genomic classification consisting of 11 major molecular subtypes [118]. Lung squamous, head and neck, and a subset of bladder cancers coalesced into a unique subtype characterized by TP53 alteration, TP63 amplification and overexpression of immune-related genes. Notably, pathways related with T cell activity such as PD-1 and CTLA4 were strongly correlated with increased overall survival highlighting the prognostic role of immune signatures across cancer of different tissue origin. In view of the analogies between prognostic and predictive signatures the results of the studies mentioned above have important implication for cancer immunotherapy.

A direct correlation between somatic mutations and response to immunotherapy has been recently demonstrated by an exome-wide study of melanoma metastases in the context of anti-CTLA4 therapy. In a breakthrough report, Snyder et al. showed that there was a significant difference in mutational load between patients with a long-term clinical benefit and those with a minimal benefit or no benefit. Indeed, the mutational load was significantly correlated with increased overall survival [100]. Using a refined neoepitope analysis, the authors identified a signature of mutated antigens shared by responding patients that was able to precisely predict clinical outcome (i.e. prolonged survival) to CTLA-4 blockade [100]. This study reinforces the observation that mutated epitopes can be recognized by T cells and potentially targeted by mutation-specific T cell therapy [119, 120]. It also suggests that mutations of genes coding for proteins that are not implicated in oncogenic progression—and therefore classically considered passenger mutations—play a driving role in dictating the final cancer phenotype [100]. Future studies should elucidate whether these concepts could be applied to other cancer types and in the context of other immunotherapeutic approaches. Despite the progresses made in the last few years, whether and in which degree genomic alterations of tumor and immune genes impact on tumor responsiveness remains largely obscure.

Genetics of the Host

Genome-wide association studies (GWAS) have enabled the identification of more than 85 loci that contribute to susceptibility to autoimmune diseases [121].

In addition, the study of families affected by inherited immune-related diseases through next generation exome sequencing led to the identification of critical genes involved in the immune homeostasis such as ADA2 [122], PLCG2 [123, 124], HOIL1 [125], and PIK3CD [126].

However, only very recently investigators began to assess the contribution of genetic determinants on immune-cell level and function beyond the setting of pathologic conditions [127]. The correlation between genotype and immune cell phenotype presupposes the collection of viable cells paired to deep phenotyping approaches, which is challenging when the study is conducted in a large number of subjects. A recent research has significantly contributed to our understanding of the relationship between genetic makeup and variability of immune-cell populations. By analyzing the frequency of 95 immune-cell population (i.e. 272 immune traits) in more than 1500 individuals, Orru and coworkers estimated that trait-heritability could account for about 40 % of the observed variance of the immune traits, therefore indicating a large genetic effect size [127]. By profiling and sequencing about 8 millions of single nucleotide polymorphisms (SNPs) in an additional cohort of more than 2500 individuals, the authors mapped relevant SNP associated with different traits. This unbiased analysis revealed 23 variants at 13 loci linked to multiple immune-cell subtypes resulting in 180 SNP-trait associations. Interestingly, the greatest estimated heritability was observed in immune populations implicated in sophisticated immunological functions such as the T regulatory cells. The strongest association with immune-traits was found for GALM/HNRPLL, CD8A/CD8B, COQ2/PLAC8, HLAs, IL2RA, ENTDP1, NCAM1, CD4, TNFS13B, SH2B3, CIITA and FCG3 and SLFN clusters. Importantly, variants identified in three loci (HLA, IL2RA, and SH2B3/ATXN2) overlapped with those identified by GWAS in auto-immunity disease [127]. These notions could have important implications in cancer immunotherapy if we consider that the stable baseline differences of immune-cells traits across individuals are associated with distinct response to immune perturbations (i.e. influenza vaccination) [128].

As for immunotherapy, large GWAS studies have identified a strong association between IL28B gene, encoding interferon-λ-3 (IFN-λ-3), and response to IFN-α treatment in patients affected by chronic hepatitis C (HCV) infection [129–131]. Such large studies have been facilitated by the fact that IFN-α has represented for years the gold standard for the treatment of HCV infection. In cancer immunotherapies setting, however, these kinds of studies are lacking. In fact, immunotherapy joined the club of the FDA-approved treatment only 4 years ago, and studies have been historically small in term of sample size, therefore limiting the possibility of performing large-scale genotyping approaches. We are currently applying a GWAS approach to define the influence of germline polymorphisms on the likelihood to respond to adoptive therapy in a large casuistic of metastatic melanoma patients. In the preliminary analysis, comparisons between patients experiencing a complete response versus those who progressed, we identified several independent association SNPs covering 14 functional known genes. Several genes at those loci have known immunological functions, and include TLR5, MIA3, TAF1A, DUSP10, DISP1, HHIPL2, EIF3a, FAM45A, SFXN4, GRK5, RGS10, TIAL1, BAG3, INPPF5, and SEC23IP (Sommariva et al. SITC 2013).

Because of the aforementioned reasons researchers have been focusing on specific genes. Genes tested by such approaches include CTLA4, IRF5, and CCR5 and HLAs. In melanoma patients treated with ipilimumab, a single study has assessed 20 single nucleotide polymorphisms (SNPs) and two deletions in 10 immune-related genes (BTNL2,CCR5, CD86, CTLA-4, IFNAR1, IFNAR2, IFNG, IL23R, NOD2, and PTPN22) as well as HLA-A and HLA-B genotypes without observing any correlation between those genetic variants and clinical outcome [82]. Other studies have investigated HLA patterns in relation to treatment responsiveness. HLA genotype (HLA-A, HLA-B, and HLA-class II) was not associated to response in a large cohort of metastatic melanoma patients treated with IL-2 [132], while Gorgas et al. described a marginally significant association between HLA-DRB1*15, Cw6, Cw7, and B44 and survival in melanoma patients treated with adjuvant IFN-α ($P=0.028$, 0.029, 0.030, and 0.040, respectively) [133, 134]. Studies that have evaluated CTLA4, IRF5, and CCR5 polymorphisms are summarized below.

CTLA4 polymorphisms—CTLA4 is a key player in establishing immune tolerance and a critical regulator of T-cell-mediated antitumor immune responses. The major function of CTLA-4 is to shape T cell response at the time of their initial response to antigen. Although CTLA-4 is expressed by activated CD8+ effector T cells, its key role consists in the regulation of T CD4+ populations through down modulation of helper T-cell function and enhancement of regulatory T-cell immunosuppressive activity [135, 136]. CTLA-4 polymorphisms have been implicated in the susceptibility of organ-specific autoimmune diseases, but results are conflicting [126, 137]. However, germline CTLA4 heterozygous mutations resulting in quantitative reductions of CTLA-4 expression have been recently identified in subjects with severe immune dysregulation [126].

A total of 12 CTLA4 single nucleotide polymorphisms (SNPs) have been assessed by three studies in metastatic melanoma treaded with anti-CTLA4 mAbs [82, 138, 139] and in one study in high-risk melanoma treated with adjuvant IFNα therapy [140]. Some significant associations have been found, but results are overall inconclusive. Data are difficult to compare as studies have assessed different SNPs and statistical analysis has been performed using different statistical models looking at allele frequency or genotype frequency. A summary of these studies is reported in Table 5.2.

Table 5.2 CTLA-4 single nucleotide polymorphisms and responsiveness to immunotherapy

References	CTLA4 SNP tested	Setting	Findings
Gogas et al. (2010)	**rs3087243**[a] **rs231775** **rs5742909** **rs7565213** rs11571297 rs11571302	Melanoma stage II-III, treated with adjuvant interferon (N=286)	No significant association or trend with relapse free survival or overall survival were found in a univariate analysis (P value range: 0.37–0.74)
Breunis et al. [138]	**rs4553808** **rs11571317** **rs231775** **rs5742909** **rs3087243** **rs7565213** rs733618	Metastatic Melanoma treated with iIpilimumab (N=152)	rs4553808 (−1660 G vs A allele); rs11571317 (−657 T vs G allele), and rs231775 (49 A vs G allele) were associated with overall response[b] (P value: 0.002, 0.02, and 0.009, respectively)
Hamid et al. [82]	**rs11571317** **rs3087243** **rs4553808** rs1863800 **rs231775**	Metastatic Melanoma treated with Ipilimumab (N=55–57)[c]	No significant associations with clinical benefit[d] were found; a trend was observed for rs1863800, rs231775, rs3087243, and rs4553808 (P value: 0.06, 0.05, 0.07, and 0.08, respectively)
Queirolo et al. [139]	**rs5742909** **rs231775** **rs3087243** **rs4553808** **rs11571317** rs11571316	Metastatic Melanoma treated with ipilimumab or tremelimumab (N=14)	rs11571316 was associated with clinical benefit[d] (−1577 G/A vs G/G; P value: 0.041); a trend was observed for rs3087243 (CT60 G/A vs G/G; P value: 0.072); both polymorphisms were associated with overall survival (P value: <0.006)

Single nucleotide polymorphisms (SNPs) tested in more than one study are in bold; polymorphisms significantly associated with clinical outcome are underlined

[a]In a subsequent analysis rs3087243 (CT60) correlates with overall survival in a multi-variable Cox regression model including HLA-B38, HLA-C15, HLA-C3, DRB1*15, and CT60*G/G. CT G/G was associated with shorter overall survival (P=0.082, and P=0.019 in non-stage corrected and stage corrected analysis, respectively)

[b]Overall response (complete response + partial response)

[c]Genotype data not available in some patients

[d]Clinical benefit: Partial response + complete response + stable disease

In the context of metastatic melanoma treated with ipilimumab, Breunis et al. found three polymorphisms (rs4553808, rs11571317, and rs231775) to be significantly associated with response (P=0.002, 0.02, and 0.009, respectively) [138]. In the same setting, a second study by Hamid et al. assessed five SNPs, including the significant ones described by Breunis et al. The investigators did not find any significant association between the CTLA-4 polymorphisms and response to ipilimumab, even though a trend was observed for rs4553808, rs11571317, and rs231775. By analyzing a small cohort of patients treated with anti-CTLA4 mAbs (ipilimumab or Tremelimumab), Queirolo et al. showed an association between rs11571316 (G/A vs G/G, odds ratio, [OR] 12.5 P=0.041) and clinical benefit; a trend was also observed for rs3087243 (G/A vs G/G, [OR] 6.8, P=0.072) [139]. Despite the small sample size, both polymorphisms were significantly associated with overall survival (P=0.006). The polymorphism rs11571316 was not evaluated in previous studies. Although rs3087243 was not associated with clinical response in the Breunis' study (P=0.62), an inverse borderline association was found in the Hamid's study (G/A vs G/G, odds ratio [OR] 0.39, P=0.07), suggesting cautions in the interpretation of these results [138]. Six CTLA4 polymorphisms in part overlapping with those investigated by the studies mentioned above were also investigated in patients treated with adjuvant IFN-α by Gogas et al. A first analysis did not find any association between CTLA-4 polymorphisms and survival or risk of relapse [140]. Interestingly, when those polymorphisms were re-analyzed using a multi-factorial approach together with HLAs polymorphisms, rs3087243 significantly contributed to the predictive survival models which included HLA-B38, C15, C3, DRB1*15 and rs3087243 [134]. The rs3087243 GG genotype correlated with shorter overall survival (P=0.082, and P=0.019 in non-stage corrected and stage-corrected analysis, respectively), similarly with what was observed by Queirolo et al. in patients receiving ipilimumab [139]. The analysis of a significantly higher number of patients through high-throughput genotyping approaches paired with functional validations will likely help to elucidate the role of CTLA4 on the responsiveness to anti-CTLA4 therapy in the near future.

IRF5 polymorphisms—IRF5 is implicated in host defense against pathogens by inducing transcription of IFN-α and the expression of genes involved in apoptosis [141, 142]. Variants of IRF5 have been associated with susceptibility to systemic lupus erythematosus [143] and to several other autoimmune diseases, including rheumatoid arthritis [144], multiple sclerosis [145] and inflammatory bowels diseases [146].

Intriguingly, treatment-induced manifestation of autoimmunity (e.g., vitiligo, thyroiditis, enterocolitis) are observed in metastatic melanoma patients who better respond to ipilimumab [147, 148], high-dose IL-2 [2] and vaccination [149]. In patients treated with high dose IFN-α, post-treatment clinical and/or molecular manifestation of autoimmunity correlate with reduced risk of deaths or relapse in the Hellenic IFN trial [150], although the integrated analysis of the EORTC and the Nordic IFN trials failed to reproduce this results [151, 152]. In view of the analogies between the molecular mechanisms responsible for the development of autoimmunity and those implicated in the immune-mediated rejection we studied IRF5 polymorphisms in metastatic melanoma patients treated with adoptive therapy [153]. All the IRF5 variants tested (rs10954213, rs11770589 and rs6953165, rs2004640) but rs2004640 were in linkage disequilibrium and associated with response to therapy. For example, the lack of the A allele in rs10954213, which confer protection to lupus erythematous, was predominant in non-responders (P=0.005. Therefore, IRF-5 polymorphisms associated with the development of lupus erythematous influence the strength of the anti-tumor response to adoptive therapy, underlining the genetic link between the predispositions to develop different immune-mediated tissue destruction processes [153]. Even more, the observation that IRF-5 polymorphisms specific signatures detected in melanoma cell lines could be used re-classify parental metastases according to their responsiveness to immunotherapy, suggests that host's genetics can directly modulate the intrinsic cancer biology besides modulating the reaction of host's immune cells to immunotherapy [153].

CCR5 and CXCR3 polymorphisms—The CC chemokine receptor C (CCR5) is expressed, together with CXC chemokine receptor 3 (CXCR3) by activated Th-1, cytotoxic T, and NK cells. As mentioned before, several studies suggested that the recruitment of activated T lymphocytes through CXCR3 and CCR5 chemokine ligands plays a key role in immune-mediated tissue destruction, including tumor rejection [17, 24, 37, 45, 154]. However, we observed that systemically administered IL-2 induces inflammation within tumors leading to production of lymphocyte chemo attractants including CXCR3 and CCR5 ligands [19, 36, 46, 155]. It could be hypothesized that polymorphisms and/or expression of CXCR3 and CCR5 may quantitatively and/or qualitatively influence chemokine receptor expression, and therefore influence migration of TILs to tumors and affect immune-mediated tumor rejection. The CCR5Δ32 mutation/polymorphism consists of a 32-base deletion encoding a protein not expressed on the cell surface. Heterozygosity results in decreased, and homozygosity absent receptor expression (Huang et al. 1996; Carrington et al. 1999). Homozygous carriers of this mutation are resistant HIV-1, which uses CCR5 as co-receptor to enter is targeted cells, while heterozygous state is associated with up to 2–4 years delay in disease progression [156]. Data on CCR5Δ32 polymorphism in autoimmune disease are conflicting as it seems to be protective against the development of some autoimmune diseases (i.e. rheumatoid arthritis) but not in others such as systemic lupus erythematosus and sclerosing cholangitis, in which a detrimental effect has been suggested [157–160]. An inverse association between this polymorphism and the occurrence of allograft rejection has also been described [161]. In the context of immunotherapy, reports focusing on metastatic melanoma are discordant. A retrospective study reported decreased survival of patients carrying this polymorphism (either in heterozygous or homozygous state) treated with immunotherapy or immunochemotherapy [162]. In two preliminary reports, Essner et al. described that individuals who were hetero- or homozygous carriers for CCR5Δ32 polymorphism experienced significantly decreased overall survival compared to wild-type patients within an adjuvant therapeutic vaccination protocol (Essner et al. ASCO 2005) or after surgical resection (Essner et al. ASCO 2006). Conversely, Hamid et al. did not find any association between CCR5Δ32 or CCR5 rs1799987 polymorphisms and responsiveness to ipilimumab [82]. As for CXCR3, the rs2280964 polymorphism has been associated with altered receptor expression, lymphocyte chemotactic activity, and risk of developing asthma [163]. We recently explored the effect of CCR5Δ32 and CXCR3 rs2280964 polymorphism in the context of metastatic melanoma patients treated with adoptive therapy and high dose IL-2 [40, 41]. Twenty-five out of 141 patients carried the CCR5Δ32 mutation (24 heterozygous and 1 homozygous). As single factors, CCR5Δ32 was slightly associated with increased overall response rate ($P=0.043$), while CXCR3 had no impact on treatment responsiveness. We also observed that the concomitant down-regulation of CXCR3 and CCR5 receptor in TILs due to the down-regulation of the corresponding genes and/or to the presence of CCR5Δ32 mutation, strongly correlate with the frequency and the degree of response [164] ($P<0.001$). These counterintuitive and surprising results could be explained by the dynamics of TIL migration to the tumor in relation to concomitant IL-2 administration. The migration of TILs at tumor site does not follow a linear kinetic [165]. Two hours after infusion, immediately followed by administration of IL-2, TILs massively localize in lung, spleen and liver but not at tumor sites. TILs' migration into tumor sites is detectable 24–48 h after infusion, in concomitance with a partial clearance of the TILs from the lung [165]. However, concentrations of CCR5 and CXCR3 ligands (e.g., CCL3, CCL4, CXCL9, CXCL10 and CXCL11) increase immediately after IL-2 administration [155]. It is possible that, following IL-2 administration, early compartmentalization is mediated by the release of specific chemokines (primarily CXCR3 and CCR5 ligands) by resident immune-cells and stromal cells from peripheral organs (e.g., spleen, lung and liver). It is tempting to hypothesize that a low expression of CXCR3 and CCR5 chemokine receptors by TILs might prevent their sequestration by extratumoral tissues and paradoxically allow their subsequent migration to the tumor when the cytokine storm has receded and the tumor become the only tissue maintaining expression of chemokines [40, 41]. This explanation would be consistent with the observation that high levels of CXCL9, CXCL10, CXCL11, and CCL5 in pre-treatment tumors are associated with an increased response rate in this setting [40, 41]. It is possible that during other treatments (e.g., combination of immunochemotherapy [162], or vaccine therapy), the proportional induction of the CXCR3 and CCR5 ligands following treatment is not as unbalanced as it is in during administration of high-dose IL-2, resulting in different kinetic modulation of T cells. Even though adoptive therapy studies in mouse models have emphasized a key role of the upregulation of CCR5 in TILs in mediating tumor rejection (Gonzalez-Martin et al. 2012), our findings do not support the notion that a maximal expression of CCR5 (and CXCR3) by TILs is a critical factor in the context of adoptive T cell therapy in humans. The role of these receptors in other types of immunotherapy such as dendritic cell vaccination needs to be further investigated.

In conclusion, some studies have assessed the contribution of germline variants on the outcome of cancer immunotherapy, but none of them have been confirmed in an independent cohort and functional validations are lacking. In the last 5 years, the remarkable success of refined immunotherapies has made this approach available to a large number of patients. At the same time, technology has advanced and large scale high-throughput genotyping and deep sequencing can now be performed with sustainable costs. Altogether, these factors offer the opportunity to explore the scope of human and tumor genetic variants in relation to immune responsiveness. Data from these studies are expected to prompt the development of personalized immunotherapy in the near future.

References

1. Atkins MB, Lotze MT, Dutcher JP et al (1999) High-dose recombinant interleukin 2 therapy for patients with metastatic melanoma: analysis of 270 patients treated between 1985 and 1993. J Clin Oncol 17(7):2105–2116
2. Atkins MB, Mier JW, Parkinson DR, Gould JA, Berkman EM, Kaplan MM (1988) Hypothyroidism after treatment with interleukin-2 and lymphokine-activated killer cells. N Engl J Med 318(24):1557–1563. doi:10.1056/NEJM198806163182401

3. Kirkwood JM, Ernstoff MS, Davis CA, Reiss M, Ferraresi R, Rudnick SA (1985) Comparison of intramuscular and intravenous recombinant alpha-2 interferon in melanoma and other cancers. Ann Intern Med 103(1):32–36
4. Kirkwood JM, Tarhini AA, Panelli MC et al (2008) Next generation of immunotherapy for melanoma. J Clin Oncol 26(20):3445–3455. doi:10.1200/JCO.2007.14.6423
5. Mazumder A, Rosenberg SA (1984) Successful immunotherapy of natural killer-resistant established pulmonary melanoma metastases by the intravenous adoptive transfer of syngeneic lymphocytes activated in vitro by interleukin 2. J Exp Med 159(2):495–507
6. Traversari C, van der Bruggen P, Luescher IF et al (1992) A nonapeptide encoded by human gene MAGE-1 is recognized on HLA-A1 by cytolytic T lymphocytes directed against tumor antigen MZ2-E. J Exp Med 176(5):1453–1457
7. van der Bruggen P, Traversari C, Chomez P et al (1991) A gene encoding an antigen recognized by cytolytic T lymphocytes on a human melanoma. Science 254(5038):1643–1647
8. Belli F, Testori A, Rivoltini L et al (2002) Vaccination of metastatic melanoma patients with autologous tumor-derived heat shock protein gp96-peptide complexes: clinical and immunologic findings. J Clin Oncol 20(20):4169–4180
9. Marincola FM, Ferrone S (2003) Immunotherapy of melanoma: the good news, the bad ones and what to do next. Semin Cancer Biol 13(6):387–389. doi:10.1016/j.semcancer.2003.09.002. pii: S1044579X03000695
10. Marincola FM, Jaffee EM, Hicklin DJ, Ferrone S (2000) Escape of human solid tumors from T-cell recognition: molecular mechanisms and functional significance. Adv Immunol 74:181–273
11. Marincola FM, Wang E, Herlyn M, Seliger B, Ferrone S (2003) Tumors as elusive targets of T-cell-based active immunotherapy. Trends Immunol 24(6):335–342. pii: S1471490603001169
12. Schwartzentruber DJ, Lawson DH, Richards JM et al (2011) gp100 peptide vaccine and interleukin-2 in patients with advanced melanoma. N Engl J Med 364(22):2119–2127. doi:10.1056/NEJMoa1012863
13. Bedognetti D, Balwit JM, Wang E et al (2011) SITC/iSBTc Cancer immunotherapy biomarkers resource document: online resources and useful tools – a compass in the land of biomarker discovery. J Transl Med 9
14. Klebanoff CA, Acquavella N, Yu Z, Restifo NP (2011) Therapeutic cancer vaccines: are we there yet? Immunol Rev 239(1):27–44. doi:10.1111/j.1600-065X.2010.00979.x
15. Ji RR, Chasalow SD, Wang L et al (2012) An immune-active tumor microenvironment favors clinical response to ipilimumab. Cancer Immunol Immunother 61(7):1019–1031. doi:10.1007/s00262-011-1172-6
16. Ulloa-Montoya F, Louahed J, Dizier B et al (2013) Predictive gene signature in MAGE-A3 antigen-specific cancer immunotherapy. J Clin Oncol 31(19):2388–2395. doi:10.1200/JCO.2012.44.3762
17. Wang E, Bedognetti D, Marincola FM (2013) Prediction of response to anticancer immunotherapy using gene signatures. J Clin Oncol. doi:10.1200/JCO.2013.49.2157
18. Wang E, Miller LD, Ohnmacht GA et al (2002) Prospective molecular profiling of melanoma metastases suggests classifiers of immune responsiveness. Cancer Res 62(13):3581–3586
19. Weiss GR, Grosh WW, Chianese-Bullock KA et al (2011) Molecular insights on the peripheral and intratumoral effects of systemic high-dose rIL-2 (aldesleukin) administration for the treatment of metastatic melanoma. Clin Cancer Res 17(23):7440–7450. doi:10.1158/1078-0432.CCR-11-1650
20. Galon J, Costes A, Sanchez-Cabo F et al (2006) Type, density, and location of immune cells within human colorectal tumors predict clinical outcome. Science 313(5795):1960–1964. doi:10.1126/science.1129139
21. Mlecnik B, Tosolini M, Kirilovsky A et al (2011) Histopathologic-based prognostic factors of colorectal cancers are associated with the state of the local immune reaction. J Clin Oncol 29(6):610–618. doi:10.1200/JCO.2010.30.5425

22. Ascierto ML, De Giorgi V, Liu Q et al (2011) An immunologic portrait of cancer. J Transl Med 9:146. doi:10.1186/1479-5876-9-146
23. Fridman WH, Pages F, Sautes-Fridman C, Galon J (2012) The immune contexture in human tumours: impact on clinical outcome. Nat Rev Cancer 12(4):298–306. doi:10.1038/nrc3245
24. Galon J, Angell Helen K, Bedognetti D, Marincola Francesco M (2013) The continuum of cancer immunosurveillance: prognostic, predictive, and mechanistic signatures. Immunity 39(1):11–26
25. Hodi FS, O'Day SJ, McDermott DF et al (2010) Improved survival with ipilimumab in patients with metastatic melanoma. N Engl J Med 363(8):711–723. doi:10.1056/NEJMoa1003466
26. Robert C, Thomas L, Bondarenko I et al (2011) Ipilimumab plus dacarbazine for previously untreated metastatic melanoma. N Engl J Med 364(26):2517–2526. doi:10.1056/NEJMoa1104621
27. Brahmer JR, Tykodi SS, Chow LQ et al (2012) Safety and activity of anti-PD-L1 antibody in patients with advanced cancer. N Engl J Med 366(26):2455–2465. doi:10.1056/NEJMoa1200694
28. Postow MA, Callahan MK, Wolchok JD (2015) Immune checkpoint blockade in cancer therapy. J Clin Oncol. doi:10.1200/jco.2014.59.4358
29. Robert C, Long GV, Brady B et al (2015) Nivolumab in previously untreated melanoma without BRAF mutation. N Engl J Med 372(4):320–330. doi:10.1056/NEJMoa1412082
30. Tomei S, Wang E, Delogu LG, Marincola FM, Bedognetti D (2014) Non-BRAF-targeted therapy, immunotherapy, and combination therapy for melanoma. Expert Opin Biol Ther. doi:10.1517/14712598.2014.890586
31. Topalian SL, Hodi FS, Brahmer JR et al (2012) Safety, activity, and immune correlates of anti-PD-1 antibody in cancer. N Engl J Med 366(26):2443–2454. doi:10.1056/NEJMoa1200690
32. Lee DW, Kochenderfer JN, Stetler-Stevenson M et al (2014) T cells expressing CD19 chimeric antigen receptors for acute lymphoblastic leukaemia in children and young adults: a phase 1 dose-escalation trial. Lancet. doi:10.1016/s0140-6736(14)61403-3
33. Restifo NP, Dudley ME, Rosenberg SA (2012) Adoptive immunotherapy for cancer: harnessing the T cell response. Nat Rev Immunol 12(4):269–281. doi:10.1038/nri3191
34. Robbins PF, Kassim SH, Tran TL et al (2014) A pilot trial using lymphocytes genetically engineered with an NY-ESO-1-reactive T-cell receptor: long-term follow-up and correlates with response. Clin Cancer Res. doi:10.1158/1078-0432.ccr-14-2708
35. Ruella M, Kalos M (2014) Adoptive immunotherapy for cancer. Immunol Rev 257(1):14–38. doi:10.1111/imr.12136
36. Bedognetti D, Wang E, Sertoli MR, Marincola FM (2010) Gene-expression profiling in vaccine therapy and immunotherapy for cancer. Expert Rev Vaccines 9(6):555–565. doi:10.1586/erv.10.55
37. Spivey TL, Uccellini L, Ascierto ML et al (2011) Gene expression profiling in acute allograft rejection: challenging the immunologic constant of rejection hypothesis. J Transl Med 9:174. doi:10.1186/1479-5876-9-174
38. Wang E, Bedognetti D, Tomei S, Marincola FM (2013) Common pathways to tumor rejection. Ann N Y Acad Sci 1284(1):75–79. doi:10.1111/nyas.12063
39. Wang E, Worschech A, Marincola FM (2008) The immunologic constant of rejection. Trends Immunol 29(6):256–262. doi:10.1016/j.it.2008.03.002. pii: S1471-4906(08)00111-7
40. Bedognetti D, Wang E, Sato-Matsushita M, Marincola FM, Ascierto ML (2013) Molecular profiling of immunotherapeutic resistance. In: Cancer immunotherapy: immune suppression and tumor growth, 2nd edn. Elsevier, pp 373–394
41. Bedognetti D, Spivey TL, Zhao Y et al (2013) CXCR3/CCR5 pathways in metastatic melanoma patients treated with adoptive therapy and interleukin-2. Br J Cancer 109(9):2412–2423. doi:10.1038/bjc.2013.557
42. Marincola FM (2007) In support of descriptive studies; relevance to translational research. J Transl Med 5:21. doi:10.1186/1479-5876-5-21
43. Marincola FM (2011) The trouble with translational medicine. J Intern Med 270(2):123–127. doi:10.1111/j.1365-2796.2011.02402.x

44. Carretero R, Wang E, Rodriguez AI et al (2011) Regression of melanoma metastases after immunotherapy is associated with activation of antigen presentation and interferon-mediated rejection genes. Int J Cancer. doi:10.1002/ijc.26471
45. Panelli MC, Stashower ME, Slade HB et al (2007) Sequential gene profiling of basal cell carcinomas treated with imiquimod in a placebo-controlled study defines the requirements for tissue rejection. Genome Biol 8(1):R8. doi:10.1186/gb-2007-8-1-r8
46. Panelli MC, Wang E, Phan G et al (2002) Gene-expression profiling of the response of peripheral blood mononuclear cells and melanoma metastases to systemic IL-2 administration. Genome Biol 3(7):RESEARCH0035
47. Trinchieri G (2012) Cancer and inflammation: an old intuition with rapidly evolving new concepts. Annu Rev Immunol 30:677–706. doi:10.1146/annurev-immunol-020711-075008
48. Fridman WH, Galon J, Pages F, Tartour E, Sautes-Fridman C, Kroemer G (2011) Prognostic and predictive impact of intra- and peritumoral immune infiltrates. Cancer Res 71(17): 5601–5605. doi:10.1158/0008-5472.CAN-11-1316
49. Ascierto PA, De Maio E, Bertuzzi S et al (2011) Future perspectives in melanoma research. Meeting report from the "Melanoma Research: a bridge Naples-USA. Naples, December 6th–7th 2010". J Transl Med 9:32. doi:10.1186/1479-5876-9-32. pii: 1479-5876-9-32
50. Azimi F, Scolyer RA, Rumcheva P et al (2012) Tumor-infiltrating lymphocyte grade is an independent predictor of sentinel lymph node status and survival in patients with cutaneous melanoma. J Clin Oncol 30(21):2678–2683. doi:10.1200/JCO.2011.37.8539
51. Mahmoud SM, Paish EC, Powe DG et al (2011) Tumor-infiltrating CD8+ lymphocytes predict clinical outcome in breast cancer. J Clin Oncol 29(15):1949–1955. doi:10.1200/JCO.2010.30.5037
52. Salgado R, Denkert C, Demaria S et al (2015) The evaluation of tumor-infiltrating lymphocytes (TILs) in breast cancer: recommendations by an International TILs Working Group 2014. Ann Oncol 26(2):259–271. doi:10.1093/annonc/mdu450
53. Hwang WT, Adams SF, Tahirovic E, Hagemann IS, Coukos G (2012) Prognostic significance of tumor-infiltrating T cells in ovarian cancer: a meta-analysis. Gynecol Oncol 124(2):192–198. doi:10.1016/j.ygyno.2011.09.039
54. Pages F, Berger A, Camus M et al (2005) Effector memory T cells, early metastasis, and survival in colorectal cancer. N Engl J Med 353(25):2654–2666. doi:10.1056/NEJMoa051424
55. Denkert C, Loibl S, Noske A et al (2010) Tumor-associated lymphocytes as an independent predictor of response to neoadjuvant chemotherapy in breast cancer. J Clin Oncol 28(1):105–113. doi:10.1200/JCO.2009.23.7370
56. Imanguli MM, Swaim WD, League SC, Gress RE, Pavletic SZ, Hakim FT (2009) Increased T-bet+cytotoxic effectors and type I interferon-mediated processes in chronic graft-versus-host disease of the oral mucosa. Blood 113(15):3620–3630. doi:10.1182/blood-2008-07-168351
57. Mantovani A, Romero P, Palucka AK, Marincola FM (2008) Tumour immunity: effector response to tumour and role of the microenvironment. Lancet 371(9614):771–783. doi:10.1016/S0140-6736(08)60241-X. pii: S0140-6736(08)60241-X
58. Ascierto ML, Idowu MO, Zhao Y et al (2013) Molecular signatures mostly associated with NK cells are predictive of relapse free survival in breast cancer patients. J Transl Med 11:145. doi:10.1186/1479-5876-11-145
59. Ascierto ML, Kmieciak M, Idowu MO et al (2012) A signature of immune function genes associated with recurrence-free survival in breast cancer patients. Breast Cancer Res Treat 131(3):871–880. doi:10.1007/s10549-011-1470-x
60. Curtis C, Shah SP, Chin SF et al (2012) The genomic and transcriptomic architecture of 2,000 breast tumours reveals novel subgroups. Nature 486(7403):346–352. doi:10.1038/nature10983
61. Desmedt C, Haibe-Kains B, Wirapati P et al (2008) Biological processes associated with breast cancer clinical outcome depend on the molecular subtypes. Clin Cancer Res 14(16):5158–5165. doi:10.1158/1078-0432.CCR-07-4756

62. Leffers N, Fehrmann RS, Gooden MJ et al (2010) Identification of genes and pathways associated with cytotoxic T lymphocyte infiltration of serous ovarian cancer. Br J Cancer 103(5):685–692. doi:10.1038/sj.bjc.6605820
63. Verhaak RG, Tamayo P, Yang JY et al (2013) Prognostically relevant gene signatures of high-grade serous ovarian carcinoma. J Clin Invest 123(1):517–525. doi:10.1172/JCI65833
64. Zhang L, Conejo-Garcia JR, Katsaros D et al (2003) Intratumoral T cells, recurrence, and survival in epithelial ovarian cancer. N Engl J Med 348(3):203–213. doi:10.1056/NEJMoa020177348/3/203
65. Mann GJ, Pupo GM, Campain AE et al (2013) BRAF mutation, NRAS mutation, and the absence of an immune-related expressed gene profile predict poor outcome in patients with stage III melanoma. J Invest Dermatol 133(2):509–517. doi:10.1038/jid.2012.283
66. Messina JL, Fenstermacher DA, Eschrich S et al (2012) 12-Chemokine gene signature identifies lymph node-like structures in melanoma: potential for patient selection for immunotherapy? Sci Rep 2:765. doi:10.1038/srep00765
67. Sivendran S, Chang R, Pham L et al (2014) Dissection of immune gene networks in primary melanoma tumors critical for antitumor surveillance of patients with stage II-III resectable disease. J Invest Dermatol 134(8):2202–2211. doi:10.1038/jid.2014.85
68. Moran CJ, Arenberg DA, Huang CC et al (2002) RANTES expression is a predictor of survival in stage I lung adenocarcinoma. Clin Cancer Res 8(12):3803–3812
69. Chew V, Chen J, Lee D et al (2012) Chemokine-driven lymphocyte infiltration: an early intratumoural event determining long-term survival in resectable hepatocellular carcinoma. Gut 61(3):427–438. doi:10.1136/gutjnl-2011-300509
70. Jiang Z, Xu Y, Cai S (2010) CXCL10 expression and prognostic significance in stage II and III colorectal cancer. Mol Biol Rep 37(6):3029–3036. doi:10.1007/s11033-009-9873-z
71. Mlecnik B, Tosolini M, Charoentong P et al (2010) Biomolecular network reconstruction identifies T-cell homing factors associated with survival in colorectal cancer. Gastroenterology 138(4):1429–1440. doi:10.1053/j.gastro.2009.10.057. pii: S0016-5085(09)01960-X
72. Tosolini M, Kirilovsky A, Mlecnik B et al (2011) Clinical impact of different classes of infiltrating T cytotoxic and helper cells (Th1, th2, treg, th17) in patients with colorectal cancer. Cancer Res 71(4):1263–1271. doi:10.1158/0008-5472.CAN-10-2907
73. Bedognetti D, Wang E, Marincola FM (2014) Meta-analysis and metagenes: CXCL-13-driven signature as a robust marker of intratumoral immune response and predictor of breast cancer chemotherapeutic outcome. Oncoimmunology 3, e28727. doi:10.4161/onci.28727
74. Ignatiadis M, Singhal SK, Desmedt C et al (2012) Gene modules and response to neoadjuvant chemotherapy in breast cancer subtypes: a pooled analysis. J Clin Oncol 30(16):1996–2004. doi:10.1200/JCO.2011.39.5624
75. Stoll G, Enot D, Mlecnik B, Galon J, Zitvogel L, Kroemer G (2014) Immune-related gene signatures predict the outcome of neoadjuvant chemotherapy. Oncoimmunology 3(1), e27884. doi:10.4161/onci.27884
76. Perez EA, Thompson EA, Ballman KV et al (2015) Genomic analysis reveals that immune function genes are strongly linked to clinical outcome in the North Central Cancer Treatment Group N9831 Adjuvant Trastuzumab Trial. J Clin Oncol. doi:10.1200/jco.2014.57.6298
77. Harlin H, Meng Y, Peterson AC et al (2009) Chemokine expression in melanoma metastases associated with CD8+ T-cell recruitment. Cancer Res 69(7):3077–3085. doi:10.1158/0008-5472.CAN-08-2281
78. Gajewski TF, Louahed J, Brichard VG (2010) Gene signature in melanoma associated with clinical activity: a potential clue to unlock cancer immunotherapy. Cancer J 16(4):399–403
79. Ruiz R, Hunis B, Raez LE (2014) Immunotherapeutic agents in non-small-cell lung cancer finally coming to the front lines. Curr Oncol Rep 16(9):400. doi:10.1007/s11912-014-0400-6
80. Herbst RS, Soria JC, Kowanetz M et al (2014) Predictive correlates of response to the anti-PD-L1 antibody MPDL3280A in cancer patients. Nature 515(7528):563–567. doi:10.1038/nature14011
81. Sznol M (2011) Molecular markers of response to treatment for melanoma. Cancer J 17(2):127–133

82. Hamid O, Schmidt H, Nissan A et al (2011) A prospective phase II trial exploring the association between tumor microenvironment biomarkers and clinical activity of ipilimumab in advanced melanoma. J Transl Med 9:204. doi:10.1186/1479-5876-9-204
83. Muller AJ, Mandik-Nayak L, Prendergast GC (2010) Beyond immunosuppression: reconsidering indoleamine 2,3-dioxygenase as a pathogenic element of chronic inflammation. Immunotherapy 2(3):293–297. doi:10.2217/imt.10.22
84. Nagalla S, Chou JW, Willingham MC et al (2013) Interactions between immunity, proliferation and molecular subtype in breast cancer prognosis. Genome Biol 14(4):R34. doi:10.1186/gb-2013-14-4-r34
85. Bindea G, Mlecnik B, Tosolini M et al (2013) Spatiotemporal dynamics of intratumoral immune cells reveal the immune landscape in human cancer. Immunity 39(4):782–795. doi:10.1016/j.immuni.2013.10.003
86. Schmidt M, Hellwig B, Hammad S et al (2012) A comprehensive analysis of human gene expression profiles identifies stromal immunoglobulin kappa C as a compatible prognostic marker in human solid tumors. Clin Cancer Res 18(9):2695–2703. doi:10.1158/1078-0432.CCR-11-2210
87. Alistar A, Chou JW, Nagalla S, Black MA, D'Agostino R Jr, Miller LD (2014) Dual roles for immune metagenes in breast cancer prognosis and therapy prediction. Genome Med 6(10):80. doi:10.1186/s13073-014-0080-8
88. Disis ML (2010) Immune regulation of cancer. J Clin Oncol 28(29):4531–4538. doi:10.1200/jco.2009.27.2146
89. Prendergast GC, Chang MY, Mandik-Nayak L, Metz R, Muller AJ (2011) Indoleamine 2,3-dioxygenase as a modifier of pathogenic inflammation in cancer and other inflammation-associated diseases. Curr Med Chem 18(15):2257–2262
90. Scott GN, DuHadaway J, Pigott E et al (2009) The immunoregulatory enzyme IDO paradoxically drives B cell-mediated autoimmunity. J Immunol 182(12):7509–7517. doi:10.4049/jimmunol.0804328
91. Correale P, Rotundo MS, Del Vecchio MT et al (2010) Regulatory (FoxP3+) T-cell tumor infiltration is a favorable prognostic factor in advanced colon cancer patients undergoing chemo or chemoimmunotherapy. J Immunother 33(4):435–441. doi:10.1097/CJI.0b013e3181d32f01
92. Frey DM, Droeser RA, Viehl CT et al (2010) High frequency of tumor-infiltrating FOXP3(+) regulatory T cells predicts improved survival in mismatch repair-proficient colorectal cancer patients. Int J Cancer 126(11):2635–2643. doi:10.1002/ijc.24989
93. Salama P, Phillips M, Grieu F et al (2009) Tumor-infiltrating FOXP3+ T regulatory cells show strong prognostic significance in colorectal cancer. J Clin Oncol 27(2):186–192. doi:10.1200/JCO.2008.18.7229
94. Ladoire S, Martin F, Ghiringhelli F (2011) Prognostic role of FOXP3+ regulatory T cells infiltrating human carcinomas: the paradox of colorectal cancer. Cancer Immunol Immunother 60(7):909–918. doi:10.1007/s00262-011-1046-y
95. Roncador G, Brown PJ, Maestre L et al (2005) Analysis of FOXP3 protein expression in human CD4+CD25+ regulatory T cells at the single-cell level. Eur J Immunol 35(6):1681–1691. doi:10.1002/eji.200526189
96. Walker MR, Kasprowicz DJ, Gersuk VH et al (2003) Induction of FoxP3 and acquisition of T regulatory activity by stimulated human CD4+CD25- T cells. J Clin Invest 112(9):1437–1443. doi:10.1172/JCI19441
97. Le DT, Ladle BH, Lee T et al (2011) CD8(+) Foxp3(+) tumor infiltrating lymphocytes accumulate in the context of an effective anti-tumor response. Int J Cancer 129(3):636–647. doi:10.1002/ijc.25693
98. Taube JM, Klein A, Brahmer JR et al (2014) Association of PD-1, PD-1 ligands, and other features of the tumor immune microenvironment with response to anti-PD-1 therapy. Clin Cancer Res 20(19):5064–5074. doi:10.1158/1078-0432.ccr-13-3271
99. Tumeh PC, Harview CL, Yearley JH et al (2014) PD-1 blockade induces responses by inhibiting adaptive immune resistance. Nature 515(7528):568–571. doi:10.1038/nature13954

100. Snyder A, Makarov V, Merghoub T et al (2014) Genetic basis for clinical response to CTLA-4 blockade in melanoma. N Engl J Med 371(23):2189–2199. doi:10.1056/NEJMoa1406498
101. Wang E, Uccellini L, Marincola FM (2012) A genetic inference on cancer immune responsiveness. Oncoimmunology 1(4):520–525
102. Iida N, Dzutsev A, Stewart CA et al (2013) Commensal bacteria control cancer response to therapy by modulating the tumor microenvironment. Science 342(6161):967–970. doi:10.1126/science.1240527
103. Viaud S, Saccheri F, Mignot G et al (2013) The intestinal microbiota modulates the anticancer immune effects of cyclophosphamide. Science 342(6161):971–976. doi:10.1126/science.1240537
104. Zitvogel L, Galluzzi L, Viaud S et al (2015) Cancer and the gut microbiota: an unexpected link. Sci Transl Med 7(271):271ps1. doi:10.1126/scitranslmed.3010473
105. Balkwill FR (2012) The chemokine system and cancer. J Pathol 226(2):148–157. doi:10.1002/path.3029
106. Jin P, Civini S, Zhao Y et al (2014) Direct T cell-tumour interaction triggers TH1 phenotype activation through the modification of the mesenchymal stromal cells transcriptional programme. Br J Cancer 110(12):2955–2964. doi:10.1038/bjc.2014.235
107. Zeimet AG, Reimer D, Wolf D et al (2009) Intratumoral interferon regulatory factor (IRF)-1 but not IRF-2 is of relevance in predicting patient outcome in ovarian cancer. Int J Cancer 124(10):2353–2360. doi:10.1002/ijc.24214
108. Callahan MJ, Nagymanyoki Z, Bonome T et al (2008) Increased HLA-DMB expression in the tumor epithelium is associated with increased CTL infiltration and improved prognosis in advanced-stage serous ovarian cancer. Clin Cancer Res 14(23):7667–7673. doi:10.1158/1078-0432.CCR-08-0479
109. Monsurro V, Beghelli S, Wang R et al (2010) Anti-viral state segregates two molecular phenotypes of pancreatic adenocarcinoma: potential relevance for adenoviral gene therapy. J Transl Med 8:10. doi:10.1186/1479-5876-8-10
110. Spivey TL, De Giorgi V, Zhao YD et al (2012) The stable traits of melanoma genetics: an alternate approach to target discovery. BMC Genomics 13. doi:10.1186/1471-2164-13-156
111. Bindea G, Mlecnik B, Angell HK, Galon J (2014) The immune landscape of human tumors: implications for cancer immunotherapy. Oncoimmunology 3(1), e27456. doi:10.4161/onci.27456
112. Grivennikov SI, Wang K, Mucida D et al (2012) Adenoma-linked barrier defects and microbial products drive IL-23/IL-17-mediated tumour growth. Nature 491(7423):254–258. doi:10.1038/nature11465
113. Tomei S, Bedognetti D, De Giorgi V et al (2014) The immune-related role of BRAF in melanoma. Mol Oncol. doi:10.1016/j.molonc.2014.07.014
114. Liu Q, Tomei S, Ascierto ML et al (2014) Melanoma NOS1 expression promotes dysfunctional IFN signaling. J Clin Invest 124(5):2147–2159. doi:10.1172/JCI69611
115. Linsley PS, Speake C, Whalen E, Chaussabel D (2014) Copy number loss of the interferon gene cluster in melanomas is linked to reduced T cell infiltrate and poor patient prognosis. PLoS One 9(10), e109760. doi:10.1371/journal.pone.0109760
116. Gu-Trantien C, Loi S, Garaud S et al (2013) CD4(+) follicular helper T cell infiltration predicts breast cancer survival. J Clin Invest 123(7):2873–2892. doi:10.1172/JCI67428
117. Mlecnik B, Bindea G, Angell HK et al (2014) Functional network pipeline reveals genetic determinants associated with in situ lymphocyte proliferation and survival of cancer patients. Sci Transl Med 6(228):228ra37. doi:10.1126/scitranslmed.3007240
118. Hoadley KA, Yau C, Wolf DM et al (2014) Multiplatform analysis of 12 cancer types reveals molecular classification within and across tissues of origin. Cell 158(4):929–944. doi:10.1016/j.cell.2014.06.049
119. Tran E, Turcotte S, Gros A et al (2014) Cancer immunotherapy based on mutation-specific CD4+ T cells in a patient with epithelial cancer. Science 344(6184):641–645. doi:10.1126/science.1251102

120. van Rooij N, van Buuren MM, Philips D et al (2013) Tumor exome analysis reveals neoantigen-specific T-cell reactivity in an ipilimumab-responsive melanoma. J Clin Oncol 31(32):e439–e442. doi:10.1200/jco.2012.47.7521
121. Rai E, Wakeland EK (2011) Genetic predisposition to autoimmunity—what have we learned? Semin Immunol 23(2):67–83. doi:10.1016/j.smim.2011.01.015
122. Zhou Q, Yang D, Ombrello AK et al (2014) Early-onset stroke and vasculopathy associated with mutations in ADA2. N Engl J Med 370(10):911–920. doi:10.1056/NEJMoa1307361
123. Ombrello MJ, Remmers EF, Sun G et al (2012) Cold urticaria, immunodeficiency, and autoimmunity related to PLCG2 deletions. N Engl J Med 366(4):330–338. doi:10.1056/NEJMoa1102140
124. Zhou Q, Lee GS, Brady J et al (2012) A hypermorphic missense mutation in PLCG2, encoding phospholipase Cgamma2, causes a dominantly inherited autoinflammatory disease with immunodeficiency. Am J Hum Genet 91(4):713–720. doi:10.1016/j.ajhg.2012.08.006
125. Boisson B, Laplantine E, Prando C et al (2012) Immunodeficiency, autoinflammation and amylopectinosis in humans with inherited HOIL-1 and LUBAC deficiency. Nat Immunol 13(12):1178–1186. doi:10.1038/ni.2457
126. Lucas CL, Kuehn HS, Zhao F et al (2014) Dominant-activating germline mutations in the gene encoding the PI(3)K catalytic subunit p110delta result in T cell senescence and human immunodeficiency. Nat Immunol 15(1):88–97. doi:10.1038/ni.2771
127. Orru V, Steri M, Sole G et al (2013) Genetic variants regulating immune cell levels in health and disease. Cell 155(1):242–256. doi:10.1016/j.cell.2013.08.041
128. Tsang JS, Schwartzberg PL, Kotliarov Y et al (2014) Global analyses of human immune variation reveal baseline predictors of postvaccination responses. Cell 157(2):499–513. doi:10.1016/j.cell.2014.03.031
129. O'Brien TR (2009) Interferon-alfa, interferon-lambda and hepatitis C. Nat Genet 41(10):1048–1050. doi:10.1038/ng.453
130. Suppiah V, Moldovan M, Ahlenstiel G et al (2009) IL28B is associated with response to chronic hepatitis C interferon-alpha and ribavirin therapy. Nat Genet 41(10):1100–1104. doi:10.1038/ng.447
131. Tanaka Y, Nishida N, Sugiyama M et al (2009) Genome-wide association of IL28B with response to pegylated interferon-alpha and ribavirin therapy for chronic hepatitis C. Nat Genet 41(10):1105–1109. doi:10.1038/ng.449
132. Marincola FM, Shamamian P, Rivoltini L et al (1995) HLA associations in the antitumor response against malignant melanoma. J Immunother Emphasis Tumor Immunol 18(4): 242–252
133. Gogas H, Kirkwood JM, Falk CS et al (2010) Correlation of molecular human leukocyte antigen typing and outcome in high-risk melanoma patients receiving adjuvant interferon. Cancer 116(18):4326–4333. doi:10.1002/cncr.25211
134. Wang E, Zhao Y, Monaco A et al (2012) A multi-factorial genetic model for prognostic assessment of high risk melanoma patients receiving adjuvant interferon. PLoS One 7(7), e40805. doi:10.1371/journal.pone.0040805
135. Pardoll DM (2012) The blockade of immune checkpoints in cancer immunotherapy. Nat Rev Cancer 12(4):252–264. doi:10.1038/nrc3239
136. Wang XY, Zuo D, Sarkar D, Fisher PB (2011) Blockade of cytotoxic T-lymphocyte antigen-4 as a new therapeutic approach for advanced melanoma. Expert Opin Pharmacother 12(17):2695–2706. doi:10.1517/14656566.2011.629187
137. Ueda H, Howson JM, Esposito L et al (2003) Association of the T-cell regulatory gene CTLA4 with susceptibility to autoimmune disease. Nature 423(6939):506–511. doi:10.1038/nature01621
138. Breunis WB, Tarazona-Santos E, Chen R, Kiley M, Rosenberg SA, Chanock SJ (2008) Influence of cytotoxic T lymphocyte-associated antigen 4 (CTLA4) common polymorphisms on outcome in treatment of melanoma patients with CTLA-4 blockade. J Immunother 31(6):586–590

139. Queirolo P, Morabito A, Laurent S et al (2013) Association of CTLA-4 polymorphisms with improved overall survival in melanoma patients treated with CTLA-4 blockade: a pilot study. Cancer Invest 31(5):336–345. doi:10.3109/07357907.2013.793699
140. Gogas H, Dafni U, Koon H et al (2010) Evaluation of six CTLA-4 polymorphisms in high-risk melanoma patients receiving adjuvant interferon therapy in the He13A/98 multicenter trial. J Transl Med 8:108. doi:10.1186/1479-5876-8-108
141. Barnes BJ, Kellum MJ, Pinder KE, Frisancho JA, Pitha PM (2003) Interferon regulatory factor 5, a novel mediator of cell cycle arrest and cell death. Cancer Res 63(19):6424–6431
142. Barnes BJ, Moore PA, Pitha PM (2001) Virus-specific activation of a novel interferon regulatory factor, IRF-5, results in the induction of distinct interferon alpha genes. J Biol Chem 276(26):23382–23390. doi:10.1074/jbc.M101216200
143. Graham RR, Kozyrev SV, Baechler EC et al (2006) A common haplotype of interferon regulatory factor 5 (IRF5) regulates splicing and expression and is associated with increased risk of systemic lupus erythematosus. Nat Genet 38(5):550–555. doi:10.1038/ng1782
144. Dawidowicz K, Allanore Y, Guedj M et al (2011) The interferon regulatory factor 5 gene confers susceptibility to rheumatoid arthritis and influences its erosive phenotype. Ann Rheum Dis 70(1):117–121. doi:10.1136/ard.2010.129171
145. Vandenbroeck K, Alloza I, Swaminathan B et al (2011) Validation of IRF5 as multiple sclerosis risk gene: putative role in interferon beta therapy and human herpes virus-6 infection. Genes Immun 12(1):40–45. doi:10.1038/gene.2010.46
146. Dideberg V, Kristjansdottir G, Milani L et al (2007) An insertion-deletion polymorphism in the interferon regulatory Factor 5 (IRF5) gene confers risk of inflammatory bowel diseases. Hum Mol Genet 16(24):3008–3016. doi:10.1093/hmg/ddm259
147. Beck KE, Blansfield JA, Tran KQ et al (2006) Enterocolitis in patients with cancer after antibody blockade of cytotoxic T-lymphocyte-associated antigen 4. J Clin Oncol 24(15):2283–2289. doi:10.1200/JCO.2005.04.5716
148. Phan GQ, Yang JC, Sherry RM et al (2003) Cancer regression and autoimmunity induced by cytotoxic T lymphocyte-associated antigen 4 blockade in patients with metastatic melanoma. Proc Natl Acad Sci U S A 100(14):8372–8377. doi:10.1073/pnas.1533209100153320910
149. Phan GQ, Attia P, Steinberg SM, White DE, Rosenberg SA (2001) Factors associated with response to high-dose interleukin-2 in patients with metastatic melanoma. J Clin Oncol 19(15):3477–3482
150. Gogas H, Ioannovich J, Dafni U et al (2006) Prognostic significance of autoimmunity during treatment of melanoma with interferon. N Engl J Med 354(7):709–718. doi:10.1056/NEJMoa053007
151. Bouwhuis MG, Suciu S, Collette S et al (2009) Autoimmune antibodies and recurrence-free interval in melanoma patients treated with adjuvant interferon. J Natl Cancer Inst 101(12):869–877. doi:10.1093/jnci/djp132
152. Bouwhuis MG, Suciu S, Testori A et al (2010) Phase III trial comparing adjuvant treatment with pegylated interferon Alfa-2b versus observation: prognostic significance of autoantibodies—EORTC 18991. J Clin Oncol 28(14):2460–2466. doi:10.1200/JCO.2009.24.6264
153. Uccellini L, De Giorgi V, Zhao Y et al (2012) IRF5 gene polymorphisms in melanoma. J Transl Med 10(1):170. doi:10.1186/1479-5876-10-170
154. Worschech A, Haddad D, Stroncek DF, Wang E, Marincola FM, Szalay AA (2009) The immunologic aspects of poxvirus oncolytic therapy. Cancer Immunol Immunother 58(9):1355–1362. doi:10.1007/s00262-009-0686-7
155. Panelli MC, White R, Foster M et al (2004) Forecasting the cytokine storm following systemic interleukin (IL)-2 administration. J Transl Med 2(1):17. doi:10.1186/1479-5876-2-171479-5876-2-17
156. Liu S, Kong C, Wu J, Ying H, Zhu H (2012) Effect of CCR5-Delta32 heterozygosity on HIV-1 susceptibility: a meta-analysis. PLoS One 7(4), e35020. doi:10.1371/journal.pone.0035020
157. Eri R, Jonsson JR, Pandeya N et al (2004) CCR5-Delta32 mutation is strongly associated with primary sclerosing cholangitis. Genes Immun 5(6):444–450. doi:10.1038/sj.gene.63641136364113

158. Gomez-Reino JJ, Pablos JL, Carreira PE et al (1999) Association of rheumatoid arthritis with a functional chemokine receptor, CCR5. Arthritis Rheum 42(5):989–992. doi:10.1002/1529-0131(199905)42:5<989::AID-ANR18>3.0.CO;2-U
159. Melum E, Karlsen TH, Broome U et al (2006) The 32-base pair deletion of the chemokine receptor 5 gene (CCR5-Delta32) is not associated with primary sclerosing cholangitis in 363 Scandinavian patients. Tissue Antigens 68(1):78–81. doi:10.1111/j.1399-0039.2006.00604.x
160. Prahalad S (2006) Negative association between the chemokine receptor CCR5-Delta32 polymorphism and rheumatoid arthritis: a meta-analysis. Genes Immun 7(3):264–268. doi:10.1038/sj.gene.6364298
161. Fischereder M, Luckow B, Hocher B et al (2001) CC chemokine receptor 5 and renal-transplant survival. Lancet 357(9270):1758–1761. pii: S0140673600048984
162. Ugurel S, Schrama D, Keller G et al (2008) Impact of the CCR5 gene polymorphism on the survival of metastatic melanoma patients receiving immunotherapy. Cancer Immunol Immunother 57(5):685–691. doi:10.1007/s00262-007-0407-z
163. Choi JW, Park CS, Hwang M et al (2008) A common intronic variant of CXCR3 is functionally associated with gene expression levels and the polymorphic immune cell responses to stimuli. J Allergy Clin Immunol 122(6):1119–1126 e7. doi:10.1016/j.jaci.2008.09.026. pii: S0091-6749(08)01721-1
164. Bedognetti D, Uccellini L, Wang E et al (2010) Evaluation of CXCR3 and CCR5 polymorphisms and gene-expression as predictive biomarkers of clinical response to adoptive therapy in melanoma patients. J Immunother 33(8):860
165. Fisher B, Packard BS, Read EJ et al (1989) Tumor localization of adoptively transferred indium-111 labeled tumor infiltrating lymphocytes in patients with metastatic melanoma. J Clin Oncol 7(2):250–261

Chapter 6
Production of Clinical T Cell Therapies

David Stroncek, Jianjian Jin, Virginia David-Ocampo, Vicki Fellowes, Larry Moses, and Marianna Sabatino

Abstract The improving effectiveness of adoptive T cell therapies has led to their increased clinical application. Most of these adoptive T cell therapies are being produced in small lots in cell therapy centers affiliated with or located within academic health centers. Typically, the cells are produced from autologous or HLA compatible donors and one lot is used for a single patient. As part of early phase clinical trials, the best available methods and devices for the manufacture of clinical grade T cell therapies are described. For most adoptive T cell therapies the starting material is a peripheral blood mononuclear cell (PBMC) product that is collected by apheresis using closed system blood cell separators. Many manufacturing processes require that red blood cells be removed from the PBMCs or that T cells or T cell subsets are isolated. Classically, T cells have been cultured in flasks, but culture in closed systems which reduces the risk of microbial contamination is desirable and bags and bioreactors are often used for T cell culture and expansion. T cell culture involves growth and expansion in media supplemented with serum, cytokines and feeder cells or other artificial stimulators, i.e. anti-CD3/28 beads or K562 cell line. Recently, closed system transduction methods have been developed that can be used to produce genetically engineered T cells. Automated instruments are available to wash and concentrate products. The final product is assessed for the quantity of cells present, purity, sterility and potency. The use of these best practices is allowing for the consistent manufacturing of high quality cellular therapies to support early phase clinical trials.

Keywords Adoptive cell therapies • Good manufacturing practices • Cell culture • Cell selection • Tumor infiltrating leukocytes • Genetically engineered T cells • Cell processing • Chimeric antigen receptor T cells • High affinity receptor T cells

D. Stroncek, M.D. (✉) • J. Jin • V. David-Ocampo • V. Fellowes • L. Moses
Cell Processing Section, Department of Transfusion Medicine, Clinical Center, National Institutes of Health, 10 Center Drive – MSC-1184, Building 10, Room 1C711, Bethesda, MD 20892-1184, USA
e-mail: dstroncek@cc.nih.gov

M. Sabatino
Kite Pharma, Inc., Santa Monica, CA, USA

P.A. Ascierto et al. (eds.), *Developments in T Cell Based Cancer Immunotherapies*, Cancer Drug Discovery and Development,
DOI 10.1007/978-3-319-21167-1_6

Introduction

Adoptive cellular therapy using T cells is becoming more effective and its use is growing. For many years tumor infiltrating lymphocytes (TIL) have been isolated from metastatic melanoma lesions, expanded and given as an effective anti-tumor autologous therapy [1]. The discovery that the administration of leukocyte reductive chemotherapy and irradiation therapy prior to TIL infusion enhances their clinical effectiveness is leading to the wide spread use of this therapy [2, 3].

The use of other adoptive T cell therapies for cancer is also growing rapidly. Autologous peripheral blood T cells are being genetically engineered to produce potent anti-cancer cells. T cells can be genetically engineered to express high affinity T cell receptors (TCR) and chimeric antigen receptors (CAR) that are specific for antigens expressed by tumors [4]. The transduction of autologous T cells with high affinity TCRs reactive with melanoma specific antigens allows for the treatment of melanoma patients when they do not have metastatic lesions that can resected for TIL production or when TIL cannot be cultured from a resected lesion. In addition, autologous lymphocytes engineered to express high affinity TCRs are being used to treat types of metastatic cancer which have not been treated with TIL. T cells expressing TCRs specific for the fetal embryonic antigen NY-ESO-1 are being used to treat patients with metastatic melanoma and metastatic synovial cell sarcoma [5] and T cells with TCRs specific for MAGE-A3 are being used to treat patients with metastatic melanoma, synovial cell sarcoma and esophageal cancer [6]. CAR T cells are engineered to express a vector encoding the zeta chain of CD3, the single change variable region of an antibody directed to a tumor antigen and a co-stimulatory molecule. CAR T cells specific for CD19 antigen have proven to be very effective for treating B cell malignancies including: acute lymphocytic leukemia, chronic lymphocytic leukemia and lymphoma. CAR T cells specific for CD20 have been shown to have clinical activity in patients with B cell lymphomas [7] and those specific for GD2 in patients with neuroblastoma [8]. CAR T cells directed to other antigens are also being developed including CD22 [9], CD23 [10], CD70 [11], immunoglobulin kappa light chain [12], B-cell maturation antigen (BCMA) [13] and erythropoietin-producing hepatocellular carcinoma A (EphA2) [14].

An important aspect of producing cell and gene therapies involves the use of good manufacturing practices (GMP). GMP is a system of evolving practices that have been found to yield safe, high quality products. GMPs involves the entire process of producing cell therapies from the screening, testing, and selection of the donor; to cell collection and processing; and to cell administration and recipient follow-up. It is a common misconception that GMP only means manufacturing cells in a highly controlled facility specifically designed for this purpose. However, it also involves maintaining a group of qualified and properly trained staff; developing and adhering to standard operating procedures; appropriate evaluation and testing of the starting materials, reagents, intermediate products and the final product; qualifying the donor and vendors supplying materials and much more. Many aspects of GMP production of cell and gene therapies are well worked out. For example to qualify that the donor of starting materials there are screening questions that the donor must answer and blood tests that must be performed to minimize the possibility that the donor may have an infectious disease that could be present in the starting material and which could be transmitted to the recipient. These requirements have been standardized and tests are readily available. However, the requirements concerning GMP manufacturing of cellular therapies are less clear because the best methods of manufacturing have yet to be defined. This is due in part to the evolving nature of cellular therapies. New cell therapies are being rapidly developed and existing therapies are constantly being modified.

Most T cell therapies produced at academic centers are manufactured under INDs for phase I and phase II clinical trials. Most of these cell therapies never make it past phase I or II trials since they are either not effective or not safe and as a result, academic centers are constantly developing new therapies. Since many new cell therapies will fail, many laboratories at academic centers do not spend as much resources and time to develop manufacturing processes for products used in a phase I/II clinical trials as for products used for later phase trials. As a result many protocols taken to phase I and II trials use the best available but not necessarily the best possible reagents and methods. Using the best available processes and reagents allows a protocol to be taken from the research laboratory to the clinic in a reasonable amount of time and allows for the timely evaluation of new therapies. Using the best possible methods for phase I and II trials may require a very lengthy duration of time to develop a cell therapy that in the end is not effective. For early phase trials it would be ideal to use the best possible methods, but it is reasonable to use the best available manufacturing processes that will allow the production of cells to begin in a timely manner so subject accrual can begin in order to determine if the product is safe and has any clinical efficacy. However, if the results of early clinical trials are promising, it's important to quickly convert the methods used for manufacturing the phase I/II products to those most compliant with GMPs and those that are appropriate for the higher expectations associated with late phase products. If necessary, efforts should be made to develop new GMP compliant methods, devices and reagents since any products that will go on to phase III trials and commercialization are expected to be manufactured using methods that yield the safest, most consistent and most effective products.

There are problems associated with not using the best possible methods for the production of early phase clinical trials. The re-engineering of early phase cellular therapy manufacturing processing procedures to meet the more demanding requirement of late phase products is generally very time consuming and costly since cell therapies are very complex. Cell therapies make use of difficult to obtain starting materials such as cells and vectors, include many steps and often require prolonged cell culture and expansion. T cell therapy manufacturing protocols generally take 1 or 2 weeks or longer to complete. Due to the complex nature and long duration of the manufacturing protocols it often takes a year or more to re-engineer and validate a cell production processes and multiple cycles of changes may be required to convert a manufacturing process used in phase I/II trails to a more robust, reliable and cost effective manufacturing processing.

This chapter will describe the various steps involved in manufacturing T cell therapies and the current best practices. The chapter is focused on the manufacturing of cells for a single patient; either for autologous use or allogeneic cells collected for administration to a specific patient. These patient specific allogeneic cells are generally from HLA-compatible relatives or unrelated donors. This type of manufacturing is usually preformed at academic health centers and typically involves phase I/II clinical trials. The size of each lot of autologous or allogeneic directed donor product is generally small, only large enough to treat one patient. Methods to manufacture consistently high quality autologous and directed donor T cell therapies that are compliant with GMP and that make use of resources available at cell processing laboratories in academic centers are described. The use of these practices helps academic cellular therapy centers produce consistently high quality products which allows for the better assessment of the efficacy of early phase products and allows for a better transition to phase III trials and licensure.

Supplies and Reagents

A wide variety of media, media supplements, cytokines and growth factors are used to produce cellular and gene therapies. The quality of these reagents has an important impact on the consistency and safety of the final cellular or gene therapy product. For example, many research laboratories use fetal bovine serum (FBS) as a media supplement to support the growth and expansion of cells in culture, but FBS can negatively impact the safety and quality of cellular therapies. There is considerable amount of lot to lot variability in the ability of FBS to support cell proliferation. Even if a laboratory goes through great lengths to select lots of FBS with similar ability to support the cell of interest, considerable differences may occur in the cultured cells when a new lot of FBS is used. The use of FBS also presents a potential safety issues. The use of animal products and animal derived reagents in cell manufacturing exposes the product recipient to risk of acquiring a xenogeneic infection. In addition, it is difficult to entirely remove foreign proteins from cells cultured in FBS and some recipients will develop immune responses including anaphylaxis due to IgE antibodies to bovine proteins if they are repeatedly exposed to cells cultured in FBS [15–17]. As a result, it is best to avoid the use of all animal derived reagents in the manufacturing of cellular therapies.

Media, media supplements, cytokines and growth factors that are xeno-free are available. From among the xeno-free reagents it is important to select the highest quality reagents. Most reagents are available in several different grades ranging from pharmaceutical grade to research grade. The highest quality reagents are of the highest purity and contain the least contaminants. They are also manufactured under the most stringent conditions and tested more rigorously to ensure that they are consistently of a very high quality. Some cytokines and growth factors such as IL-2, interferon-γ, granulocyte colony-stimulating factor and granulocyte macrophage-colony stimulating factor are available as pharmaceutical grade reagents. Many other cytokines and growth factors are available as GMP grade reagents made specifically for the manufacturing of clinical cellular therapies. Some cellular therapy laboratories use cytokines and growth factors that are less than GMP grade since they are less costly, but this practice is not recommended since it could lead to cellular and gene therapy products of inferior quality.

Closed System Processing

One of the most important concepts of manufacturing cellular therapies is to maintain the cells in a system that is closed to the external environment. Since cell and gene therapies cannot be sterilized at the end of manufacturing, it is important to maintain the sterility of the products throughout the collection and manufacturing process. In order to prevent microbial contamination of cell therapies, as much as possible, closed systems should be used in the collection and manufacturing process. However, most early phase cellular therapies produced at academic centers often move directly from research laboratories to cell processing laboratories and research laboratories almost always use systems that are open to the environment for cell culture: T flasks, multi-well plates, and tubes. Therefore, one of the challenges for clinical cell processing laboratories is to convert open culture systems into closed systems.

Culture bags can be used to create closed systems. These sterile, plastic, one time use bags are manufactured with tubular tails (Fig. 6.1). Sterile tube welders are commercially available which sterilely connect tubing from one bag to tubing from another bag (Fig. 6.2). This technology allows cellular therapy laboratories to construct customized networks of bags for cell expansion, separation, washing and concentration. It has been used widely in blood centers and blood banks for collecting and processing whole blood, red blood cells, platelets and plasma. To separate bags the tubing connecting the bags is heat sealed to create two small segments, gentle pulling of the tubing (where the two segments meet) separates the tubing and the bags while maintaining a closed system (Fig. 6.3). Bags for cell culture are available in different sizes ranging for as small as 7 mL to 5 L.

As cells in culture proliferate media must often be changed and the culture volume expanded. For cells cultured in up to 600 mL bags, media can be changed by simply centrifuging the bags in floor model centrifuges specifically designed for processing whole blood. After centrifugation the cell-free supernatant can be expressed through the bag's tubing into another bag that has been sterilely connected to the culture bag (Fig. 6.4). After the bag containing the cell-free supernatant is heat sealed and removed, another bag containing fresh media is connected to the culture bag and the media is added to the cell pellet. Alternatively, as cells proliferate they can be transferred into progressively larger bags and fresh media can be added. For cultures requiring 3 or more liters of volume, the contents of the culture bags can be spilt into multiple bags. Bag cultures have been used to expand TIL cultures up to 60 L. When multiple bags are used, at the end of the culture period, the bags can be sterilely linked and the contents can be combined, concentrated and washed using automated closed system instruments as described below.

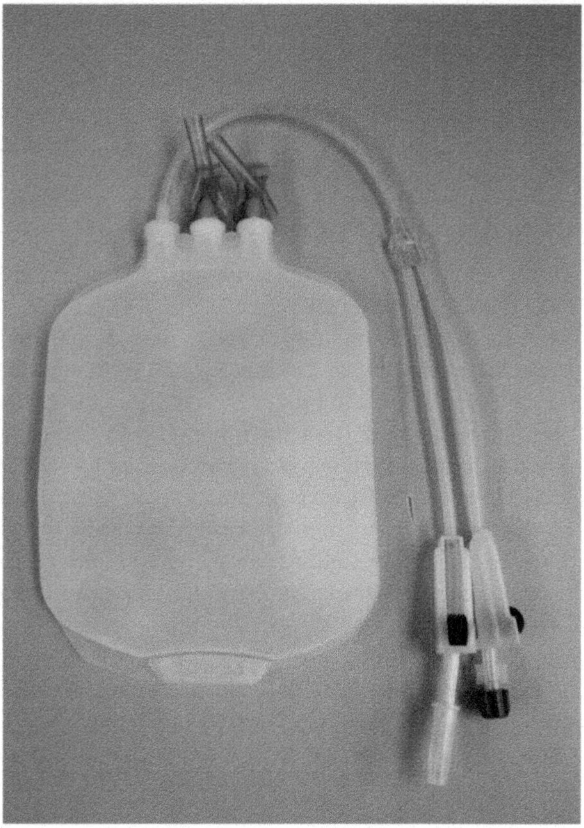

Fig. 6.1 Bag specifically designed for cell culture. This cell culture bag (Lifecell Tissue Culture Flask, Baxter Healthcare Corporation, Deerfield, IL) is made with gas permeable plastic and has two tubing leads which can be connected to other bags or containers using the ports or by sterile connection of the tubing to the tubing of another bag or vessel

One limitation of bag culture is that bags do not lend themselves to automated filling and drainage. Other closed system devices and reactors are available for T cell culture and they will be discussed later in this chapter.

Many automated devices that are used to collect and process blood cells also make use of sterile disposable plastic liners to maintain a closed system during processing. Bags are integrated into these systems that are used to hold the collected cells or solutions for washing. These disposable kits, which are used only once, are sterilized during the manufacturing process and they prevent the cellular product from becoming contaminated microbes and with cells from other donors whose cells have been processed using the same instrument.

Donor Screening and Testing

The first step in the production of cellular therapies is to identify and evaluate the person who will donate the cellular starting material. Donors of cellular therapy products are screened and tested, much like blood donors, to determine if they have

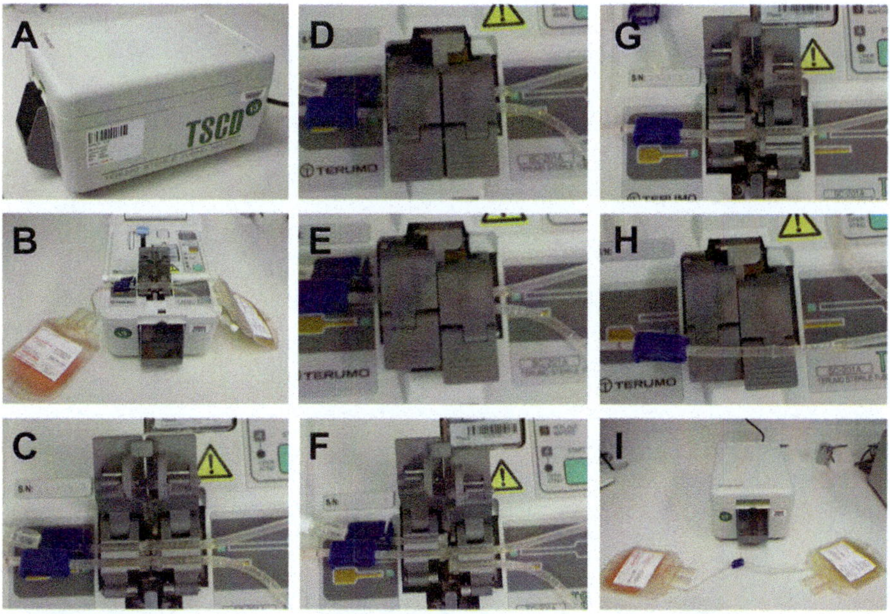

Fig. 6.2 Sterile connection of tubes. This sterile tubing welder (Terumo BCT, Lakewood, CO, USA) (**A**) is used to sterilely connect two separate tubes. It is used to sterilely link bags containing media, cells, reagents or wash solutions. To connect tubing from two bags the instrument's clamps are opened and the tubing to be welded are inserted into the clamps (**B** and **C**). To hold the tubing is place, the clamps are closed and the tubes are welded (**D** and **E**). The connected tubes (**F**, **G** and **H**) and bags (**I**) are shown

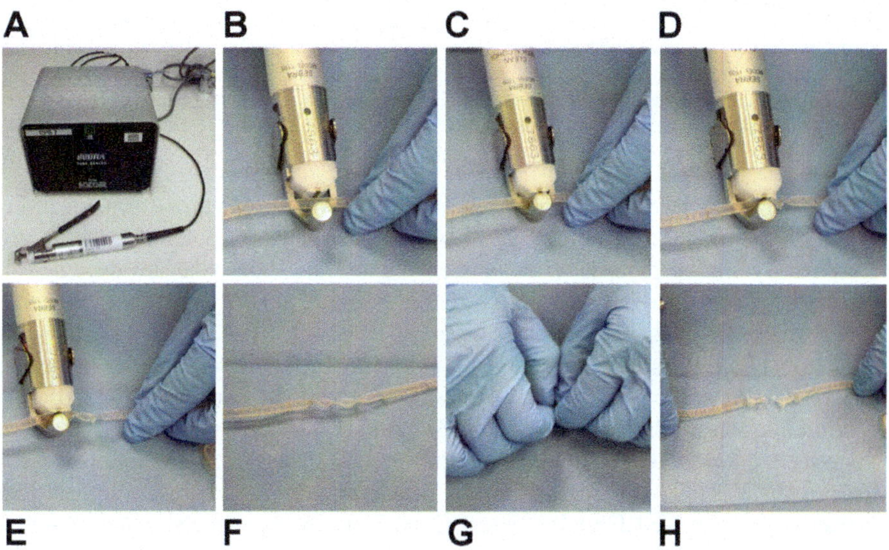

Fig. 6.3 Sealing and separation of tubing. Instruments have been specifically designed to seal tubing and separate culture bags connected by tubing (**A**). The tube is first sealed at the site where the tubing and bags are to be separated (**B** and **C**). After the first seal has been completed a second (**D**) and third seal (**E**) are made within a few centimeters of the first seal (**F**). At the site of the middle seal gentle pressure is used to pull the tubes apart (**G**). The tubes remain sealed at each end and the addition seal in each tube is to ensure that the cultures are not contaminated by a leak at the site of separation (**H**). A Sebra tube sealing device (Haemonetics Corporation, Braintree, MA) is shown

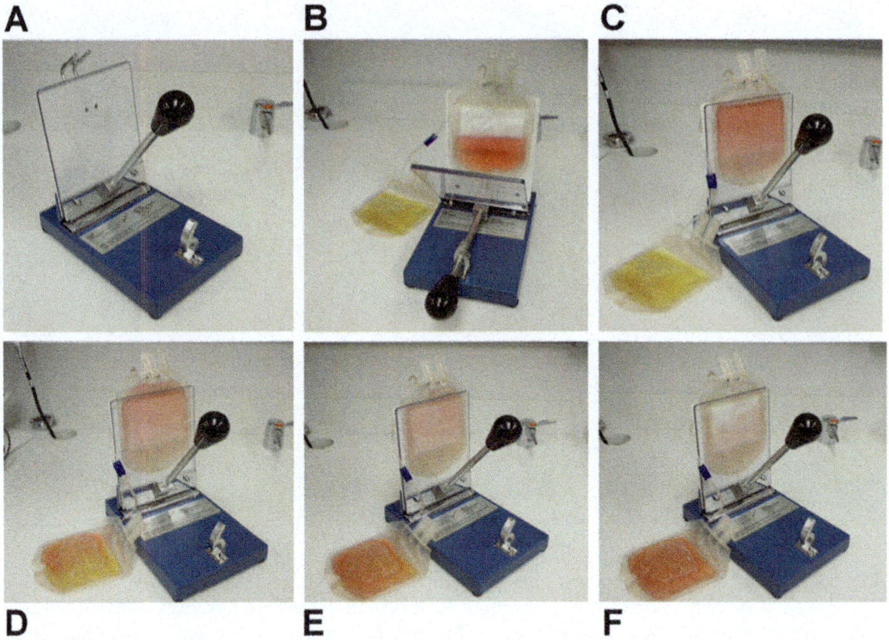

Fig. 6.4 Expression of supernatant from one bag to another. Plasma extractors can be used to separate supernatant from cells that have been pelleted in the bottom of a bag by centrifugation (**A**). The bag with the pelleted cells is placed in the plasma extractor (Fenwal Inc., Lake Zurich, IL, USA) (**B**). The clear spring-loaded plate is released and allowed to slowly force the supernatant from the bag. The speed of remove is controlled using a clamp on the bag's tubing (**C**). The supernatant gradually flows into a waste bag (**D**, **E** and **F**). At the end of the process the cells are resuspended and new media is added to the culture bag. The bags have colored media to better visual the bags

a history of past behavior and experiences that may have exposed them to an infectious agent that could be transmitted by blood cells. Typically, potential donors are screened by asking them a series of well-defined questions that are designed to determine if they have risk factors associated exposure to HIV, hepatitis B, hepatitis C and other infectious agents. The content of these questions are outlined by standards for cell and gene therapies prepared by some professional societies including the AABB (formally the American Association of Blood Banks). Blood from the potential donors is also tested for anti-HIV, anti-HTLV1/2, anti-HCV, HBsAg and Syphilis. Donors of leukocyte rich products such as T cell therapies are also tested for anti-CMV. This screening and testing process is not required when the product collected will be used for autologous therapy, but many centers screen and test autologous donors since the processing of products from patients with infectious diseases may require special precautions to prevent the cross contamination of products intended for other patients that are being processed at the same time. In addition, after the processing of an infectious product is complete, the facility may need more rigorous cleaning to prevent the contamination of products subsequently processed in the laboratory.

Collecting the Cellular Starting Material

For the production of genetically engineered autologous T cell therapies the T cell rich starting material is collected by apheresis using a blood cell separator. Blood cell separators were originally designed to collect neutrophils for transfusion and are now most commonly used to collect platelets for transfusion and peripheral blood mononuclear cell (PBMC) products for cellular and gene therapy.

Blood cell separators can collect large quantities of PBMCs quickly and sterilely. Enough cells for an entire autologous T cell therapy, 5 to 10×10^9 cells, can be collected in a couple of hours. Most of the red blood cells and plasma is returned back to the donor during the process. The blood cell separators make use of one time use sterile disposable kits in which blood is collected from the donor's peripheral vein in one arm and is directed through the instrument where cells and plasma are separated. The blood cell separator collects the desired cell type in to a bag and the cells and plasma that are not needed are returned to a peripheral vein in the donor's other arm. The closed disposable collection system ensures that the products are free from microbial contamination. There is a very small risk that the cells become contaminated during the collection procedure. When blood cell separators are used for the collection of platelet products for transfusion it has been found that only about 1 in 5000 products are contaminated with bacteria [18]. The contamination events are typically due to bacteria from the donor's skin or donor bacteremia.

The most frequently used blood cell separators make use of differences in density to separate cells; consequently, when the instruments are set up to collect lymphocytes, they also collect monocytes which are of a similar density. The lymphocyte-rich PBMC products collected by apheresis are also contaminated with small quantities of granulocytes, red blood cells and platelets. Blood cells separators are very reliable, but the composition of the final PBMC product may vary among donors. The quantities of T cells collected as well as the quantities of contaminating cells can vary. Since the composition of the cells is highly dependent on the donor's blood counts, the composition of products collected from autologous donors who have received prior therapies can be especially variable.

For many T cell therapies the starting material is autologous mononuclear cells collected by apheresis. T cells may then be enriched by elutriation or purified by selection or simply by stimulating T cells which may outgrow other cells in culture. For TIL therapy the starting material is obtained from surgically resected metastatic lesions. TIL can be obtained from small pieces for tumors or from tumors digested with proteolytic enzymes. The use of tumor pieces rather than digests avoids the need to obtain GMP grade enzymes. Traditionally, TIL are obtained for the tumor pieces or digests by culture in plates with IL-2 over several weeks. However, these tissue culture plates are open to the environment and are subject to contamination with bacteria and fungus. It has recently been shown that initial TIL culture can be performed in G-Rex10 gas permeable flasks (Wilson Wolf Manufacturing, New Brighton, MN, USA) at higher cell concentration than classical T flasks or culture bags (Fig. 6.5) [19]. Culture in G-Rex flasks is less susceptible to contamination since these flasks allow T cells to grow at higher densities therefore fewer manipulation are required.

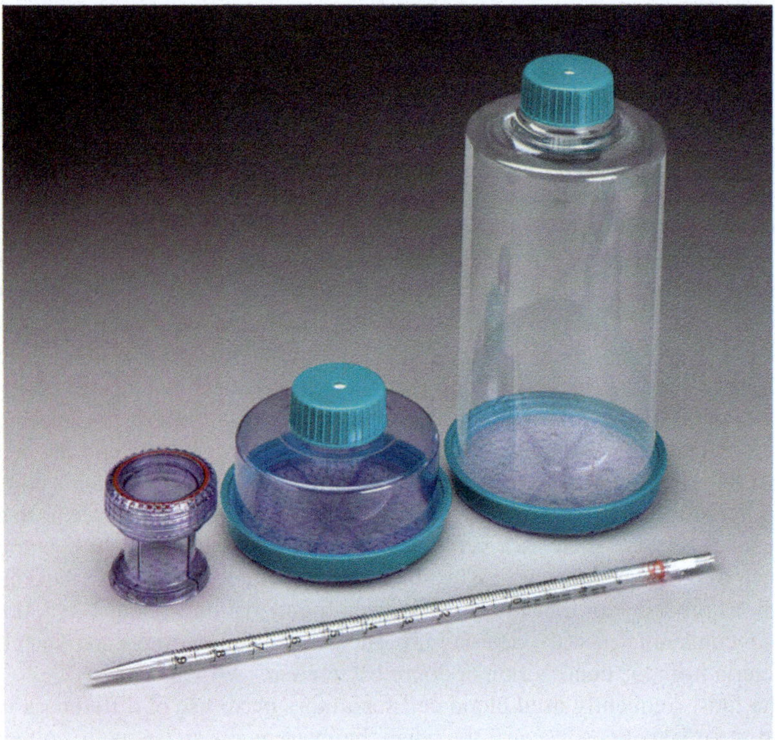

Fig. 6.5 Gas permeable flasks that have been used for T cell culture. These flasks have a gas permeable member at the bottom (G-Rex, Wilson Wolf Manufacturing, New Brighton, MN, USA). Three different size flasks are shown. The smallest flask, G-Rex10, has a gas permeable membrane of 10 cm^2 and a capacity of 40 mL. The middle and largest size flasks both have membranes of 100 cm^2; the middle flask, G-Rex100, has a capacity of 500 mL and the largest flask, G-Rex100L, has a capacity of 2000 mL

Product Segregation

It is important to take measures to ensure that products from different patients are not mixed up or cross contaminated. The vessels containing the cells are, of course, labeled with the recipient's name and hospital number, but specific laboratory practices should be in place to reduce the possibility of misidentifying products. Products manufactured for different patients should be keep separate and laboratory staff should work on products from only one patient. If products from multiple patients are to be processed simultaneously, each product should be processed in a separate biosafety cabinet, they should be cultured in a separate incubator and by separate staff. Strict control of product labels is useful in preventing mislabeling of products. At receipt of the starting material, release of the product from inventory and at other critical points, two processing staff should confirm that the appropriate product has been received, issued or selected for the next processing step or operation.

Cell Isolation

For many cell manufacturing protocols contaminating cells must be removed from PBMC products collected by apheresis before cell culture or further manufacturing can begin. Most often red blood cells must be removed. For some manufacturing protocols highly enriched T cells must be obtained from the starting PBMC product. A variety of methods can be used for RBC removal and lymphocyte enrichment. To obtain highly enriched populations of T cells selection processes using monoclonal antibodies must be used.

Removal of Contaminating RBCs

Several methods are available to remove RBCs. If large quantities of RBCs are present in the product and if some leukocyte loss can be tolerated, the product can be centrifuged and the leukocyte rich buffy coat can be isolated. Up to 80 % of RBCs can be removed by isolating the buffy coat with a loss of less than 20 % of the leukocytes. For products of approximately 250–500 mL total volume, the product can be placed in a bag and centrifuged in a floor model centrifuge to isolate the buffy coat. For products of larger volume, automated instruments made for cell washing such as the Cobe 2991 cell processor (Terumo BCT, Lakewood, CO, USA) can be used to isolate buffy coats. The Cobe 2991 is automated, but less so than other instruments [20, 21].

Another simple method to remove RBCs involves sedimentation. When 6 % hetastarch in 0.9 % sodium chloride, an infusible grade solution that is used for volume expansion, is added to a cell suspension containing RBCs, the RBCs form rouleaux and settle quickly. After a volume of hetastarch equal to approximately 25 % of the final product volume is added to the RBC-containing product and the suspension is mixed, the RBCs settle over 30–90 min and the leukocyte rich supernatant can be removed. RBC sedimentation with hetastarch is simple, but some leukocytes are lost with the RBCs. RBC sedimentation with hetastarch works well for products that are up to 400 mL in volume. The sedimentation process can be performed in a 600 mL bag and devices designed to express plasma from one unit of centrifuged whole blood (Fig. 6.2) can be used to express the leukocyte rich supernatant into a culture bag.

Ammonium chloride lysis is another method for RBC removal. Sterile solutions for RBC lysis such as ACK Lysing Buffer (Lonza, Allendale, NJ) are commercially available. The PBMCs are incubated with a solution of ammonium chloride until RBC lysis is complete, but the time of incubation should not exceed 10 min. The cells are washed to remove free hemoglobin and ammonium chloride at the conclusion of the process. Lysis allows for greater removal of RBC without loss of leukocytes. This reagent is not approved specifically for human use, but it is used by some clinical laboratories safely as an ancillary reagent, i.e. used in production, but not infused.

Density gradient sedimentation can also be used to separate lymphocytes from RBCs. Density gradient separation has an advantage over other methods in that it not only removes RBCs but it separates the lymphocyte-rich mononuclear cells from granulocytes. This method involves centrifuging the cellular product over ficoll. During centrifugation the lymphocytes and monocytes remain at the top of the ficoll gradient, while the granulocytes and RBCs go to the bottom of the gradient. For small volume products ficoll separation can be performed in 50 mL conical tubes. For larger volume products, ficoll density gradient separation can be performed in bags, blood cell separators or cells washers such as the Cobe 2991 [22]. Another instrument that is available for ficoll density gradient separation is the Sepax 2 (Biosafe SA, Eysins, Switzerland). This instrument has been specifically designed to remove plasma and RBCs from umbilical cord blood components or isolate mononuclear cells from marrow aspirates. It can also be used to isolate buffy coats from PBMCs. One advantage of this instrument is that it is highly automated and provides documentation of the RBC depletion process for the product processing record.

Cell Enrichment

For the production of some T cell therapies it is desirable to use highly enriched T cell, B cell or T cell subset populations as the starting material. For many of these protocols, PBMCs products collected by apheresis are the starting material. The PBMC products can either be enriched or depleted of specific cell populations using commercially available GMP quality monoclonal antibodies conjugated magnetic beads. Antibodies conjugated to magnetic beads are incubated with the PBMCs product and the bound cells are removed with an external magnet. Commercial GMP magnetic beads conjugated to antibodies with many different specificities including CD3, CD4, CD8, CD14, CD19, CD25, CD34, CD56 and others are available from Miltenyi Biotech, Bergisch Gladback, Germany [23–27]. Miltenyi also manufactures the CliniMACS which is an instrument fitted with sterile plastic closed system disposables that can used to isolate cells from clinical PBMC apheresis products. This system can be used for positive selection to isolate a specific cell type such as CD3 or CD4 cells or negative selection to deplete a PBMC product of a specific cell type. When used for positive selection a final product that contains approximately 90 % of the desired type of cells can usually be obtained and typically, approximately 70 % of the desired cell type can be recovered. With this system of monocloncal antibodies and paramagnetic beads multiple negative cell selections, but only one positive cell selection can be performed at one time due to the high affinity of the antibodies and the permanent nature of the conjugation to the magnetic beads. While the CliniMACS system is semi-automated, a considerable amount of labor is involved. A new system has recently been developed that allows for the more automated separation of cells, the Prodigy. This system can also be used for cell incubation and washing, however, it has not yet been adapted for T cell applications.

Another system that makes use of magnetic beads uses monoclonal antibodies bound to *Strep*tamers@ which are conjugated to magnetic beads using a streptavidin derivative, Strep-Tactin@ (IBA GmbH). After the cells are selected, they can be released from the beads using a biotin solution. When the cells are released, the antibody *strep*tamer complexes break up and the low affinity antibodies are released from the cells. This ability to separate the cells, antibodies and the magnetic beads allows for multiple positive selections.

Cell Culture

Culture Vessels

Classically lymphocytes have been cultured in T flasks. However, T flask culture has some limitations because they are open systems. Culturing cells in T flasks is especially problematic when large quantities of cells must be grown. In order to culture large quantities of cells, the number of flasks used must be increased which increases the possibility of microbial contamination. When large quantities of T cells must be produced they are usually grown in multiple culture bags. Culturing of cells in bags is now common for phase I and phase II manufacturing of T cell products. However, bag cultures have some limitations, one being that cells can only be grown to a concentration of approximately $2.0–3.0 \times 10^6$ cells per mL. Growing cells at a higher concentration has the advantage of requiring less cytokines, less media and less media additives reducing the cost of the culture. In addition, smaller culture volumes make washing and concentrating the cells easier. Smaller volume cultures also require less labor for feeding, media changes, cell concentration and cell washing and more instruments are available for washing and concentrating smaller volumes.

An alternative to bag culture which allows T cells to be grown at greater concentrations are bioreactors. Closed system hollow-fiber cartridge [28] or circular chamber [29] bioreactors have been developed and have been demonstrated to support T cell growth, but they have limited by their lack of scalability. Typically they are automated, but have a very limited availability of cartridge sizes. Scaling up the production of cells generally requires growing cells in more than one bioreactor which increases the complexity and cost of the process. In addition, these bioreactors require the purchase of expensive hardware. The validation and maintenance of the hardware can also be time consuming and expensive. The disposable hollow fiber and circular cartridges can also be costly. One bioreactor, the wave, uses bags rather than cartridges and can grow cells in a wide range of bag sizes, making it very scalable. The wave gentle rocks the culture bag while gradually adding media and oxygen and it has been used to expand TIL [30]. While it is effective, it still requires capital investment in equipment, validation of the system and staff training.

Another alternative to bag culture is the growth of T cells in gas permeable flasks. Flasks with gas permeable membrane bottoms, G-Rex flasks (Fig. 6.5), allow T cells to be grown up to approximately 1×10^7 cells per mL. These gas permeable flasks have been used to grow TIL [3, 19] and viral specific cytotoxic T cells [31]. A number of modifications have been made to the flasks to facilitate GMP cell manufacturing. Caps for the flasks are available that include tubes that allow for a

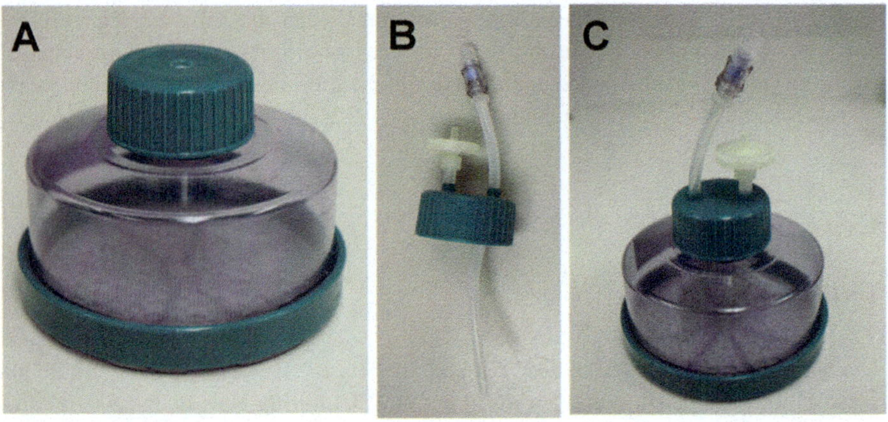

Fig. 6.6 Gas permeable flask and cap with tubes for closed system media addition and removal. Gas permeable flasks (G-Rex, Wilson Wolf Manufacturing, New Brighton, MN, USA) with 500 mL capacity and 100 cm^2 gas permeable membrane, G-Rex100 (**A**) are available with a cap that contains a tube and a vent that allow filling or emptying of the flask without opening the cap (**B**). This creates a more closed system that reduces the risk of microbial contamination (**C**)

more closed system and for easier filling and emptying of the flasks (Fig. 6.6). An instrument has recently become available that allow the automated sterile removal and addition of fluid and sterile cell harvesting (Fig. 6.7). Removing fluid from the gas permeable flasks with a repeater pump is problematic because the vacuum generated in the flask by the pump causes the flexible gas permeable membrane on the bottom of the flask to rise and could rupture the membrane. The device used to remove fluid from gas permeable flasks pushes sterile air into the flasks to displace fluid through tubing and into a bag(s). One limitation of these flasks is that they are only available in sizes that hold 40 to 2000 mL. However, flasks are being developed that will hold 4–5 L of media.

T Cell Expansion

Several media have been designed specifically for T cell culture. However, to support T cell expansion these media must be supplemented with human serum. Group AB serum is used since it lacks anti-A and anti-B which could lyse RBCs present in the cultured cells from blood group A, B or AB subjects or react with group A or B antigens adsorbed by other cells. Most T cell expansion protocols also involve the culture of the cells with a cytokine that acts as a growth factor; IL-2 is most commonly used. While IL-2 promotes T cells proliferation, it also induces T cell maturation. In addition, the culture of antigen- or anti-CD3-stimulated T cells in the presence of IL-2 induces an increase in the number of T cells expressing FOXP3 which are know as T regulatory cells (Treg) [32]. Treg cells have a potent immune

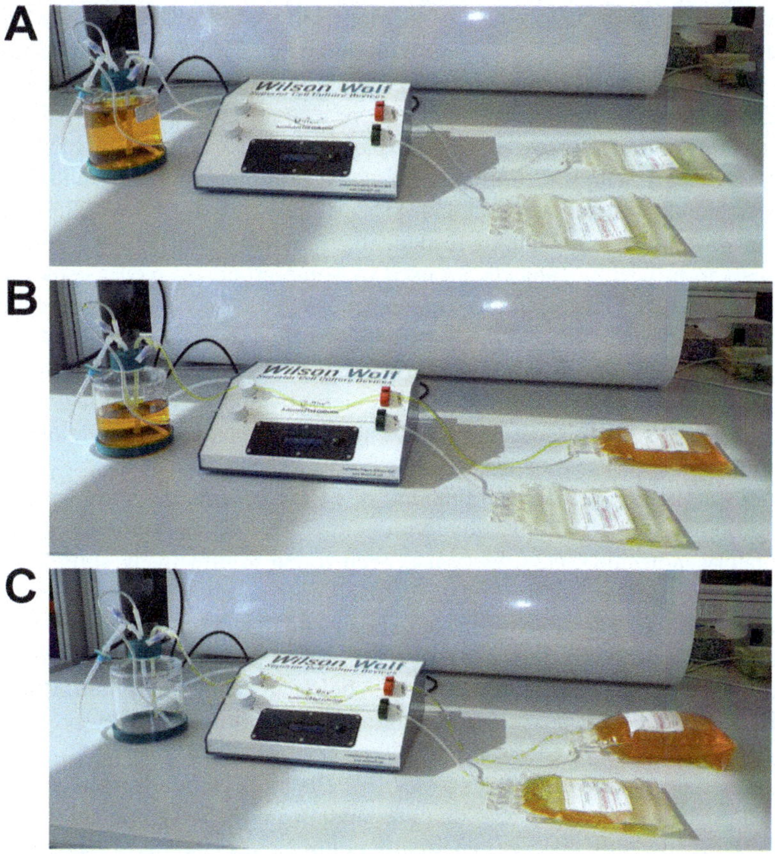

Fig. 6.7 Closed system removal of media from gas permeable flasks. Removing media from the closed system gas permeable flasks using negative pressure generated with a syringe or repeater pump can cause displacement of the flexible gas permeable membrane off the bottom of the flask which could damage the membrane. The automated cell recovery device (Wilson Wolf Manufacturing) displaces fluid from the flasks using positive pressure (**A**). Sterile filtered air displaces media without displacing the gas permeable membrane (**B** and **C**)

suppressive function and can inhibit in vivo the effects of adoptively transferred T cells [33]. Other cytokines are being tested for T cells culture that do not expand regulatory T cells and which promote a different and possibly more potent T cell phenotype. For example the culture of cytotoxic T cells in IL-21 increases the quantity of antigen-specific cytotoxic lymphocytes (CTLs) in culture while reducing the quantity of FOXP3 expressing suppressor cells [34, 35]. In addition IL-21 produces more CTLs with a central memory phenotype [36]. IL-7 and IL-15 can also be used for T cell expansion. The culture of T cells in IL-7 does not increase the number of Tregs, however, culture in IL-15 does [32, 33]. Culture of naïve T cells in IL-7 plus IL-15 allows the cells to maintain a memory T cell phenotype [37].

Antibodies, antibodies conjugated to beads and feeder cells are often added to T cell cultures in order to increase proliferation and expansion. TIL can be obtained from melanoma digests or fragments by culture in IL-2 alone; this is known as initial TIL culture [38]. However, for the more extensive expansion of isolated TIL required for clinical therapy, a "rapid expansion" protocol is used which involves the culture of TIL with IL-2, antibodies directed to the T cell receptor, anti-CD3, and allogeneic leukocytes as feeder cells [38]. GMP grade IL-2 and anti-CD3 are commercially available, but obtaining feeder cells is more difficult. PBMCs collected by apheresis from healthy subjects are used by some groups as feeder cells for rapid expansion of TIL. These cells are gamma irradiated to a sufficient degree to prevent their proliferation in vitro and in vivo. PBMCs pooled from several healthy subjects are typically used. The disadvantage of using PBMCs is that a mechanism must be in place to collect PBMCs from healthy subjects by apheresis and the donors must undergo health history screening and be tested for markers of infectious agents. As a result this process is expensive and not available to all centers.

There are some alternatives to the use of PBMCs feeder cells for T cell expansion. Magnetic beads conjugated with anti-CD3 and an antibody to the co-stimulatory molecule CD28, have also been found to effectively stimulate T cell growth in the presence of IL-2 [39, 40]. GMP grade beads conjugated to anti-CD3 and anti-CD28 are available [40]. These beads are magnetic which allows for their removal with a magnet external to the culture vessel at the end of the culture period (Fig. 6.8). While these anti-CD3/anti-CD28 beads are being use for the GMP manufacture of some T cell products, some groups have found that they preferentially induce the expansion of CD4+ T cells over CD8+ T cells [41].

Another alternative to PBMCs as feeder cells for T cell expansion are artificial antigen presenting cells (APCs) that are made from the leukemia cell line, K562 cells. The K562 cells that are genetically engineered to express Fc receptors and costimulatory ligands, such as 4-1BBL [42]. The Fc receptors are loaded with anti-CD3 and anti-CD28. The presence costimulatory ligands along with anti-CD3 and anti-CD28 allow these cells to function as APCs and support the expansion of T cells [43]. These artificial APCs have been used to expand TIL from melanoma and ovarian cancer. The degree of expansion and cell characteristic are similar to those expanded with allogeneic PBMCs but at lower TIL:APC ratios, meaning fewer feeder cells are required. In addition, these artificial APCs maintained a favorable CD8/CD4 ratio and FOXP3+ CD4+ cell frequency. Artificial APCs represent a more standardized "off-the-shelf" cellular platform for TIL and T cell expansion [43].

Genetically Engineering T Cells

Many clinical T cell therapy clinical protocols now involve the genetic engineering of T cells in order to express high affinity T cell receptors specific for tumor antigens or CAR specific for tumor antigens. Many of these studies make

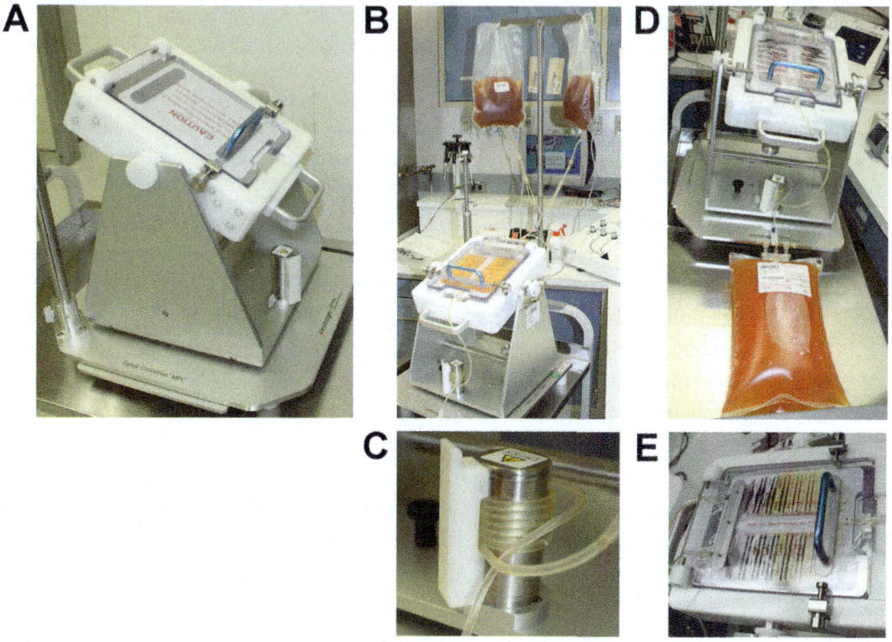

Fig. 6.8 Removal from cultured T cells of magnetic beads conjugated with CD3 and CD28 antibodies. A device (Dynal ClinExVivo MCP, Life Technologies AS, Oslo, Norway) with two magnets is used to remove magnetic beads from the cultured cells (**A**). The primary magnet is encased in the white plastic platform. A smaller secondary cylindrical secondary magnetic is below the primary magnetic. T cells cultured in bags with anti-CD3/CD28 magnetic beads flow through a bag held against the large primary magnet with a clear plastic plate (**B**), and into a collection bag (**C**) through tubing wrapped around the secondary magnetic (**D**). Most of the anti-CD3/CD28 beads are removed by the primary magnetic (**E**)

use of retroviral vector and transduction is carried out in open vessels. Typically, T flasks or 6-well plates are coated with retronectin, vector is added and the flasks are incubated and then centrifuged. After the supernatant is removed the T cells are added to the retroviral vector preloaded plates [44]. Recently, simpler, closed system methods for the transduction of T cells with retroviral vectors encoding anti-CD19 CAR have been developed [45, 46]. The two closed system methods involve the transduction and culture of T cells in culture bags. The bags are coated with retronectin and vector is added. One method involves the centrifugation of the bags containing the vector and T cells, spinoculation [46], while another simply adds the T cells to retronectin coated bags that have incubated for 2 h with the vector [45]. Both of these methods result in similar transduction efficiency to that obtained with multi-well plates and both are being used in clinical trials.

Concentrating and Washing

At the end of the culture period cellular therapies must be concentrated and washed. Often the culture media contains antibodies, cytokines and growth factors in quantities that are too great to be safely administered with the cellular therapy and the final product must be washed before it is given to the recipient. Typically, after washing, the cells are resuspended in an infusible fluid such as saline or Plasma-Lyte A (Baxter Healthcare Corporation, Deerfield, IL, USA) supplemented with human serum albumin. In addition, the final volume of the cells in culture is often several liters or greater, but the ideal volume for cellular therapies that are given intravenously is only 100–1000 mL. Consequently, the volume of the product must often be reduced. For products whose final volume is a liter or less the product can easily be concentrated and washed using bags and a floor model centrifuge. For products whose volume is more than 1–1.5 L it is desirable to use a more automated system for washing and concentrating the cells. Automated systems allow for the more rapid concentration and washing of large volume cultures. Rapid washing and concentration is important because of the limited stability of the T cell products suspended at high concentration in infusible media. Blood cell processors such as the Cobe 2991 cell processor (Terumo BCT) have been used for this application. Fenwal, is working to develop a new spinning membrane cell washer. The Cobe 2991 processes cells discontinuously (Fig. 6.9). It can concentrate up to 600 mL in one concentration/wash cycle and each cycle requires approximately 10 min to complete. The size of product that can be concentrated and washed using a Cobe 2991 over 2–4 h is limited to 5–10 L. Some laboratories are using instruments designed to recover autologous blood during surgery for washing cellular products. In addition some companies are working on instruments that make use of tangential flow to concentrate and wash cellular products.

Product Testing

It's important to evaluate the cellular product at several time points during the manufacturing process. The starting material is tested to be sure that a sufficient quantity of the desired cells are present and to ensure that quantities of contaminating cells are below critical levels. The starting material is usually tested for sterility and identity. The final product should be tested for quantities of desired and contaminating cells, endotoxin level and sterility. For cultured products, in addition to sterility, mycoplasma and endotoxin testing is required. It is also desired to measure the critical biological function or potency of the final product.

Conclusions

Many procedures have been described that allow for the manufacturing of high quality cellular therapies to support phase I/II clinical trials. Instruments, devices and reagents are available which help improve the manufacturing process. The growing

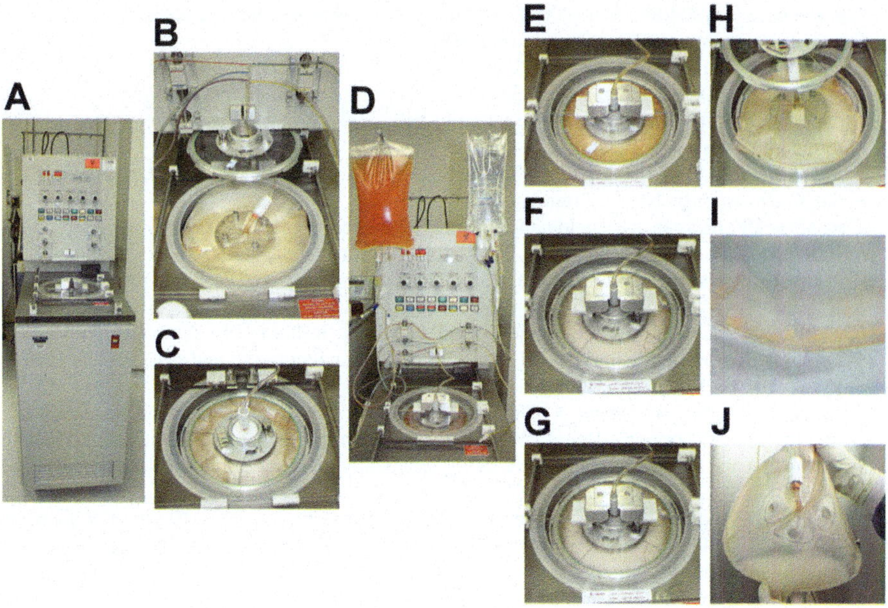

Fig. 6.9 An automated cell washing device and it's sterile, closed, one-use, plastic liner. The Cobe 2991 cell processer (Terumo BMT, Lakewood, CO, USA) can be used to wash cultured T cells (**A**). The sterile plastic circular disposable liner is placed into the instrument (**B**) and the clear cover is locked into place (**C**). Cultured cells are loaded into the circular plastic container as the container spins (**D** and **E**). Approximately 600 mL of cultured cells are added. After approximately 6 min of centrifugation, the supernatant is removed and the pelleted cells are resuspended by agitation (**F**). The process of adding 600 mL of the cultured cells, centrifugation, supernatant removal and resuspension continues until all the cultured cells have been concentrated. Each cycle takes about 10 min. The pelleted cells are then washed by using several cycles of resuspension in 600 mL of saline, centrifugation and supernatant removal (**G**). The cells are then washed with the solution that will be used for infusion or for cryopreservation. After the washes are complete the pelleted cells (**H** and **I**) are resuspended in infusion or cryopreservation solution (**J**)

success of T cell therapies has resulted in investment by academic investigators and industry in methods to further improve the manufacturing processes by creating new reagents and devices. Combining these procedures, instrument and reagents in novel ways is allowing for the rapid development methods for manufacture of emerging cell therapies that are sterile, consistent and potent.

References

1. Dudley ME, Wunderlich J, Nishimura MI, Yu D, Yang JC, Topalian SL, Schwartzentruber DJ, Hwu P, Marincola FM, Sherry R et al (2001) Adoptive transfer of cloned melanoma-reactive T lymphocytes for the treatment of patients with metastatic melanoma. J Immunother 24(4):363–373

2. Rosenberg SA, Dudley ME (2009) Adoptive cell therapy for the treatment of patients with metastatic melanoma. Curr Opin Immunol 21(2):233–240
3. Besser MJ, Shapira-Frommer R, Itzhaki O, Treves AJ, Zippel DB, Levy D, Kubi A, Shoshani N, Zikich D, Ohayon Y et al (2013) Adoptive transfer of tumor-infiltrating lymphocytes in patients with metastatic melanoma: intent-to-treat analysis and efficacy after failure to prior immunotherapies. Clin Cancer Res 19(17):4792–4800
4. Restifo NP, Dudley ME, Rosenberg SA (2012) Adoptive immunotherapy for cancer: harnessing the T cell response. Nat Rev Immunol 12(4):269–281
5. Robbins PF, Morgan RA, Feldman SA, Yang JC, Sherry RM, Dudley ME, Wunderlich JR, Nahvi AV, Helman LJ, Mackall CL et al (2011) Tumor regression in patients with metastatic synovial cell sarcoma and melanoma using genetically engineered lymphocytes reactive with NY-ESO-1. J Clin Oncol 29(7):917–924
6. Morgan RA, Chinnasamy N, Abate-Daga D, Gros A, Robbins PF, Zheng Z, Dudley ME, Feldman SA, Yang JC, Sherry RM et al (2013) Cancer regression and neurological toxicity following anti-MAGE-A3 TCR gene therapy. J Immunother 36(2):133–151
7. Till BG, Jensen MC, Wang J, Chen EY, Wood BL, Greisman HA, Qian X, James SE, Raubitschek A, Forman SJ et al (2008) Adoptive immunotherapy for indolent non-Hodgkin lymphoma and mantle cell lymphoma using genetically modified autologous CD20-specific T cells. Blood 112(6):2261–2271
8. Louis CU, Savoldo B, Dotti G, Pule M, Yvon E, Myers GD, Rossig C, Russell HV, Diouf O, Liu E et al (2011) Antitumor activity and long-term fate of chimeric antigen receptor-positive T cells in patients with neuroblastoma. Blood 118(23):6050–6056
9. James SE, Greenberg PD, Jensen MC, Lin Y, Wang J, Till BG, Raubitschek AA, Forman SJ, Press OW (2008) Antigen sensitivity of CD22-specific chimeric TCR is modulated by target epitope distance from the cell membrane. J Immunol 180(10):7028–7038
10. Giordano Attianese GM, Marin V, Hoyos V, Savoldo B, Pizzitola I, Tettamanti S, Agostoni V, Parma M, Ponzoni M, Bertilaccio MT et al (2011) In vitro and in vivo model of a novel immunotherapy approach for chronic lymphocytic leukemia by anti-CD23 chimeric antigen receptor. Blood 117(18):4736–4745
11. Shaffer DR, Savoldo B, Yi Z, Chow KK, Kakarla S, Spencer DM, Dotti G, Wu MF, Liu H, Kenney S et al (2011) T cells redirected against CD70 for the immunotherapy of CD70-positive malignancies. Blood 117(16):4304–4314
12. Vera J, Savoldo B, Vigouroux S, Biagi E, Pule M, Rossig C, Wu J, Heslop HE, Rooney CM, Brenner MK et al (2006) T lymphocytes redirected against the kappa light chain of human immunoglobulin efficiently kill mature B lymphocyte-derived malignant cells. Blood 108(12):3890–3897
13. Carpenter RO, Evbuomwan MO, Pittaluga S, Rose JJ, Raffeld M, Yang S, Gress RE, Hakim FT, Kochenderfer JN (2013) B-cell maturation antigen is a promising target for adoptive T-cell therapy of multiple myeloma. Clin Cancer Res 19(8):2048–2060
14. Chow KK, Naik S, Kakarla S, Brawley VS, Shaffer DR, Yi Z, Rainusso N, Wu MF, Liu H, Kew Y et al (2013) T cells redirected to EphA2 for the immunotherapy of glioblastoma. Mol Ther 21(3):629–637
15. Selvaggi TA, Walker RE, Fleisher TA (1997) Development of antibodies to fetal calf serum with arthus-like reactions in human immunodeficiency virus-infected patients given syngeneic lymphocyte infusions. Blood 89(3):776–779
16. Macy E, Bulpitt K, Champlin RE, Saxon A (1989) Anaphylaxis to infusion of autologous bone marrow: an apparent reaction to self, mediated by IgE antibody to bovine serum albumin. J Allergy Clin Immunol 83(5):871–875
17. Mackensen A, Drager R, Schlesier M, Mertelsmann R, Lindemann A (2000) Presence of IgE antibodies to bovine serum albumin in a patient developing anaphylaxis after vaccination with human peptide-pulsed dendritic cells. Cancer Immunol Immunother 49(3):152–156
18. Eder AF, Kennedy JM, Dy BA, Notari EP, Weiss JW, Fang CT, Wagner S, Dodd RY, Benjamin RJ (2007) Bacterial screening of apheresis platelets and the residual risk of septic transfusion reactions: the American Red Cross experience (2004–2006). Transfusion 47(7):1134–1142

19. Jin J, Sabatino M, Somerville R, Wilson JR, Dudley ME, Stroncek DF, Rosenberg SA (2012) Simplified method of the growth of human tumor infiltrating lymphocytes in gas-permeable flasks to numbers needed for patient treatment. J Immunother 35(3):283–292
20. Shu Z, Heimfeld S, Gao D (2013) Hematopoietic SCT with cryopreserved grafts: adverse reactions after transplantation and cryoprotectant removal before infusion. Bone Marrow Transplant 30
21. Ayello J, Hesdorffer C, Reiss RF (1995) A semiautomated technique for volume reduction of stem cell suspensions for autotransplantation. J Hematother 4(6):545–549
22. Przepiorka D, Van VP, Huynh L, Durett A, Agbor P, Lauppe J, Valone F, Champlin R, Korbling M (1996) Rapid debulking and CD34 enrichment of filgrastim-mobilized peripheral blood stem cells by semiautomated density gradient centrifugation in a closed system. J Hematother 5(5):497–502
23. Schumm M, Lang P, Taylor G, Kuci S, Klingebiel T, Buhring HJ, Geiselhart A, Niethammer D, Handgretinger R (1999) Isolation of highly purified autologous and allogeneic peripheral CD34+ cells using the CliniMACS device. J Hematother 8(2):209–218
24. Campbell JD, Piechaczek C, Winkels G, Schwamborn E, Micheli D, Hennemann S, Schmitz J (2005) Isolation and generation of clinical-grade dendritic cells using the CliniMACS system. Methods Mol Med 109:55–70
25. Adamson L, Palma M, Choudhury A, Eriksson I, Nasman-Glaser B, Hansson M, Hansson L, Kokhaei P, Osterborg A, Mellstedt H (2009) Generation of a dendritic cell-based vaccine in chronic lymphocytic leukaemia using CliniMACS platform for large-scale production. Scand J Immunol 69(6):529–536
26. Hannon M, Lechanteur C, Lucas S, Somja J, Seidel L, Belle L, Bruck F, Baudoux E, Giet O, Chantillon AM et al (2014) Infusion of clinical-grade enriched regulatory T cells delays experimental xenogeneic graft-versus-host disease. Transfusion 54:353–363
27. Koehl U, Brehm C, Huenecke S, Zimmermann SY, Kloess S, Bremm M, Ullrich E, Soerensen J, Quaiser A, Erben S et al (2013) Clinical grade purification and expansion of NK cell products for an optimized manufacturing protocol. Front Oncol 3:118
28. Malone CC, Schiltz PM, Mackintosh AD, Beutel LD, Heinemann FS, Dillman RO (2001) Characterization of human tumor-infiltrating lymphocytes expanded in hollow-fiber bioreactors for immunotherapy of cancer. Cancer Biother Radiopharm 16(5):381–390
29. Klapper JA, Thomasian AA, Smith DM, Gorgas GC, Wunderlich JR, Smith FO, Hampson BS, Rosenberg SA, Dudley ME (2009) Single-pass, closed-system rapid expansion of lymphocyte cultures for adoptive cell therapy. J Immunol Methods 345(1-2):90–99
30. Somerville RP, Devillier L, Parkhurst MR, Rosenberg SA, Dudley ME (2012) Clinical scale rapid expansion of lymphocytes for adoptive cell transfer therapy in the WAVE(R) bioreactor. J Transl Med 10:69
31. Vera JF, Brenner LJ, Gerdemann U, Ngo MC, Sili U, Liu H, Wilson J, Dotti G, Heslop HE, Leen AM et al (2010) Accelerated production of antigen-specific T cells for preclinical and clinical applications using gas-permeable rapid expansion cultureware (G-Rex). J Immunother 33(3):305–315
32. Ahmadzadeh M, Antony PA, Rosenberg SA (2007) IL-2 and IL-15 each mediate de novo induction of FOXP3 expression in human tumor antigen-specific CD8 T cells. J Immunother 30(3):294–302
33. Perna SK, Pagliara D, Mahendravada A, Liu H, Brenner MK, Savoldo B, Dotti G (2014) Interleukin-7 mediates selective expansion of tumor-redirected cytotoxic T lymphocytes without enhancement of regulatory T-cell inhibition. Clin Cancer Res 20:131–139
34. Li Y, Bleakley M, Yee C (2005) IL-21 influences the frequency, phenotype, and affinity of the antigen-specific CD8 T cell response. J Immunol 175(4):2261–2269
35. Li Y, Yee C (2008) IL-21 mediated Foxp3 suppression leads to enhanced generation of antigen-specific CD8+ cytotoxic T lymphocytes. Blood 111(1):229–235
36. Kaka AS, Shaffer DR, Hartmaier R, Leen AM, Lu A, Bear A, Rooney CM, Foster AE (2009) Genetic modification of T cells with IL-21 enhances antigen presentation and generation of central memory tumor-specific cytotoxic T-lymphocytes. J Immunother 32(7):726–736

37. Cieri N, Camisa B, Cocchiarella F, Forcato M, Oliveira G, Provasi E, Bondanza A, Bordignon C, Peccatori J, Ciceri F et al (2013) IL-7 and IL-15 instruct the generation of human memory stem T cells from naive precursors. Blood 121(4):573–584
38. Dudley ME, Wunderlich JR, Shelton TE, Even J, Rosenberg SA (2003) Generation of tumor-infiltrating lymphocyte cultures for use in adoptive transfer therapy for melanoma patients. J Immunother 26(4):332–342
39. Levine BL, Bernstein WB, Connors M, Craighead N, Lindsten T, Thompson CB, June CH (1997) Effects of CD28 costimulation on long-term proliferation of CD4+ T cells in the absence of exogenous feeder cells. J Immunol 159(12):5921–5930
40. Kalamasz D, Long SA, Taniguchi R, Buckner JH, Berenson RJ, Bonyhadi M (2004) Optimization of human T-cell expansion ex vivo using magnetic beads conjugated with anti-CD3 and Anti-CD28 antibodies. J Immunother 27(5):405–418
41. Li Y, Kurlander RJ (2010) Comparison of anti-CD3 and anti-CD28-coated beads with soluble anti-CD3 for expanding human T cells: differing impact on CD8 T cell phenotype and responsiveness to restimulation. J Transl Med 8:104
42. Suhoski MM, Golovina TN, Aqui NA, Tai VC, Varela-Rohena A, Milone MC, Carroll RG, Riley JL, June CH (2007) Engineering artificial antigen-presenting cells to express a diverse array of co-stimulatory molecules. Mol Ther 15(5):981–988
43. Ye Q, Loisiou M, Levine BL, Suhoski MM, Riley JL, June CH, Coukos G, Powell DJ Jr (2011) Engineered artificial antigen presenting cells facilitate direct and efficient expansion of tumor infiltrating lymphocytes. J Transl Med 9:131
44. Kochenderfer JN, Feldman SA, Zhao Y, Xu H, Black MA, Morgan RA, Wilson WH, Rosenberg SA (2009) Construction and preclinical evaluation of an anti-CD19 chimeric antigen receptor. J Immunother 32(7):689–702
45. Tumaini B, Lee DW, Lin T, Castiello L, Stroncek DF, Mackall C, Wayne A, Sabatino M (2013) Simplified process for the production of anti-CD19-CAR-engineered T cells. Cytotherapy 28
46. Hollyman D, Stefanski J, Przybylowski M, Bartido S, Borquez-Ojeda O, Taylor C, Yeh R, Capacio V, Olszewska M, Hosey J et al (2009) Manufacturing validation of biologically functional T cells targeted to CD19 antigen for autologous adoptive cell therapy. J Immunother 32(2):169–180

Chapter 7
Clinical Success of Adoptive Cell Transfer Therapy Using Tumor Infiltrating Lymphocytes

Jessica Chacon, Krit Ritthipichai, Geok Sim, Michiko Harao, Jie Qing Chen, Caitlin Creasy, Chantale Bernatchez, Patrick Hwu, and Laszlo Radvanyi

Abstract One of the hallmarks of cancer is the infiltration of all tumors (both primary and metastatic sites) with immune cells composed of lymphocytes and myeloid cells to various extents. This is indicative of the intrinsic inflammatory nature of all tumors as "invaders" in resident normal tissues causing local immune activation and immune cell recruitment. In most cases, a cascade of events sets in motion immune responses that drive innate immunity at the tumor site which then drives adaptive responses mediated by antigen-specific T cells. Tumors are infiltrated to various degrees with previously activated $CD8^+$ and $CD4^+$ T cells that are products of this immune response. These cells, also called "tumor-infiltrating lymphocytes" (TIL) have emerged as critical factors controlling cancer growth at localized tumor sites by recognizing not only over-expressed self-antigens, but also mutated neo-antigens specific for each patient for which immune tolerance does not exist. These are emerging to be the most powerful tumor antigens recognized by TIL. Many studies have

J. Chacon • K. Ritthipichai • P. Hwu
Department of Melanoma Medical Oncology, The University of Texas
M. D. Anderson Cancer Center, Houston, TX, USA

G. Sim • M. Harao • J.Q. Chen • C. Creasy • C. Bernatchez
Department of Melanoma Medical Oncology, The University of Texas
M. D. Anderson Cancer Center, Houston, TX, USA

L. Radvanyi, Ph.D. (✉)
Immuno-Oncology Translational Innovation Platform, EMD Serono Research & Development Institute, 45A Middlesex Turnpike, Billerica, MA 01821, USA
e-mail: laszlo.radvanyi@lionbio.com

P.A. Ascierto et al. (eds.), *Developments in T Cell Based Cancer Immunotherapies*, Cancer Drug Discovery and Development,
DOI 10.1007/978-3-319-21167-1_7

now shown that increased T-cell infiltration into tumors at different disease stages is correlated with increased survival for many different type of cancers. Harnessing the intrinsic effector properties and antigen specificity of these TIL has emerged over a number of decades of dedicated pre-clinical and clinical research as a powerful approach to cancer immunotherapy through the adoptive transfer of TIL expanded ex vivo followed by re-infusion into the patient. This form of adoptive cell therapy has been highly developed to treat metastatic melanoma with consistent response rates ranging from 45 to 50 % and complete durable response rates as high as 20 %. These results, together with advances in methods to expand TIL ex vivo to therapeutic numbers, has set the stage now for further developing TIL adoptive cell therapy as a standard-of-care for melanoma through practical commercial manufacturing systems. In this chapter, we introduce T-cell therapy and comprehensively describe TIL therapy for metastatic melanoma, including a discussion of how TIL are expanded for therapy, outstanding technical and biological questions relating to T-cell differentiation being addressed in the field, and the growing area of predictive biomarker research that is revealing new mechanistic insights into how TIL work and opening up the possibility to select patients for T-cell therapy by interrogating factors within the tumor microenvironment. Overall, TIL therapy for melanoma has proven to be an effective regimen to treat melanoma in multiple clinical centers now across the world, especially in a salvage setting when other front-line therapies have failed, including T-cell checkpoint blockade, such as anti-CTLA-4, anti-PD-1, and anti-PD-L1. A number of cytokine signaling mechanisms and genes regulating the differentiation of TIL in culture towards end-stage, senescent cells, especially $CD8^+$ T cells, have been elucidated allowing us to manipulate these pathways during ex vivo cell expansion to keep the cells "younger" and less differentiated when infused to ensure improved persistence in vivo. A number of promising biomarkers are being discovered in the original tumors used to expand TIL for therapy. These markers, together with other patient-specific biomarkers can be incorporated into biomarker signatures allowing for accurate selection of patients most likely to respond to therapy. All these developments, together with newer ways to selectively expand more tumor-specific TIL, will push this cell therapy more and more into the mainstream cancer care as part of growing "immunological toolbox" for cancer.

Keywords Adoptive cell therapy • Immunotherapy • Tumor-infiltrating lymphocyte • Melanoma • T-Lymphocyte • CD8 • Clinical trials • Interleukin-2 • Biomarkers • Costimulation

Introduction

One of the newer hallmarks of cancer is avoiding immune destruction and there is a renewed interest currently in developing novel therapeutics that modulate the anti-tumor immune response including anti-tumor vaccines, immune modulators, innate immune system activators, cytokines, and adoptive cell therapy.

Emerging evidence suggests that the tumor microenvironment (TME) is a very heterogeneous entity and involves a very complex interaction of multiple cell types. In most of tumors, distinct populations of adaptive and innate immune cells are commonly found infiltrated in addition to the cancer cells and resident stromal cells such as fibroblast and tumor-associated endothelial cells. These tumor infiltrating immune cells often comprise of not only lymphocytes such as cytotoxic $CD8^+$ T cells, NK cells and NKT cells that confer anti-tumor responses but also antigen presenting cells such as macrophages and dendritic cells, which play a critical role in activating effector cytotoxic cells and in promoting anti-tumor responses, and granulocytes (i.e. neutrophils, eosinophils) [1]. However, the local cytokine milieu and various growth factors that are pro-inflammatory in the TME also recruit negative regulatory cells or condition pre-existing macrophages, myeloid cells, and T cells into M2 macrophages, myeloid derived suppressor cells (MDSC) and T regulatory cells that exhibit immunosuppressive functions. These regulatory cells suppress anti-tumor responses by limiting efficacy of CD8- and NK cell-mediated anti-tumor responses via multiple mechanisms including cell-cell contact or secretion of immunosuppressive cytokines. For instance, tumor infiltrating-$CD4^+CD25^{hi}Foxp3^+$ Tregs have been shown to suppress TILs in head and neck squamous cell carcinoma (HNSCC) patients via secretion of IL-10 and TGF-β [2]. Tregs may also promote tumor progression by direct killing of tumor-reactive cells T and NK in a granzyme dependent mechanism [3]. Furthermore, evidence from several studies show that Tregs express high levels of ectoenzymes (CD39 and CD73) that hydrolyse ATP and ADP to AMP and adenosine. Activation of adenosine signaling pathway on T cells directly inhibits their proliferation [4, 5]. In addition, the presence of MDSC in TME suppress effector cell function via production of Indoleamine 2,3-dioxygenase (IDO) [6] and reactive oxygen species (ROS) [7], depletion of metabolites (arginine and cysteine) by up-regulation of arginase-1 (ARG1) [8] and promoting the activation and development of Ag-specific regulatory T cells within tumors [9]. Despite these immunosuppressive features in the TME and the active suppression of tumor-infiltrating T cells by factors from the tumor cells, Tregs, and MDSC, in most patients a significant and local anti-tumor cell-mediated immune response occurs in an ongoing fashion that needs to be boosted by a "push" and "pull" approach in which agonists (e.g., tumor vaccines and factors further activating T cells) are needed to further activate the response and antagonists are needed to block immunosuppressive factors (e.g., blockers of negative co-stimulatory molecules like PD-1 in T cells).

Attempts to augment the immune system to treat cancer began over a century ago. In the nineteenth and twentieth century, bacterial culture products (supernatants) were injected into tumor of cancer patients in studies mainly conducted by Coley [10–14]. This work was later extended to include treatment of tumor-bearing individuals with injected killed bacteria into the tumor and eradication of the tumor was a result of Tumor Necrosis Factor (TNF) production in response to the bacterial endotoxins [10–14].

In 1926, J.B. Murphy proposed that lymphoid cells played a key role in eradicating solid tumors that had previously been transplanted in animal models [15].

In 1958, Sir Peter Medawar coined the term "*immunological competent cell*" to describe a cell that is "*fully qualified to undertake an immunological response* [15, 16]." In the mid-1960s, Alexander and associates treated mice that had sarcomas using lymphocytes from immunized syngeneic animals [16, 17]. Alex Fefer and colleagues demonstrated in 1969 that mice containing lymphomas that were virally induced could be treated using chemotherapy and infusion of lymphocytes [18, 19]. In the mid-1970s, Eberlein and colleagues demonstrated that immune cells could be expanded using the growth factor Interleukin-2 (IL-2), be injected intravenously into a tumor-bearing mice and eradicate the implanted tumors [19–23]. In the early 1990s, Boon and colleagues discovered (from MAGE-3) the first tumor-associated Human Leukocyte Antigen (HLA)-restricted T-cell epitope [24].

The immune system is our first line of defense to help protect us from foreign pathogens. Knowledge regarding the immune system has greatly increased over the years, paving the way for immunotherapy to become a novel way to treat diseases, including cancer. Many cancer cells become resistant to conventional therapy, such as chemotherapy and radiation, resulting in large majority patients relapsing, therefore immunotherapies have been developed to overcome this resistance [19, 25–27]. Passive and active immunotherapies have been developed over the years. Passive immunotherapies consist of antibodies, cytokines or other elements of the immune system that are constructed in a laboratory and subsequently administered to patients to provide immunity against a disease. On the other hand, active immunotherapy involves the direct stimulation of the host immune system, usually through the use of vaccines. Another powerful way to augment the immune system is through adoptive T-cell therapy (ACT) using tumor-infiltrating lymphocytes (TIL) expanded ex vivo outside the TME. This allows dysfunctional tumor-specific T cells to numerically expand while escaping the ongoing suppression they faced while in the TME. These "rejuvenated" T cells can then be adoptively transferred back into patients in great numbers to overcome the previous suppression where they are more highly active and can home back into the TME and mediate anti-tumor responses.

In this chapter, we will specifically focus on the type of passive immunotherapy called Adoptive T-cell therapy.

Adoptive T-Cell Therapy

Interleukin-2 (IL-2) was FDA approved for the treatment of late-stage melanoma in 1998 [28, 29]. The treatment of late-stage melanoma patients consisted of intravenously (I.V.) giving patients high dose infusion (600,000–720,000 IU/kg) IL-2 every 8 h, with several rounds of these high dose-IL-2 infusions given 3–5 weeks apart [28, 29]. Using high-dose IL-2 to treat melanoma patients has resulted in varied response rates. In patients containing one or more visceral lesions, this high-dose IL-2 resulted in clinical response rates of only 10–15 %, while patients that contained cutaneous melanoma response rates reached 50 % objective response rates [28, 29]. In patients with visceral metastasis, treatment with high-dose IL-2 resulted in 5–6 % of patients undergoing complete remission that has lasted over 10 years [28, 29].

Results from using high-dose IL-2 to treat melanoma patients prompted researchers at the National Cancer Institute (NCI), Bethesda, Maryland to investigate the role of combining high-dose IL-2 with adoptive T-cell therapy (ACT) [20, 30]. ACT is a type of passive immunotherapy that acquires either cancer patient peripheral blood or lymphocytes obtained within a tumor and expands these cells into large numbers using growth factors (such as IL-2) in a laboratory [25, 26, 31]. These cells are then re-infused into the patient. In the following section, we will discuss the different types of adoptive T-cell therapy, which vary in the site where the lymphocytes are collected and the methods used to expand the cells. A major advantage of using ACT to treat cancer patients is the opportunity for researchers to select or engineer in vitro the desired or optimal T-cell phenotype, function, anti-tumor response, antigen specificity, and expand this selected population prior to infusing the cells back into the patients.

Tumor-Infiltrating Lymphocytes

When ACT was first combined with high-dose IL-2 at the NCI by Grimm and colleagues in 1982, they developed a novel cytolytic cell system using lymphokine activated killer (LAK) cells [30, 32, 33]. LAK cells were generated from cancer patient peripheral blood lymphocytes and normal donors, using high-dose IL-2 to grow the cells [30, 32, 33]. Using animal models, infusion of LAK cells and high-dose IL-2 was able to eradicate metastatic tumors [30, 32–35]. In 1985, Rosenberg and colleagues demonstrated in a clinical trial that LAK cells infused with high-dose IL-2 had an impact on metastatic melanoma patients resulting in a 21 % response rate [34–36]. In a follow-up study conducted in 1987, it was demonstrated that combining high-dose IL-2 with LAK cells was more effective in treating cancer patients than high-dose IL-2 alone [34–36]. However, in the clinical trial conducted by Rosenberg and colleagues, the majority of the responses were partial responses and this initiated the investigators to inquire about other cells that may induce anti-tumor responses.

It was demonstrated in 1986, using mice models, that tumor-infiltrating lymphocytes (TIL) obtained from melanoma tumors from the mouse could be expanded in vitro in IL-2 and subsequently eradicate the tumors when adoptively transferred [19, 37, 38]. In a phase II trial, when metastatic melanoma patients were treated with TIL and high-dose IL-2, the response rate was 39 % [19, 37]. However, in a groundbreaking Phase II clinical trial in 2002, conducted at the NCI, Dudley and colleagues demonstrated that combining cyclophosphamide and fludarabine (a lymphodepleting chemotherapy regimen) prior to infusing the TIL with high-dose IL-2 resulted in 50 % response rates in metastatic melanoma patients [25, 39, 40]. Lymphodepletion allowed the transferred TIL to persist longer in the patient and rid any cells that may be in competition with the infused TIL for homeostatic cytokines, such as Interleukin-7 (IL-7) and Interleukin-15 (IL-15) [25, 39–42]. In addition, lymphode-

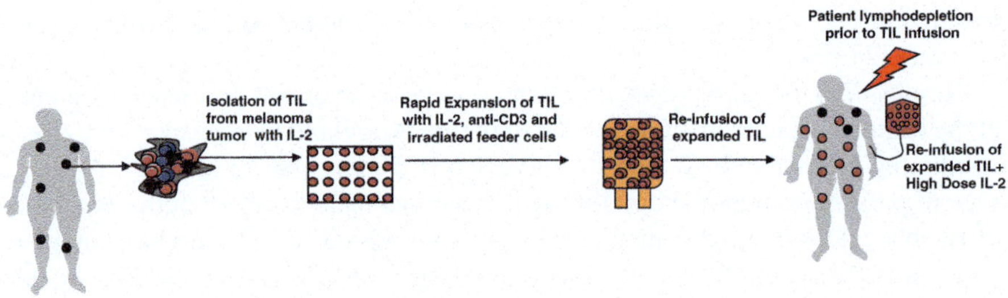

Fig. 7.1 Adoptive T-cell therapy using tumor-infiltrating lymphocytes. Tumors are surgically extracted from patients. The tumors are subsequently cut up into multiple fragments or enzymatically digested and the TIL are initially expanded over a 3–5 week period in media and IL-2. In some cases, TIL can be initially expanded from core or punch biopsies of accessible tumors. The TIL then undergo the Rapid Expansion Protocol 'REP' in which the TIL are expanded over a 2 week period with media, IL-2, anti-CD3, and irradiated PBMCs (feeder cells). The post-REP TIL are then re-infused into a lymphodepleted patient and high-dose IL-2 is given to help the transferred TIL persist

pletion also eliminates endogenous suppressor cells such as T regulatory cells (Tregs) that may inhibit the infused T cells functions [39–44]. However, the re-appearance of Tregs and other suppressor cells and factors that may inhibit T-cell function is a concern even after lymphodepletion since the lymphodepeltion is non-myeloablative and transient [39–44]. Therefore, new approaches using total body irradiation (TBI) of 2Gy or 12Gy, in addition to the traditional cyclophosphamide and fludarabine lymphodepletion have been developed at the NCI [25, 45]. Using the 12Gy TBI plus chemotherapy, the NCI found an impressive objective clinical response rate of 72 % and a complete response rate of up to 40 % [25, 45].

The standard protocol for ACT using TIL for the treatment of metastatic melanoma patients is represented schematically in Fig. 7.1. The melanoma tumor is surgically resected from the patient. The tumor is then subsequently cut up into multiple 3–5 mm^2 fragments or enzymatically digested using collagenase and hyaluronidase to isolate the TIL and placed in a 24-well plate with TIL-media and high-dose IL-2 (up to 6000 IU/ml) for a period of 4–5 weeks [25, 31, 46]. After this period, the TIL are referred as 'pre-Rapid Expansion' TIL or pre-REP TIL. The pre-REP TIL are then subjected to undergo a "rapid expansion protocol" (REP). The REP is a 2 week expansion period in which the TIL are placed in flasks with media, high-dose IL-2, anti-CD3, and irradiated peripheral blood mononuclear cells (PBMCs) that we refer to as feeder cells [26, 31, 47]. After the 2 week expansion period, the TIL are referred to as 'post-REP' TIL. During the expansion period, the TIL numerically expand to billions of cells [26, 31, 47]. The billions of post-REP TIL are then infused into a lymphodepleted patient after harvesting and formulating and high-dose IL-2 is subsequently given in order to help the transferred TIL persist [26, 31, 47]. Using this TIL to treat metastatic melanoma patients has resulted in clinical response rates of up to 50 % in Phase II clinical trials at NCI, as well as other centers around the world, including at MD Anderson Cancer Center [26, 46, 48]. Although TIL therapy was pioneered and has been mainly studied in melanoma,

TIL therapy has also been developed to treat patients with breast cancer, renal cancer, ovarian cancer, lung cancer, cervical cancer, gastric cancer, and head and neck cancers. Particularly, the presence of TIL has been correlated with overall survival in these cancers.

Ruffini et al. found that in stage I squamous-cell carcinomas patients, the presence of TIL was associated with significant survival [49]. Horne et al. also found that in stage IA non-small cell lung carcinoma, the overall 5 year survival was affected by presence of TIL. High levels of intratumoral TIL was associated with improved recurrence-free survival in Stage IA [50]. In a study conducted by Kilic and colleagues [51], it was also discovered that in stage IA-IB non-small cell lung cancer, a higher density of TIL associated with lower disease recurrence and improved 5 year disease-free survival. A higher degree of TILs within large node-negative non-small-cell lung cancer correlates with decreased risk of disease recurrence and improved disease-free survival [51].

In a study conducted by Kopecký et al. [52], the investigators found that in renal cell carcinoma, T cells are the most prominent lymphocyte population in the tumor-infiltrating lymphocytes. $CD3^+CD8^+$ lymphocytes were the prominent population in the TIL whereas $CD3^+CD4^+$ lymphocytes were the prominent population in the peripheral blood [52]. Wu, Kuo, and Ho found that cervical cancer tumors contain a higher proportion of the $CD4^+CD25^+FoxP3^+$ Tregs than in the cervical intraepithelial neoplasia [53]. When patients had lymph node metastasis, Tregs accumulated more within the tumor than when patients did not have lymph node metastasis [53].

Coukos group demonstrated that intraepithelial TIL correlated with improved survival and should be considered the most important immune biomarker for ovarian cancer [54]. Freedman and colleagues found that in epithelial ovarian carcinoma (EOC), TIL can be isolated and expanded with 200–600 IU/ml IL-2 [55] and HLA-class I expression on the EOC cells correlated with infiltration of T cells [55]. In addition, Webb demonstrated that the presence of $CD8^+$ TIL associated with prolonged survival in high-grade serous ovarian cancer (HGSC) [56]. Tumors containing $CD8^+$ TIL that were $CD103^+$ showed poor prognosis equivalent to tumors lacking $CD8^+$ TIL altogether [56]. In addition to $CD8^+$ T cells, $CD20^+$ Tumor-infiltrating B cells also associated with survival in HGSC [57].

A study conducted by Dirican et al. [58] found that in 236 gastric adenocarcinoma patients, the neutrophil-lymphocyte ratio (NLR) was independently associated with survival. No significant advantages were detected for overall survival and increased TIL number [58]. Feichtenbeiner et al. demonstrated in 50 gastric patient tumor tissue microarrays that high intraepithelial infiltration of $CD8^+$ and $FoxP3^+$ TIL was associated with 10-year metastasis-free survival [59]. However, prognostic influence of tumor-infiltrating immune cells in gastric cancer critically depends on their cell-to-cell distance. $FoxP3^+$ TIL must be located within a distance between 30 and 110 μm of $CD8^+$ T cells to positively impact positive prognosis [59]. In a study conducted by Shen and colleagues, in 133 gastric cancer patients, $CD4^+$ and $CD8^+$ TIL were not associated with overall survival [60]. However, increased $Foxp3^+$ Tregs/$CD8^+$ ratio was an independent factor for worse overall survival in addition to the presence of intratumoral high Foxp3⁺ Tregs which was also associated with worse overall survival [60].

Balermpas et al. found that in 101 head and neck squamous cell carcinoma patients, patients with high CD3 and CD8 expression had significantly increased overall survival progression free survival and distant metastasis free survival [61]. Junker and colleagues demonstrated that TIL isolated from head and neck squamous cell carcinoma can be expanded up to 3500-fold within 17 days in 80 % of patients [62, 63]. In addition, 60 % of patient TIL demonstrated tumor reactivity and were T-effector memory $CD4^+$ and $CD8^+$ subset cells [62, 63]. The TIL were tumor reactive against NY-ESO-1, cyclin B1, and Bcl-x derived peptides [62, 63].

The importance of TIL has also been investigated in breast cancer. We will discuss in the next section, the possible use of using TIL for the treatment of triple negative breast cancer based on unpublished results from our group recently.

Isolating and Expanding TIL from Triple Negative Breast Cancer

TIL have been isolated and expanded from breast cancer tissue for the past few decades. Breast cancer is mainly divided into three subtypes according to receptor expression based on hormone receptor (estrogen receptor (ER) and progesterone receptor (PR)) and HER2. The three subtypes are Luminal (ER^+ and or PR^+), $HER2^+$ and triple negative (ER^-, PR^- and $HER2^-$) subtype. Hormone receptor positive breast cancer breast is characterized as a slower growing "chronic" disease. On the other hand, most of $HER2^+$ and triple negative subtypes have aggressive characteristics.

Studies have been conducted to determine if there is a correlation between specific immune cell subsets and/or chemokines and overall survival in breast cancer patients. Tsang et al. found that the chemokine fractalkine CX3CL1 is associated with poor clinical outcome in breast cancer patients as increased CX3CL1 expression was detected in 33.3 % of primary invasive breast cancers [64]. West et al. determined in ER-breast cancer patients, high expression of $Foxp3^+$ TIL strongly correlated with anti-tumor immunity and prolonged recurrence-free survival [65]. Seo et al. found that $CD4^+$ and $CD8^+$ T-cell infiltration was closely correlated with breast cancer stem cell phenotype and epithelial-mesenchymal transition (EMT) [66]. Also, high levels of CD4 and CD8 and $FoxP3^+$ TIL were associated with complete response [66].

During our studies at MDA, breast cancer TIL have been generated using enzymatic digestion and mechanical disruption with the addition of cytokines. Although TIL successfully grew from breast cancer tissue, the number of TIL expanded is not adequate to apply adoptive T cell transfer to treat cancer patients. Hudson et al. successfully isolated TIL from 23 of 30 patients using IL-2, TNF alpha and OKT3 monoclonal antibody stimulation and the yield of total TIL growth was as much as 34 million in 21 day culture. TIL from 66 % of patients were predominantly $CD8^+$ phenotype, however, only one of four of these TIL lysed their autologous tumor [67]. Swartzentruber et al. have shown that TIL generated with high dose IL-2 (6000 IU/ml) grew to a median of 6,700-fold in 65 days, however, predominantly

$CD4^+$ cells expanded in the culture (73 %) and lysis of autologous tumor occurred in only one of 12 expanded TIL [68].

The degree of TIL has shown an inverse correlation with the level of Estrogen receptor [69]. Additionally, intratumoral and stromal lymphocytic infiltration reduced risk of relapse and death in the triple negative subtype [70]. Few studies have succeeded in isolating breast cancer TIL, however the cytotoxity of these cells were very low. The reason for lack of killing is unclear but one study demonstrated lack of co-stimulatory molecules or the lack of IL-2 expression in primary breast TIL [71].

In our experience, regardless of primary systemic therapy, our group has succeeded in isolating TIL from breast cancer tissue without using enzymatic digestion or mechanical disruption. Here, TIL are isolated from fresh triple negative breast cancer tissues cut into small fragments and cultured with media and high dose IL-2 (6000 IU/ml) for 28–35 days. The total TIL yield ranged from 0.04 to 14.1 million per fragment, however, TIL grown with only high dose IL-2 did not generate a high $CD8^+$ population (18.3 ± 14.9 %). In comparison, TIL isolated and expanded with high dose IL-2 and CD137 (4-1BB) agonist, a member of the TNF receptor super family, resulted in great expansion of TIL and high percentage of CD8 population compared with only IL-2 culture condition (2.15–71.4 million per fragment, 36.9 ± 28.8 %). Furthermore, TIL expanded with IL-2 and 4-1BB agonist showed high cytotoxic capability. For generating breast cancer TIL, we may have to select suitable TIL populations (e.g. triple negative breast cancer) and use additional cytokine or co-stimulatory molecules to overcome the lack of responsiveness.

An advantage of using TIL therapy to treat cancer is that TIL have a polyclonal nature that recognize multiple tumor-associated antigens (TAA), including the well-known melanocyte/melanoma differentiation antigens as well as a wide range of unknown antigens. However, alternative approaches have been developed to expand antigen specific T cells from patient-derived PBMCs.

Chimeric Antigen Receptor

Another exciting form of ACT utilizes chimeric antigen receptors (CAR). The first generation of the CAR contained a link between light chain and heavy chain of a monoclonal antibody with variable domains [72, 73]. This was associated with transmembrane and the T-Cell Receptor (TCR) complex cytoplasmic tail of the zeta (ζ) chain [72, 73]. Engineering T cells using this approach was developed to circumvent Major Histocompatibility Complex (MHC) restriction due to the fact that tumor cells evade recognition by down-regulating the MHC. However, this first generation of the CAR led to poor proliferation of the engineered T cells and it was noted that stimulation of the TCR via the cytoplasmic tail of the zeta (ζ) chain was insufficient to help the T cells persist [72, 73]. Therefore, the second generation of the CAR was generated to provide improved co-stimulation to the T cells by adding the CD28 endo-domain, in addition to the zeta (ζ) chain that was already on the first generation CAR in order to mimic the two-signals needed for T-cell activation [74, 75]. Other endo-domains from other co-stimulatory molecules have been added to the CAR constructs, such as the endo-domains from CD137 (4-1BB) and CD134 (OX40) to generate a third generation of CAR consisting of the zeta (ζ) chain, and the endo-domains of CD28 and CD137 [76].

Adoptive transfer of autologous CAR transduced T cells induced substantial tumor regressions in leukemia and B-cell lymphomas [77, 78]. While treating leukemia and B-cell lymphomas, CAR constructs contained a CD19-specific Immunoglobulin G (IgG) single chain variable fusion (scFv) fused to the TCR ζ chain and the endo-domains of CD137 [77–79]. Treatment of leukemias and lymphomas using CAR transduced T cells have shown promising results in regards to anti-tumor activity and persistence [77, 78]. In regards to melanoma treatment, targets of CAR transduced T cells include overexpressed genes, such as gangliosides GD2, and GD3, which are over expressed in 50–80 % of metastatic melanomas [80, 81]. Using IL-2 administration and a second generation CAR that targeted GD2 with endo-domains of CD28 and CD134 significantly improved the survival of mice following transfer [82]. In addition, researchers found that administering IL-2 and using a TCR transduced CAR that targeted GD3 and contained the CD28 endo-domain resulted in complete response rate in 50 % of melanoma tumor-burdened mice [81].

TCR-Transgenic T Cells

Another type of ACT involves TCR-transduced T cells. These T cells are transduced using retroviral or lentiviral vectors containing TCR genes that encode variable regions for different melanoma-associated antigens, such as glycoprotein-100 (gp100) and melanoma antigen recognized by T cells-1 (MART-1) [83, 84].

In a pilot trial, T cells were transduced with a gp-100 specific TCR and infused with IL-2 into melanoma patients that had been previously lymphodepleted [85, 86]. The TCR-transduced T cells persisted and no toxicity was detected in the patients, however, the T cells demonstrated minimal effector function due to the low surface levels of the gp-100 TCR [85, 86]. However, when high-affinity MART-1 transduced TCR T cells were expanded and infused into melanoma patients, this resulted in a 30 % clinical response rate [87, 88]. $CD4^+$ T cells have also been transduced to target another melanoma antigen, NY-ESO-1, and have demonstrated objective response rates in 5 out of 11 patients, with 2 complete responses [89].

Antigen-Specific $CD8^+$ and $CD4^+$ T Cells

Using autologous patient PBMC, tumor-associated antigen (TAA)-specific $CD4^+$ and $CD8^+$ T cell clones have been expanded using multiple in vitro antigen stimulations. $CD8^+$ T cells are believed to be the optimal population for ACT due to their specialized ability to recognize and kill tumors via the TCR binding to the peptide:Major Histocompatibility Complex (MHC) class I [90]. Using ACT, various groups have attempted to use antigen-specific $CD8^+$ cells to treat metastatic melanoma patients [90, 91]. A key approach isolates lymphocytes from peripheral blood of patients. $CD8^+$ T cells that recognize melanoma antigens gp100, MART-1 or tyrosinase are expanded using autologous mature dendritic cells that have been pulsed with peptide [91–94]. In a phase I clinical trial, $CD8^+$ T-cell clones that recognized MART-1 and gp100 that were expanded and infused into metastatic melanoma patients were able to respond to IL-2, persist in vivo, traffic to the tumor sites and induce tumor regression [90, 91, 95]. Using this approach resulted in clinical responses of about 30 %, with patients experiencing complete responses, partial responses, or stabilization of the disease for up to 11 months [90, 91, 95]. In another clinical trial conducted by Mackensen and colleagues, a 30 % clinical response was demonstrated when melanoma patients received MART-1-specific T cells that were able to migrate to the tumor sites [91, 94]. Butler et al. conducted studies using K562 artificial antigen presenting cells (aAPCs) to expand MART-1 specific $CD8^+$ T cells [96]. The aAPCs were transduced to express MART-1, HLA-A2, CD86 and 4-1BBL and the MART-1 specific-$CD8^+$ cells were obtained from melanoma patient peripheral blood [96].

Currently, the specific role of antigen-specific $CD4^+$ used to treat metastatic melanoma patients is being investigated [97]. Studies have demonstrated that adoptively transferring tumor reactive $CD4^+$ T helper (Th)17 cells into C57Bl/6 mice resulted in tumor eradication [98, 99]. Although the use of antigen-specific $CD8^+$ T cells have been largely studied, many groups believe that polyclonal $CD4^+$ and $CD8^+$ population used for ACT for melanoma patient treatment may be necessary to induce better clinical responses [93, 100–102]. In addition to the HLA class I-restricted melanoma antigens mentioned earlier, there are also HLA class II-restricted melanoma antigens, such as tyrosinase, NY-ESO-1 and Melanoma Associate antigen-1 (MAGE-1) [92]. A clinical trial using Th1 $CD4^+$ T cells clones that were specific for tyrosinase or NY-ESO-1 antigens to treat melanoma patients demonstrated a persistence of these cells for up to 2 months [92]. In addition, four patients showed a partial response or stabilization of the disease and one patient exhibited complete durable response of more than 3 years [92].

Outstanding Issues with ACT Using TIL: Technical Aspects

Long Process of Isolating TIL for Adoptive T-Cell Therapy

Although ACT using TIL has resulted in a 50 % clinical response rate across various institutes [25, 31, 46, 48], one of the caveats for this therapy is the actual expansion process of the TIL. The process of isolating and expanding TIL from melanoma tumors can be time-consuming and labor intensive. In addition, not all TIL from all patients are able to be isolated and expanded [25, 31, 46, 48]. In order for the TIL to be eligible for the secondary expansion (REP), the pre-REP TIL must reach a minimum of 50×10^6 cells following the initial isolation and expansion period making it worthwhile to continue the process [31, 46]. However, only 65–80 % of patients pre-REP TIL reach this threshold although recently there is some improvement in this area [31, 46]. Although the expansion period induces the growth of billions of TIL, this process may also induce differentiation [31, 46]. An approach to shorten the time expanding the TIL without inducing the differentiation of the TIL is referred to as the 'young' TIL method. This approach uses enzymatic digestion, not the traditional fragment set-up to isolate the TIL from the tumor [48, 103–106]. The TIL are then expanded with high-dose IL-2 over a maximum period of 3–5 weeks. Using the enzymatic digest method to expand the TIL has shown success in generating the minimum number of pre-REP TIL needed to undergo the secondary expansion in as little as 3 weeks, increasing the number of patients eligible for secondary from 50–60 % to 80 % or greater [103–106]. In addition the 'young' TIL protocol does not test for anti-tumor reactivity in the TIL, therefore all patient TIL undergo the secondary expansion despite their anti-tumor reactivity and are infused into lymphodepleted patients, along with high dose IL-2. Besser and colleagues reported the results of their first clinical trial that utilized the 'young' TIL protocol [103, 107]. They reported a clinical response of 50 %, which was comparable to the response rates seen with using the conventional method to isolate and expand the TIL that also selected for tumor-reactive before secondary expansion [103, 107]. Although the 'young' TIL method did not increase the clinical response rates, data suggests that the 'young' TIL are less differentiated and may persist in vivo longer, and overall, this method reduces the initial expansion period making the process more practical.

Irradiated PBMC as Feeder Cells

During the secondary expansion phase of the TIL (REP), irradiated PBMCs from a mixture of 3-6 normal allogeneic donors act as 'feeder cells.' The feeder cells secrete cytokines and are a mix of different cells within the population of normal donor PBMC. However, sometimes we encounter technical issues with the feeder cells. Although we pool normal donor PBMC, and irradiate them prior to culturing these cells with the TIL, the TIL do not always expand. The exact mechanism of what exactly the feeder cells are doing to activate the TIL is still not clear. We presume that the anti-CD3 crosslinks with the Fc portions on the feeder cells and this activates the TIL. However, we do not know for sure if this is occurring. The feeder cells may also be a source for some growth factors, anti-oxidants, and co-stimulatory factors for TIL expansion; although we found that the feeder cells had very little expression of 4-1BB and 4-1BBL [108]. We have not thoroughly dissected the different subsets of cells that are within the feeder population. A group of normal donor PBMC is pooled in order to have a heterogeneous population of feeder cells to ensure adequate TIL stimulatory activity in case one single donor may be suboptimal. It would be interesting to determine what subsets within the PBMC population and/or cytokines initiate the outgrowth of the TIL. In addition, there may be some suppressor factors in the PBMC that may induce poor expansion of the TIL. This needs to be further investigated. Some preliminary unpublished data we have is that the monocytes producing the cytokine TNF-alpha in the PBMC feeders are critical for expanding $CD8^+$ TIL.

An alternative to using irradiated PBMC as feeder cells during the REP consists of utilizing a human K562 erythroleukemia cell line as an artificial antigen presenting cell (aAPC) [109]. This type of aAPC lacks endogenous MHC class I expression and has been engineered to stably express co-stimulatory molecules CD83, CD86, and/or CD137L. The use of aAPCs for the expansion of TIL has been investigated during the initial expansion of the TIL, during the pre-REP phase as well for the rapid expansion (REP) phase recently. The use of aAPC to expand melanoma TIL resulted in a less differentiated TIL state, with increased expression of CD27, CD28, and CD62L. In addition to the aAPCs expressing critical co-stimulatory molecules, they have also been engineered to express secreted or cell surface cytokines, such as IL-7, IL-15, IL-21, and IL-12.

Outstanding Issues with ACT Using TIL: Biological Aspects

Dynamics of T Cell Differentiation

After encountering tumor cells, naïve T cells undergo activation and clonal expansion. Some of those can further differentiate into T central memory (T_{CM}), T effector memory cells (T_{EM}), T effector cells (T_{EFF}) and terminally differentiated effector cells (T_{TDE}) [110, 111]. Several cell surface and intracellular molecules or markers are dynamically changed upon T cell differentiation, and are used to classify the discrete population of T cell subsets as demonstrated in Fig. 7.2. Protein tyrosine phosphatase receptor type C (PTPRC), also known as CD45 antigen, is one of the major surface markers to distinguish human T cell subsets. Naïve T cells highly

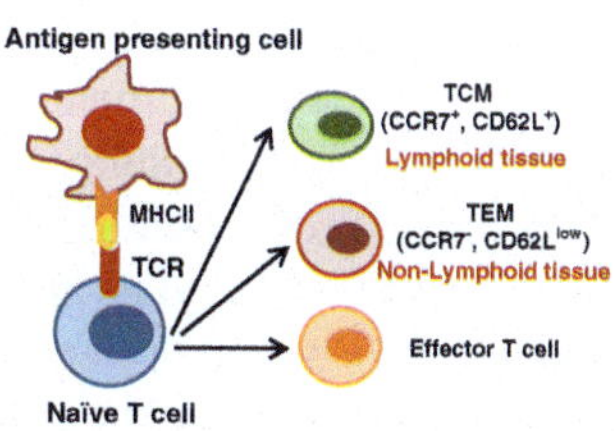

	Marker	Naïve	Central memory	Effector memory	EMRA	Memory stem cell
CD45	CD45RA	High	No	No	Intermediate	High
	CD45RO	No	High	High	High	No
Homing receptor	CCR7	High	Intermediate	No	No	High
	CD62L	High	Low	No	No	High
Co-stimulatory	CD28	High	High	Low	Low to No	High
Co-inhibitory	BTLA	High	Intermediate	Low	Low to No	N/A
Replicative senescence	CD57	No	Low	Intermediate	High	No
Cytotoxicity	Perforin/ Granzyme	Low	Low	Intermediate	High	Low
Inhibitory	KLRG1	No	Low	Intermediate	High	No

Fig. 7.2 T-cell differentiation stages. After encountering antigen, naïve $CD8^+$ T cells differentiate into different subsets including Central Memory (CM), Effector Memory (EM), and Effector Memory RA (EMRA). Several markers used to distinguish different T cell subset are shown in the table. Recently, a new memory T-cell subset with self-renewal capacity and longer persistence similar to central memory cells yet distinct in phenotype called "memory stem cells" has been described

express the full-length form of CD45, so called CD45RA [111–113]. Re-expression of CD45RA has also been demonstrated in terminally differentiated T cells (T_{TDE}) together with down-regulation of CD27 and CD28. Recent findings also showed that the expression of CD45RO, a shorter form if CD45, is highly enriched in antigen experienced T cells, both effector memory and central memory T cells. The chemokine receptor CCR7 (CC-Chemokine receptor 7) and the adhesion molecule CD62L (L-selectin) are also used to delineate central memory from effector memory on the basis of migration and lymphoid tissue homing properties. Central memory T cells (T_{CM}) highly express both CCR7 and CD62L, while Effector memory T cells (T_{EM}) have no expression of CCR7, and slightly express CD62L [110, 111, 114]. After repeated stimulation, human $CD8^+$T cells undergo differentiation and lose the expression of co-stimulatory surface molecules: CD28 and CD27. Therefore, these two markers have also been used to define differentiation stages of T cells into three categories: relatively undifferentiated ($CD28^+CD27^+$), intermediate differentiated ($CD28^-CD27^+$), and terminally differentiated ($CD28^-CD27^-$) [114]. It was reported that terminally differentiated T cell (T_{TDE}) had significantly short telomeres, loss of telomerase activity, and reduction of proliferation [115]. Defects in proliferation and telomere length in highly differentiated T cells has been shown to be associated with impairment of the CD28 signaling pathway, which promote telomerase activity via Akt phosphorylation and activation [116]. Using an in vivo LCMV-specific $CD8^+$ murine T cell model, researchers have demonstrated that sustained Akt signaling resulted in the mouse $CD8^+$ effector CTL to become short-lived effector cells (SLECs) and lose $CD8^+$ T-cell memory by inactivating the FOXO transcription factor known to enhance expression of CD62L, CCR7, and IL-7 genes [117]. Inhibition of Akt using a selective Akt inhibitor, A-443654, in LCMV-infected mice prevented SLECs from terminal differentiation and apoptosis [117]. It remains unclear whether Akt inhibition could rescue T cells from terminal differentiation. The TCR (T cell receptor) strength in a cancer model could potentially be weaker than viral infection, and sustained Akt activation might not exist in the TIL setting. Thus, further study is needed to clarify if inhibiting the Akt pathway would benefit the development of less differentiated TIL.

In addition to the malfunction in CD28 signaling, KLRG1, killer cell lectin-like receptor G1, has been shown to attenuate Akt phosphorylation by the recruitment of phosphatase (SHP-1 and SHP-2) into its immunoreceptor tyrosine-based inhibitory motif (ITIM). Interestingly, KLRG1 blockade can restore cell proliferation, but does not revive the function of telomerase. It might indicate that phosphorylation of human telomerase reverse transcriptase alone is not adequate to restore telomerase activity and telomere shortening [118]. A study in HIV infection has shown that CD57 expression was involved in the clonal exhaustion process [119]. Both $CD4^+$ and $CD8^+$ T cells expressing CD57 exhibited reduced proliferation under exogenous IL-2 and IL-15 treatment. In addition, $CD8^+CD57^+$ T cells are more susceptible to undergo apoptosis upon activation induced cell death (AICD) as compared to $CD8^+CD57^-$ counterparts [120]. Recently, perforin (Perf) and granzyme B (gB), are also considered as markers of T cell differentiation, even though the major function of these two enzymes is known to induce cell death via apoptosis of CTL target cells. Effector and effector memory T cells are the major cell type expressing perf and gB, particularly those positive for CD57 [121]. In contrast, naïve and central memory barely express these cytolytic enzymes. It must be noted that there is no single bona fide marker to define different subsets of T cells due to the dynamic changes of several molecules during T cell differentiation. Thus, the use of a combination of phenotypic characteristics and functional properties such as proliferative capacity, surface makers, cytotoxicity, apoptotic susceptibility, and telomere length are mandated for the delineation of different stages of T cell differentiation. In the past decades, we have been making progress in understanding the biological function and T-cell subsets, but the mechanistic basis involved in T cell differentiation is still largely undefined. A number of interesting questions including whether the master regulator(s) of T cell differentiation really exist, how T cells decide their fates in differentiation after chronic antigen stimulation, what exact signaling pathway(s) contribute to the development of different T cell subsets, and so forth. Unveiling the puzzle of T cell differentiation will help us to generate optimal phenotypes of TIL, which will help improve the clinical outcome of the patients treated by ACT.

Heterogeneity of TIL and Clinical Outcome

The tumor bed contains heterogeneous populations of TIL. Differences in phenotypes and differentiation status of TIL reflects their proliferation, migration, and anti-tumor function. In melanoma tumors, several infiltrating T cell subsets can be found including $CD8^+$T cells, $CD4^+$T cells, Tregs (regulatory T cells), and gamma delta T cells. Correlation between different subsets of T cells in TIL infusion products and clinical response has been investigated, but the conclusions that can be made remain unclear. Some evidence from the National Cancer Institute (NCI) and MD Anderson cancer center suggested the importance of $CD8^+$TIL [46, 122, 123]. Recent data have revealed that total number of T cells in the infused product and a high proportion of $CD8^+$ TIL favorably correlates with positive outcomes of melanoma patients treated by ACT [46]. Although $CD4^+$ TIL are not considered as an important key player in tumor eradication, their cytotoxic function and role in helping $CD8^+$ T cells to promote cancer regression cannot be overlooked. $CD4^+$ T cells have been shown to exhibit cytotoxic function in controlling poxvirus infection via perforin-dependent manner [124]. Using a mouse melanoma model, it was reported that OX40 and 4-1BB pathways are required to program naïve $CD4^+$ into cytotoxic Th1 differentiation, and is critical to maximize $CD4^+$ tumoricidal function [125]. It is still in doubt if the observation in mouse models can recapitulate the phenomenon in human settings. In fact, in our phase II clinical trial, we have noticed that some melanoma patients, whose infusion products contained a major population of $CD4^+$ TIL, somehow achieved clinical response or even complete regression in some certain cases. Possibly, cytotoxic $CD4^+$T cells might mediate cytotoxicity by using granzyme and perforin in contributing tumor regression. Perhaps, $CD4^+$ cells may target melanoma tumor cells expressing MHC class II instead of MHC class I. TIL can contain a subpopulation of gamma-delta T cells. The functional study of gamma delta TIL is still limited, even though they can occasionally expand during the TIL REP. A phase I clinical trial demonstrated that zoledronate-activated Vgamma9Vdelta2 T cells can react against a broad ranges of tumor cell types such as melanoma, ovarian cancer, breast cancer, adenocarcinoma, and colon cancer. Transfusion of expanded gamma-delta T cells was clinically proven to be safe, and these cells were able to migrate into tumor bed due to high expression of chemokine receptors including CCR5, CCR7, CXCR3, and CXCR5 [126]. Clinical outcome was tightly correlated with the number of infused T cells, which was consistent with our observations. The success of adoptive T cell in melanoma therapy relies on a number of parameters, such as the quantity of infusion product and it percentage of $CD8^+$ T cells. However, we should not simply ignore another subset of T cells such as cytotoxic $CD4^+$ and gamma delta T cells, which are occasionally found in in melanoma T cell infusion product. However, this situation may be quite different in TIL therapy for other solid tumor types.

Optimal TIL Therapy

Fine Tuning the Balance Between Anti-Tumor Function Versus Persistency in vivo

So far, it remains inconclusive which exact subset(s) of $CD8^+$ T cell can elicit long-term tumor control. However, recent evidence supports the correlation between the differentiation stage of the T cells and clinical response. It was reported that in vivo persistence of adoptively transferred T cells in peripheral blood was associated with cancer regression. A study at the NCI showed that TIL clones expressing CD28 and CD27 had long telomeres and long-term persistence in peripheral blood, whereas short-term persistent clones up-regulated CD57 and down-regulated CD28 and CD27 [115]. This suggested that "Young TIL" or a relatively undifferentiated phenotype conferred greater ability for survival and persistence in vivo as compared with more differentiated TIL". On the contrary, in our clinical trials at MD Anderson Cancer Center, we have observed that more differentiated $CD8^+CD27^-$ T cells in TIL infusion products were somehow linked with cancer regression and positive clinical response [31]. It is highly possible that more differentiated effector stage TIL, which have more cytotoxic function, could facilitate tumor regression. It should be noted that some of $CD8^+CD27^-$ still retained CD28 expression that might enhance cell survival and persistence in tumors. Although more differentiated T cells seems to be more proficient in cytolytic killing of cancer, they are also susceptible to undergoing apoptosis via the activation-induced cell death (AICD) pathway. In addition, more differentiated T cells are defective in telomerase function, which reduces proliferative capacity and eventually leads to replicative senescence. On the other hand, less differentiated T cells that have longer telomeres and proliferate better upon antigen re-stimulation can persist better after clonal expansion, but the cytotoxic function is less efficient. This fundamental concept suggests that the optimal proportion between "young" versus "old" phenotype in the TIL infusion product is a key variable to achieve both anti-tumor capacity and persistence in vivo, which is required for long-term antitumor response.

Cytokines Involved in T Cell Proliferation and Differentiation

IL-2 is known as a major cytokine to promote the growth and survival of T cells. An ideal concept in T cell expansion for ACT is to achieve the maximal expansion while maintaining a less differentiated phenotype. Through understanding the mechanistic basis of proliferation and differentiation, several efforts have been made to utilize different cytokines to steer T cells to the desired phenotypes. Some examples of cytokine such as IL-7, IL-15, and IL-21, which are known to play the role in memory T cell development, have been extensively investigated for ACT application. IL-7 was shown to enhance survival and cell proliferation by increasing anti-apoptotic protein such as Bcl-2 and MCL1 [127]. It was reported that IL-7 preferentially expanded and polarized antigen specific T cells toward a memory phenotype. Using systemic lymphoma in an immunodeficient mouse model, IL-7 and IL-21 in combination effectively promote tumor eradication [128]. In TIL ex vivo expansion using artificial APC, IL-21 was recently shown to promote greater fold expansion with enrichment of a $CD8^+$ $CD27^+CD28^+$ phenotype and high perforin and granzyme expression [129]. Although TIL expanded with IL-21 have not been utilized for ACT yet, a "young" phenotype with cytotoxic potency is expected to exhibit effective killing function and persist longer in tumor bed. In our experience, we have found that IL-21 at low dose (10 ng/ml) induced higher fold expansion and $CD8^+$, but we did not observe increased $CD28^+$ and $CD27^+$, and in some cases we found that $CD28^+$ was significantly down-regulated. On the other hand, when we used high dose IL-21 (100 ng/ml), TILs were greatly decreased in fold expansion, but had less differentiated phenotype. Although using cytokines to expand TIL seems to be promising, the further clinical studies are necessary to verify the killing function and persistency in vivo setting.

Immunomodulators in TIL Therapy

Anti-PD-1

Combining TIL therapy with blocking antibodies in vivo against negative co-stimulatory molecules, such as PD-1, which can be negatively expressed on TAA-specific T cells, represents a promising approach. B7-H1 or PD-L1, expressed on melanoma cells and other forms of cancer, resulting in tumor-induced immune suppression has been correlated to poor clinical response [130, 131]. However, more recently increased PD-L1 expression in melanoma and other solid tumors has been found to be a biomarker of increased activated T-cell infiltration, with the PD-L1 expression being an "adaptive resistance" response blocking further anti-tumor Tcell activity. A number of FDA-approved monoclonal antibodies blocking PD-1 have been recently approved for metastatic melanoma that overcome T-cell suppression and increase anti-tumor activity and progression-free and overall survival in treated patients. These block PD-1 signaling in TIL in vivo and enhancing T-cell effector function and cell division [132]. With TIL therapy showing a high rate of effectiveness and constitutive expression of PD-1 on many TAA-reactive TILs, especially $CD8^+$ T cells, a combo with of TIL with anti-PD-1 may be a highly synergistic combination to enhance clinical response [133–135]. Other checkpoint molecules, such as Lag3, TIM-3, VISTA, and others may also be targeted using an antibody blocking approach in vivo after TIL infusion. Another approach can be the use of gene editing using zonc finger nucleases, talens, or CRISPRs to eliminate PD-1 expression or the expression of other checkpoint molecules altogether. However, we should be cautious of these approaches as these checkpoint signaling pathways may have hitherto undiscovered positive functions in the anti-tumor response.

Anti-CD137 and Anti-CD134

Augmenting positive co-stimulatory pathways, such as CD134 (OX40) and CD137 (4-1BB) may also be beneficial. Agents can be infused into patients following ACT or can be used in vitro to expand T cells with improved effector and memory phenotypes. In clinical trials using CAR-transduced T cells recognizing CD19 in patients with CLL, inclusion of a CD137 signaling endodomain was found to be critical in maintaining T-cell persistence and anti-tumor cell activity [136, 137]. In our studies, we have found that expanding tumor-infiltrating lymphocytes using agonistic anti-4-1BB antibodies can enhance the proliferation and anti-tumor function of the TIL.

BRAF

Although the combination of total body irradiation (TBI) together with cyclophosphamide and fludarabine as a preconditioning regimen has further increased response rates of TIL therapy, this approach is too toxic and requires stem cell backup each time, precluding its wide range application. One potential alternative is combining TIL therapy with B-RAF inhibitors, vemurafenib. The rationale behind this is that melanoma cells can undergo experience apoptosis associated with inactivation of an oncogene such as mutated V600E B-RAF, they release TAAs, such as tissue differentiation antigens, cancer/testis antigens or products of mutated genes expressed by transformed cells [138–140]. The subsequent activation of antigen-presenting cells (APC) and antigen presentation could further stimulate adoptively transferred TIL. Administration of adoptively transferred TIL in combination with BRAF inhibitor will likely synergize effects of both immunotherapy and oncogene-targeted therapy. The combination of TIL therapy and MEK inhibitors may not be optimal as MEK inhibition may negatively impact T cell function due to ERK inactivation. However, some unpublished recent data suggests that MEK inhibition may not be toxic to effector T cells and may inhibit

AICD in the tumor microenvironment. Targeted therapy for melanoma may not work due to emergence of drug-resistant clones. By inducing long-term, polyclonal T cell response against tumors via antigen spreading induced by targeted therapy, this may significantly increase the durability of the response.

Biomarkers in TIL Therapy

An increasing amount of evidence now shows that the classification of cancer leading to an estimation of clinical outcome should include analysis of the tumor microenvironment, as this takes into consideration the interaction with the host immune response. Differences between response and resistance of melanoma patients to different immunotherapies may therefore be further elucidated by analysis of the immunological status of the tumor microenvironment that, together with other systemic biomarkers, may help predict patient prognosis or response to immunotherapy. Currently, there are multiple approaches to study and measure the effects of the immune system in melanoma.

The Nanostring nCounter analysis system can accurately quantify RNA levels from fresh-frozen and Formalin Fixed Paraffin Embedded (FFPE) samples in small amounts of total RNA for gene expression profiling [141]. This new technology could aid in the identification of potential immunotherapy targets for melanoma [142] or as biomarkers to predict melanoma patient prognosis through the use of different genes panels of interest, such as an immunology panel, an inflammatory panel, or a cancer related gene panel. This could therefore help to identify patients who could benefit from immunotherapy when predictive gene expression profiles or signatures can be found. Messina et al. found that a 12-chemokine (CCL18, CCL19, CCL2, CCL21, CCL3, CCL4, CCL5, CCL8, CXCL10, CXCL11, CXCL13 and CXCL9) gene expression signature (GES) score was associated with better overall survival in metastatic melanoma [143], and Tanese et al. showed the same results. They compared the gene expression profile between iNOS-positive and iNOS-negative tumor samples and found that the upregulation of CXCL10 expression in iNOS-negative groups correlated with the most favorable prognosis [144]. Further, Tanese et al. demonstrated that inducible Nitric Oxide Synthase (iNOS) and Nitrotyrosine (NT) expression in patients with stage III melanoma strongly correlated with poor survival [145]. Epigenetic analysis found that DNA methylation is associated with melanoma progression [146, 147] even though it has been shown that BRAF V600E, KIT and NRAS mutations are the most important catalysts for melanoma development and could be influential in predicting poor outcome in melanoma patients [148].

Patients that have clinical responses to melanoma vaccines and anti-CTLA-4 monoclonal antibody (ipilimumab) therapies have a high ratio of T cell infiltration in their tumors [149, 150]. Steven A. Rosenberg's group at the NCI also found an overall increase (not statistically significant) of the total number of $CD8^+$ T cell infiltration in metastatic melanoma between the responders and the non-responders to TIL therapy [151]. One explanation for this trend could be that the infiltrations of inflammatory and lymphocytic cells were not randomly distributed in the tumor. The Rosenberg group further illustrates the idea of selective accumulation of T cells in tumors through the significant increase of FoxP3+ CD4 Treg cells in the intratumoral areas as compared to peritumoral areas [151]. Galon et al. developed the so-called Immunoscore approach by immunohistochemical technology which takes into account the location, density, and function of different immune cell types. Immunoscore is useful for routine clinical use for the classification of cancer, identification of the prognostic factors for disease-free survival (DFS) and overall survival (OS), and elucidation of the potential targets for immunotherapy [152–154].

Recent studies suggest that the immunosuppressive mechanisms in the tumor microenvironment that inhibit T cell activation [155] may also explain the cause for resistance to tumor immunotherapies. Thomas F Gajewski's group at the University of Chicago found that the inhibitory pathways for T cell activation may result from a negative feedback loop through demonstrating that CD8+ T cell infiltrated metastatic melanoma had higher expression of inhibitory factors including, indoleamine-2,3-dioeygenase (IDO), PD-L1/B7-H1, and FoxP3+ regulatory T cells (Tregs). These factors can be considered as candidates for inhibitors of T cell function at the tumor site [156].

Getting TIL "Past the Finish Line"

TIL "Tri" therapy consisting of a non myeloablative chemotherapy regimen followed by TIL infusion and High Dose IL-2 was reported to induce >50 % clinical response rates for the first time over a decade ago [39]. Since then results have been reproduced in hundreds of patients treated by different groups worldwide [40, 46, 103, 157]. The cumbersome manufacturing process has been an obstacle to bringing this therapy to a phase III clinical trial for licensing. As new technologies become available the process of expanding TIL is becoming easier and more centers are embarking on phase II TIL ACT studies such as centers in Seattle and Los Angeles in the USA and Toronto, Ontario, Canada, as well as a number of centers in Europe. TIL are now grown in more efficient devices, such as G-Rex gas permeable flasks (Wilson Wolf Manufacturing) or bioreactors such as WAVE (GE Healthcare) [158–160]. A first randomized TIL ACT trial was recently published where 69 patients were treated with either bulk TIL product (N=34) or CD8+ T-cell-enriched TIL product (N=35). The trial did not show benefit in enriching for CD8+ TIL but did demonstrate the ability to treat 69 patients within 18 month time frame at one institution [161].

The TIL product is uncharacterized inasmuch as there is no biomarker of potency available yet. Numbers of bulk TIL that currently need to be grown and infused are very large, going up to 1.5×10^{11}, making the industrialization process difficult. The future of TIL therapy lies in better characterization of the TIL product to delineate biomarkers of the most effective TIL population. If cell surface biomarkers are identified, an instrument could pull the desired T cells out to be expanded. Technology to achieve the selection step is available, with the arrival on the market of clinical grade cell sorters. Studies have hinted at the CD8+ T-cell content being critical for response to TIL and indeed infusing highly enriched (>99.6 %) $CD8^+$ TIL had the same clinical response as bulk TIL product (in this study the median CD8 content was 93 %) suggesting that tumor control is effectively performed by the $CD8^+$ T-cell population [46, 103, 161]. However the clinical response rate could not be improved by selective infusion of $CD8^+$ TIL arguing against adding a $CD8^+$ selection step to the already complex TIL manufacturing. Longer follow-up time will be needed to determine the durability of response induced by infusion of selected $CD8^+$ TIL. Moreover, the recent publication of a $CD4^+$ tumor-reactive TIL clone causing tumor regression in vivo puts $CD4^+$ TIL back at the forefront of tumor control and invalidates the use of CD8 or CD4 to positively or negatively select a TIL subpopulation for infusion [162]. This first attempt to select a TIL subpopulation, $CD8^+$, did not improve potency. More research is needed to elucidate which TIL subpopulation is beneficial or harmful to the response. Our own studies have pointed to some novel biomarkers and TIL subsets, such as those expressing B and T Lymphocyte Attenuator (BTLA) being correlated with response to TIL therapy. Research is underway to determine the functional properties of $CD8^+BTLA^+$ TIL subpopulation and its value as biomarker.

Selection approaches based on functionality of TIL are being developed at the moment. Strategies aiming at selecting tumor reactive TIL by their expression of molecules typically associated with antigen specific T cell activation such as 4-1BB (short term activation) or PD-1 (chronic antigen exposure) immediately after tumor dissociation and TIL isolation are being explored. Pre-clinical studies have demonstrated that selecting TIL based on either 4-1BB or PD-1 enriches for a more tumor-specific TIL population although admittedly very few cells can be obtained at this early stage [163, 164]. Rapidly expanding those cells immediately after selection will shorten the manufacturing to just over 2 weeks (from 5 to 7 weeks as it stands now) but produce fewer cells for infusion. Conceivably a selected tumor-specific product could have improved potency and fewer cells might be needed for tumor control.

The TIL population of interest can be sorted out or alternatively the desired population can be selectively stimulated in vitro through unique markers to favor their specific expansion. Our group and others have applied this concept and added anti-41BB in early TIL cultures or during the Rapid Expansion Protocol to favor the growth of tumor antigen-specific TIL [165, 166]. Although this strategy does not appreciably shorten the manufacturing time, it increases the proportion of tumor reactive $CD8^+$ TIL in the final product, and therefore has the potential of increasing the potency of the TIL product.

The activity of tumor-specific TIL could likely be potentiated by combining TIL therapy with the systemic administration of immunomodulatory agents belonging to the checkpoint blockade category such as anti-CTLA4 or anti-PD-1. These agents “release the breaks on T cells” and could directly help maintain the function of activated TIL post antigen exposure in vivo. Pre-clinical mouse models have shown that combination with checkpoint blockade could enhance the therapeutic activity of TIL [133, 167]. Clinical trials exploring the combination of TIL and checkpoint blockade agents are at the planning stages.

Another drawback of TIL therapy is the toxicity of the High Dose IL-2 treatment restricting the application to patients with good performance status. The need for IL-2 to support TIL expansion and persistence in vivo has been demonstrated but the dose of IL-2 necessary to achieve this goal is not known. The use of lower doses of IL-2 administered subcutaneously every day for 14 days post TIL infusion has been tested in a pilot study of six patients and encouraging clinical responses were reported with two patients undergoing durable complete responses and no patient experiencing grade 3 or 4 toxicity [168]. Another study had also found clinical benefit of combining TIL with low dose IL-2 [169]. Studies treating larger patient cohorts will be necessary to address the potential of low dose IL-2 as supportive cytokine regimen to foster TIL growth and long term persistence in vivo. Lastly, IL-2 analogs able to support T cell growth while having a much better safety profile are in development and may greatly help disseminate TIL therapy to a broader patient population.

References

1. Schiavoni G, Mattei F, Gabriele L (2013) Type I interferons as stimulators of DC-mediated cross-priming: impact on anti-tumor response. Front Immunol 4:483
2. Strauss L, Bergmann C, Szczepanski M, Gooding W, Johnson JT, Whiteside TL (2007) A unique subset of CD4+CD25highFoxp3+ T cells secreting interleukin-10 and transforming growth factor-beta1 mediates suppression in the tumor microenvironment. Clin Cancer Res 13:4345–4354
3. Cao X, Cai SF, Fehniger TA, Song J, Collins LI, Piwnica-Worms DR, Ley TJ (2007) Granzyme B and perforin are important for regulatory T cell-mediated suppression of tumor clearance. Immunity 27:635–646
4. Kobie JJ, Shah PR, Yang L, Rebhahn JA, Fowell DJ, Mosmann TR (2006) T regulatory and primed uncommitted CD4 T cells express CD73, which suppresses effector CD4 T cells by converting 5′-adenosine monophosphate to adenosine. J Immunol 177:6780–6786
5. Borsellino G, Kleinewietfeld M, Di Mitri D, Sternjak A, Diamantini A, Giometto R, Hopner S, Centonze D, Bernardi G, Dell'Acqua ML et al (2007) Expression of ectonucleotidase CD39 by Foxp3+ Treg cells: hydrolysis of extracellular ATP and immune suppression. Blood 110:1225–1232
6. Yu J, Du W, Yan F, Wang Y, Li H, Cao S, Yu W, Shen C, Liu J, Ren X (2013) Myeloid-derived suppressor cells suppress antitumor immune responses through IDO expression and correlate with lymph node metastasis in patients with breast cancer. J Immunol 190:3783–3797
7. Corzo CA, Cotter MJ, Cheng P, Cheng F, Kusmartsev S, Sotomayor E, Padhya T, McCaffrey TV, McCaffrey JC, Gabrilovich DI (2009) Mechanism regulating reactive oxygen species in tumor-induced myeloid-derived suppressor cells. J Immunol 182:5693–5701
8. Srivastava MK, Sinha P, Clements VK, Rodriguez P, Ostrand-Rosenberg S (2010) Myeloid-derived suppressor cells inhibit T-cell activation by depleting cystine and cysteine. Cancer Res 70:68–77
9. Pan PY, Ma G, Weber KJ, Ozao-Choy J, Wang G, Yin B, Divino CM, Chen SH (2010) Immune stimulatory receptor CD40 is required for T-cell suppression and T regulatory cell activation mediated by myeloid-derived suppressor cells in cancer. Cancer Res 70:99–108
10. Coley WB (1991) The treatment of malignant tumors by repeated inoculations of erysipelas: with a report of ten original cases. 1983. Am J Med Sci (262):3–11
11. Coley WB (1896) Further observations upon the treatment of malignant tumors with the toxins of erysipelas and Bacillus prodigiosus with a report of 160 cases. Bull Johns Hopkins Hosp 65:157–162

12. Coley WB (1906) Late results of the treatment of inoperable sarcoma by the mixed toxins of erysipelas and Bacillus prodigious. Am J Med Sci 131:375–430.
13. Nauts HC, Fowler GA, Bogatko FH (1953) A review of the influence of bacterial infection and of bacterial products (Coley's toxins) on malignant tumors in man; a critical analysis of 30 inoperable cases treated by Coley's mixed toxins, in which diagnosis was confirmed by microscopic examination. Acta Med Scand Suppl 276:1–103
14. Hoffman RM (2012) The preclinical discovery of bacterial therapy for the treatment of metastatic cancer with unique advantages. Expert Opin Drug Discov 7:73–83
15. Silverstein AM (2001) The lymphocyte in immunology: from James B. Murphy to James L. Gowans. Nat Immunol 2:569–571
16. Miller JF (1999) Discovering the origins of immunological competence. Annu Rev Immunol 17:1–17
17. Baldwin RW (1966) Tumour-specific immunity against spontaneous rat tumours. Int J Cancer 1:257–264
18. Fass L, Fefer A (1972) Factors related to therapeutic efficacy in adoptive chemoimmunotherapy of a Friend virus-induced lymphoma. Cancer Res 32:2427–2431
19. Rosenberg SA, Restifo NP, Yang JC, Morgan RA, Dudley ME (2008) Adoptive cell transfer: a clinical path to effective cancer immunotherapy. Nat Rev Cancer 8:299–308
20. Eberlein TJ, Rosenstein M, Spiess P, Wesley R, Rosenberg SA (1982) Adoptive chemoimmunotherapy of a syngeneic murine lymphoma with long-term lymphoid cell lines expanded in T cell growth factor. Cancer Immunol Immunother 13:5–13
21. Kradin RL, Kurnick JT (1986) Adoptive immunotherapy of cancer with activated lymphocytes and interleukin-2. Pathol Immunopathol Res 5:193–202
22. Kradin RL, Kurnick JT, Lazarus DS, Preffer FI, Dubinett SM, Pinto CE, Gifford J, Davidson E, Grove B, Callahan RJ et al (1989) Tumour-infiltrating lymphocytes and interleukin-2 in treatment of advanced cancer. Lancet 1:577–580
23. Oliver RT (1988) The clinical potential of interleukin-2. Br J Cancer 58:405–409
24. Gaugler B, Van den Eynde B, van der Bruggen P, Romero P, Gaforio JJ, De Plaen E, Lethe B, Brasseur F, Boon T (1994) Human gene MAGE-3 codes for an antigen recognized on a melanoma by autologous cytolytic T lymphocytes. J Exp Med 179:921–930
25. Dudley ME (2011) Adoptive cell therapy for patients with melanoma. J Cancer 2:360–362
26. Rosenberg SA, Dudley ME (2009) Adoptive cell therapy for the treatment of patients with metastatic melanoma. Curr Opin Immunol 21:233–240
27. Rosenberg SA, Yang JC, Restifo NP (2004) Cancer immunotherapy: moving beyond current vaccines. Nat Med 10:909–915
28. Atkins MB, Kunkel L, Sznol M, Rosenberg SA (2000) High-dose recombinant interleukin-2 therapy in patients with metastatic melanoma: long-term survival update. Cancer J Sci Am 6(Suppl 1):S11–S14
29. Atkins MB, Lotze MT, Dutcher JP, Fisher RI, Weiss G, Margolin K, Abrams J, Sznol M, Parkinson D, Hawkins M et al (1999) High-dose recombinant interleukin 2 therapy for patients with metastatic melanoma: analysis of 270 patients treated between 1985 and 1993. J Clin Oncol 17:2105–2116
30. Rosenberg SA, Eberlein TJ, Grimm EA, Lotze MT, Mazumder A, Rosenstein M (1982) Development of long-term cell lines and lymphoid clones reactive against murine and human tumors: a new approach to the adoptive immunotherapy of cancer. Surgery 92:328–336
31. Wu R, Forget MA, Chacon J, Bernatchez C, Haymaker C, Chen JQ, Hwu P, Radvanyi LG (2012) Adoptive T-cell therapy using autologous tumor-infiltrating lymphocytes for metastatic melanoma: current status and future outlook. Cancer J 18:160–175
32. Grimm EA, Mazumder A, Zhang HZ, Rosenberg SA (1982) Lymphokine-activated killer cell phenomenon. Lysis of natural killer-resistant fresh solid tumor cells by interleukin 2-activated autologous human peripheral blood lymphocytes. J Exp Med 155:1823–1841
33. Mazumder A, Grimm EA, Zhang HZ, Rosenberg SA (1982) Lysis of fresh human solid tumors by autologous lymphocytes activated in vitro with lectins. Cancer Res 42:913–918

34. Koretz MJ, Lawson DH, York RM, Graham SD, Murray DR, Gillespie TM, Levitt D, Sell KM (1991) Randomized study of interleukin 2 (IL-2) alone vs IL-2 plus lymphokine-activated killer cells for treatment of melanoma and renal cell cancer. Arch Surg 126:898–903
35. Rosenberg SA, Lotze MT, Yang JC, Topalian SL, Chang AE, Schwartzentruber DJ, Aebersold P, Leitman S, Linehan WM, Seipp CA et al (1993) Prospective randomized trial of high-dose interleukin-2 alone or in conjunction with lymphokine-activated killer cells for the treatment of patients with advanced cancer. J Natl Cancer Inst 85:622–632
36. Atkins MB, Mier JW, Parkinson DR, Gould JA, Berkman EM, Kaplan MM (1988) Hypothyroidism after treatment with interleukin-2 and lymphokine-activated killer cells. N Engl J Med 318:1557–1563
37. Rosenberg SA, Packard BS, Aebersold PM, Solomon D, Topalian SL, Toy ST, Simon P, Lotze MT, Yang JC, Seipp CA et al (1988) Use of tumor-infiltrating lymphocytes and interleukin-2 in the immunotherapy of patients with metastatic melanoma. A preliminary report. N Engl J Med 319:1676–1680
38. Rosenberg SA, Spiess P, Lafreniere R (1986) A new approach to the adoptive immunotherapy of cancer with tumor-infiltrating lymphocytes. Science 233:1318–1321
39. Dudley ME, Wunderlich JR, Robbins PF, Yang JC, Hwu P, Schwartzentruber DJ, Topalian SL, Sherry R, Restifo NP, Hubicki AM et al (2002) Cancer regression and autoimmunity in patients after clonal repopulation with antitumor lymphocytes. Science 298:850–854
40. Dudley ME, Wunderlich JR, Yang JC, Sherry RM, Topalian SL, Restifo NP, Royal RE, Kammula U, White DE, Mavroukakis SA et al (2005) Adoptive cell transfer therapy following non-myeloablative but lymphodepleting chemotherapy for the treatment of patients with refractory metastatic melanoma. J Clin Oncol 23:2346–2357
41. Klebanoff CA, Khong HT, Antony PA, Palmer DC, Restifo NP (2005) Sinks, suppressors and antigen presenters: how lymphodepletion enhances T cell-mediated tumor immunotherapy. Trends Immunol 26:111–117
42. Wang LX, Shu S, Plautz GE (2005) Host lymphodepletion augments T cell adoptive immunotherapy through enhanced intratumoral proliferation of effector cells. Cancer Res 65:9547–9554
43. Poehlein CH, Haley DP, Walker EB, Fox BA (2009) Depletion of tumor-induced Treg prior to reconstitution rescues enhanced priming of tumor-specific, therapeutic effector T cells in lymphopenic hosts. Eur J Immunol 39:3121–3133
44. Turk MJ, Guevara-Patino JA, Rizzuto GA, Engelhorn ME, Sakaguchi S, Houghton AN (2004) Concomitant tumor immunity to a poorly immunogenic melanoma is prevented by regulatory T cells. J Exp Med 200:771–782
45. Dudley ME, Yang JC, Sherry R, Hughes MS, Royal R, Kammula U, Robbins PF, Huang J, Citrin DE, Leitman SF et al (2008) Adoptive cell therapy for patients with metastatic melanoma: evaluation of intensive myeloablative chemoradiation preparative regimens. J Clin Oncol 26:5233–5239
46. Radvanyi LG, Bernatchez C, Zhang M, Fox PS, Miller P, Chacon J, Wu R, Lizee G, Mahoney S, Alvarado G et al (2012) Specific lymphocyte subsets predict response to adoptive cell therapy using expanded autologous tumor-infiltrating lymphocytes in metastatic melanoma patients. Clin Cancer Res 18:6758–6770
47. Dudley ME, Wunderlich JR, Shelton TE, Even J, Rosenberg SA (2003) Generation of tumor-infiltrating lymphocyte cultures for use in adoptive transfer therapy for melanoma patients. J Immunother 26:332–342
48. Itzhaki O, Levy D, Zikich D, Treves AJ, Markel G, Schachter J, Besser MJ (2013) Adoptive T-cell transfer in melanoma. Immunotherapy 5:79–90
49. Ruffini E, Asioli S, Filosso PL, Lyberis P, Bruna MC, Macri L, Daniele L, Oliaro A (2009) Clinical significance of tumor-infiltrating lymphocytes in lung neoplasms. Ann Thorac Surg 87:365–371, discussion 371–362
50. Horne ZD, Jack R, Gray ZT, Siegfried JM, Wilson DO, Yousem SA, Nason KS, Landreneau RJ, Luketich JD, Schuchert MJ (2011) Increased levels of tumor-infiltrating lymphocytes are associated with improved recurrence-free survival in stage 1A non-small-cell lung cancer. J Surg Res 171:1–5

51. Kilic A, Landreneau RJ, Luketich JD, Pennathur A, Schuchert MJ (2011) Density of tumor-infiltrating lymphocytes correlates with disease recurrence and survival in patients with large non-small-cell lung cancer tumors. J Surg Res 167:207–210
52. Kopecky O, Lukesova S, Vroblova V, Vokurkova D, Moravek P, Safranek H, Hlavkova D, Soucek P (2007) Phenotype analysis of tumour-infiltrating lymphocytes and lymphocytes in peripheral blood in patients with renal carcinoma. Acta Medica (Hradec Kralove) 50:207–212
53. Wu MY, Kuo TY, Ho HN (2011) Tumor-infiltrating lymphocytes contain a higher proportion of FOXP3(+) T lymphocytes in cervical cancer. J Formos Med Assoc 110:580–586
54. Hagemann AR, Hagemann IS, Cadungog M, Hwang WT, Patel P, Lal P, Hammond R, Gimotty PA, Chu CS, Rubin SC et al (2011) Tissue-based immune monitoring II: multiple tumor sites reveal immunologic homogeneity in serous ovarian carcinoma. Cancer Biol Ther 12:367–377
55. Kooi S, Zhang HZ, Patenia R, Edwards CL, Platsoucas CD, Freedman RS (1996) HLA class I expression on human ovarian carcinoma cells correlates with T-cell infiltration in vivo and T-cell expansion in vitro in low concentrations of recombinant interleukin-2. Cell Immunol 174:116–128
56. Webb JR, Milne K, Watson P, Deleeuw RJ, Nelson BH (2014) Tumor-infiltrating lymphocytes expressing the tissue resident memory marker CD103 are associated with increased survival in high-grade serous ovarian cancer. Clin Cancer Res 20:434–444
57. Milne K, Alexander C, Webb JR, Sun W, Dillon K, Kalloger SE, Gilks CB, Clarke B, Kobel M, Nelson BH (2012) Absolute lymphocyte count is associated with survival in ovarian cancer independent of tumor-infiltrating lymphocytes. J Transl Med 10:33
58. Dirican A, Ekinci N, Avci A, Akyol M, Alacacioglu A, Kucukzeybek Y, Somali I, Erten C, Demir L, Can A et al (2013) The effects of hematological parameters and tumor-infiltrating lymphocytes on prognosis in patients with gastric cancer. Cancer Biomark 13:11–20
59. Feichtenbeiner A, Haas M, Buttner M, Grabenbauer GG, Fietkau R, Distel LV (2014) Critical role of spatial interaction between CD8(+) and Foxp3(+) cells in human gastric cancer: the distance matters. Cancer Immunol Immunother 63:111–119
60. Shen Z, Zhou S, Wang Y, Li RL, Zhong C, Liang C, Sun Y (2010) Higher intratumoral infiltrated Foxp3+ Treg numbers and Foxp3+/CD8+ ratio are associated with adverse prognosis in resectable gastric cancer. J Cancer Res Clin Oncol 136:1585–1595
61. Balermpas P, Michel Y, Wagenblast J, Seitz O, Weiss C, Rodel F, Rodel C, Fokas E (2014) Tumour-infiltrating lymphocytes predict response to definitive chemoradiotherapy in head and neck cancer. Br J Cancer 110:501–509
62. Junker N, Andersen MH, Wenandy L, Dombernowsky SL, Kiss K, Sorensen CH, Therkildsen MH, Von Buchwald C, Andersen E, Straten PT et al (2011) Bimodal ex vivo expansion of T cells from patients with head and neck squamous cell carcinoma: a prerequisite for adoptive cell transfer. Cytotherapy 13:822–834
63. Junker N, Kvistborg P, Kollgaard T, Straten P, Andersen MH, Svane IM (2012) Tumor associated antigen specific T-cell populations identified in ex vivo expanded TIL cultures. Cell Immunol 273:1–9
64. Tsang JY, Ni YB, Chan SK, Shao MM, Kwok YK, Chan KW, Tan PH, Tse GM (2013) CX3CL1 expression is associated with poor outcome in breast cancer patients. Breast Cancer Res Treat 140:495–504
65. West NR, Kost SE, Martin SD, Milne K, Deleeuw RJ, Nelson BH, Watson PH (2013) Tumour-infiltrating FOXP3(+) lymphocytes are associated with cytotoxic immune responses and good clinical outcome in oestrogen receptor-negative breast cancer. Br J Cancer 108:155–162
66. Seo AN, Lee HJ, Kim EJ, Kim HJ, Jang MH, Lee HE, Kim YJ, Kim JH, Park SY (2013) Tumour-infiltrating CD8+ lymphocytes as an independent predictive factor for pathological complete response to primary systemic therapy in breast cancer. Br J Cancer 109:2705–2713
67. Hudson JM, Castilleja A, Murray JL, Honda T, Kudelka A, Singletary E, Wharton JT, Ioannides CG (1998) Growth and antigen recognition of tumor-infiltrating lymphocytes from human breast cancer. J Interferon Cytokine Res 18:529–536

68. Deschodt-Lanckman M, Robberecht P, De Neef P, Labrie F, Christophe J (1975) In vitro interactions of gastrointestinal hormones on cyclic adenosine 3′:5′-monophosphate levels and amylase output in the rat pancreas. Gastroenterology 68:318–325
69. Whitford P, Mallon EA, George WD, Campbell AM (1990) Flow cytometric analysis of tumour infiltrating lymphocytes in breast cancer. Br J Cancer 62:971–975
70. Loi S, Sirtaine N, Piette F, Salgado R, Viale G, Van Eenoo F, Rouas G, Francis P, Crown JP, Hitre E et al (2013) Prognostic and predictive value of tumor-infiltrating lymphocytes in a phase III randomized adjuvant breast cancer trial in node-positive breast cancer comparing the addition of docetaxel to doxorubicin with doxorubicin-based chemotherapy: BIG 02-98. J Clin Oncol 31:860–867
71. Coventry BJ, Weeks SC, Heckford SE, Sykes PJ, Bradley J, Skinner JM (1996) Lack of IL-2 cytokine expression despite Il-2 messenger RNA transcription in tumor-infiltrating lymphocytes in primary human breast carcinoma: selective expression of early activation markers. J Immunol 156:3486–3492
72. Brocker T, Karjalainen K (1995) Signals through T cell receptor-zeta chain alone are insufficient to prime resting T lymphocytes. J Exp Med 181:1653–1659
73. Ramos CA, Dotti G (2011) Chimeric antigen receptor (CAR)-engineered lymphocytes for cancer therapy. Expert Opin Biol Ther 11:855–873
74. Hombach A, Wieczarkowiecz A, Marquardt T, Heuser C, Usai L, Pohl C, Seliger B, Abken H (2001) Tumor-specific T cell activation by recombinant immunoreceptors: CD3 zeta signaling and CD28 costimulation are simultaneously required for efficient IL-2 secretion and can be integrated into one combined CD28/CD3 zeta signaling receptor molecule. J Immunol 167:6123–6131
75. Maher J, Brentjens RJ, Gunset G, Riviere I, Sadelain M (2002) Human T-lymphocyte cytotoxicity and proliferation directed by a single chimeric TCRzeta/CD28 receptor. Nat Biotechnol 20:70–75
76. Finney HM, Akbar AN, Lawson AD (2004) Activation of resting human primary T cells with chimeric receptors: costimulation from CD28, inducible costimulator, CD134, and CD137 in series with signals from the TCR zeta chain. J Immunol 172:104–113
77. Kalos M, Levine BL, Porter DL, Katz S, Grupp SA, Bagg A, June CH (2011) T cells with chimeric antigen receptors have potent antitumor effects and can establish memory in patients with advanced leukemia. Sci Transl Med 3:95ra73
78. Savoldo B, Ramos CA, Liu E, Mims MP, Keating MJ, Carrum G, Kamble RT, Bollard CM, Gee AP, Mei Z et al (2011) CD28 costimulation improves expansion and persistence of chimeric antigen receptor-modified T cells in lymphoma patients. J Clin Invest 121:1822–1826
79. Kochenderfer JN, Rosenberg SA (2013) Treating B-cell cancer with T cells expressing anti-CD19 chimeric antigen receptors. Nat Rev Clin Oncol 10:267–276
80. Erfurt C, Muller E, Emmerling S, Klotz C, Hertl M, Schuler G, Schultz ES (2009) Melanoma-associated chondroitin sulphate proteoglycan as a new target antigen for CD4+ T cells in melanoma patients. Int J Cancer 124:2341–2346
81. Lo AS, Ma Q, Liu DL, Junghans RP (2010) Anti-GD3 chimeric sFv-CD28/T-cell receptor zeta designer T cells for treatment of metastatic melanoma and other neuroectodermal tumors. Clin Cancer Res 16:2769–2780
82. Yvon E, Del Vecchio M, Savoldo B, Hoyos V, Dutour A, Anichini A, Dotti G, Brenner MK (2009) Immunotherapy of metastatic melanoma using genetically engineered GD2-specific T cells. Clin Cancer Res 15:5852–5860
83. Adamina M (2010) When gene therapy meets adoptive cell therapy: better days ahead for cancer immunotherapy? Expert Rev Vaccines 9:359–363
84. Clay TM, Custer MC, Sachs J, Hwu P, Rosenberg SA, Nishimura MI (1999) Efficient transfer of a tumor antigen-reactive TCR to human peripheral blood lymphocytes confers anti-tumor reactivity. J Immunol 163:507–513
85. Morgan RA, Dudley ME, Rosenberg SA (2010) Adoptive cell therapy: genetic modification to redirect effector cell specificity. Cancer J 16:336–341

86. Morgan RA, Dudley ME, Wunderlich JR, Hughes MS, Yang JC, Sherry RM, Royal RE, Topalian SL, Kammula US, Restifo NP et al (2006) Cancer regression in patients after transfer of genetically engineered lymphocytes. Science 314:126–129
87. Jorritsma A, Gomez-Eerland R, Dokter M, van de Kasteele W, Zoet YM, Doxiadis II, Rufer N, Romero P, Morgan RA, Schumacher TN et al (2007) Selecting highly affine and well-expressed TCRs for gene therapy of melanoma. Blood 110:3564–3572
88. Johnson LA, Morgan RA, Dudley ME, Cassard L, Yang JC, Hughes MS, Kammula US, Royal RE, Sherry RM, Wunderlich JR et al (2009) Gene therapy with human and mouse T-cell receptors mediates cancer regression and targets normal tissues expressing cognate antigen. Blood 114:535–546
89. Robbins PF, Morgan RA, Feldman SA, Yang JC, Sherry RM, Dudley ME, Wunderlich JR, Nahvi AV, Helman LJ, Mackall CL et al (2011) Tumor regression in patients with metastatic synovial cell sarcoma and melanoma using genetically engineered lymphocytes reactive with NY-ESO-1. J Clin Oncol 29:917–924
90. Prieto PA, Durflinger KH, Wunderlich JR, Rosenberg SA, Dudley ME (2010) Enrichment of CD8+ cells from melanoma tumor-infiltrating lymphocyte cultures reveals tumor reactivity for use in adoptive cell therapy. J Immunother 33:547–556
91. Mackensen A, Meidenbauer N, Vogl S, Laumer M, Berger J, Andreesen R (2006) Phase I study of adoptive T-cell therapy using antigen-specific CD8+ T cells for the treatment of patients with metastatic melanoma. J Clin Oncol 24:5060–5069
92. Hunder NN, Wallen H, Cao J, Hendricks DW, Reilly JZ, Rodmyre R, Jungbluth A, Gnjatic S, Thompson JA, Yee C (2008) Treatment of metastatic melanoma with autologous CD4+ T cells against NY-ESO-1. N Engl J Med 358:2698–2703
93. Verdegaal EM, Visser M, Ramwadhdoebe TH, van der Minne CE, van Steijn JA, Kapiteijn E, Haanen JB, van der Burg SH, Nortier JW, Osanto S (2011) Successful treatment of metastatic melanoma by adoptive transfer of blood-derived polyclonal tumor-specific CD4+ and CD8+ T cells in combination with low-dose interferon-alpha. Cancer Immunol Immunother 60:953–963
94. Meidenbauer N, Marienhagen J, Laumer M, Vogl S, Heymann J, Andreesen R, Mackensen A (2003) Survival and tumor localization of adoptively transferred Melan-A-specific T cells in melanoma patients. J Immunol 170:2161–2169
95. van Oijen M, Bins A, Elias S, Sein J, Weder P, de Gast G, Mallo H, Gallee M, Van Tinteren H, Schumacher T et al (2004) On the role of melanoma-specific CD8+ T-cell immunity in disease progression of advanced-stage melanoma patients. Clin Cancer Res 10:4754–4760
96. Butler MO, Lee JS, Ansen S, Neuberg D, Hodi FS, Murray AP, Drury L, Berezovskaya A, Mulligan RC, Nadler LM et al (2007) Long-lived antitumor CD8+ lymphocytes for adoptive therapy generated using an artificial antigen-presenting cell. Clin Cancer Res 13:1857–1867
97. Friedman KM, Prieto PA, Devillier LE, Gross CA, Yang JC, Wunderlich JR, Rosenberg SA, Dudley ME (2012) Tumor-specific CD4+ melanoma tumor-infiltrating lymphocytes. J Immunother 35:400–408
98. Martin-Orozco N, Muranski P, Chung Y, Yang XO, Yamazaki T, Lu S, Hwu P, Restifo NP, Overwijk WW, Dong C (2009) T helper 17 cells promote cytotoxic T cell activation in tumor immunity. Immunity 31:787–798
99. Muranski P, Boni A, Antony PA, Cassard L, Irvine KR, Kaiser A, Paulos CM, Palmer DC, Touloukian CE, Ptak K et al (2008) Tumor-specific Th17-polarized cells eradicate large established melanoma. Blood 112:362–373
100. Evans R, Duffy TM, Kamdar SJ (1991) Differential in situ expansion and gene expression of CD4+ and CD8+ tumor-infiltrating lymphocytes following adoptive immunotherapy in a murine tumor model system. Eur J Immunol 21:1815–1819
101. Wong JT, Pinto CE, Gifford JD, Kurnick JT, Kradin RL (1989) Characterization of the CD4+ and CD8+ tumor infiltrating lymphocytes propagated with bispecific monoclonal antibodies. J Immunol 143:3404–3411

102. Xiang J, Huang H, Liu Y (2005) A new dynamic model of CD8+ T effector cell responses via CD4+ T helper-antigen-presenting cells. J Immunol 174:7497–7505
103. Besser MJ, Shapira-Frommer R, Treves AJ, Zippel D, Itzhaki O, Hershkovitz L, Levy D, Kubi A, Hovav E, Chermoshniuk N et al (2010) Clinical responses in a phase II study using adoptive transfer of short-term cultured tumor infiltration lymphocytes in metastatic melanoma patients. Clin Cancer Res 16:2646–2655
104. Besser MJ, Shapira-Frommer R, Treves AJ, Zippel D, Itzhaki O, Schallmach E, Kubi A, Shalmon B, Hardan I, Catane R et al (2009) Minimally cultured or selected autologous tumor-infiltrating lymphocytes after a lympho-depleting chemotherapy regimen in metastatic melanoma patients. J Immunother 32:415–423
105. Dudley ME, Gross CA, Langhan MM, Garcia MR, Sherry RM, Yang JC, Phan GQ, Kammula US, Hughes MS, Citrin DE et al (2010) CD8+ enriched "young" tumor infiltrating lymphocytes can mediate regression of metastatic melanoma. Clin Cancer Res 16:6122–6131
106. Tran KQ, Zhou J, Durflinger KH, Langhan MM, Shelton TE, Wunderlich JR, Robbins PF, Rosenberg SA, Dudley ME (2008) Minimally cultured tumor-infiltrating lymphocytes display optimal characteristics for adoptive cell therapy. J Immunother 31:742–751
107. Besser MJ, Schallmach E, Oved K, Treves AJ, Markel G, Reiter Y, Schachter J (2009) Modifying interleukin-2 concentrations during culture improves function of T cells for adoptive immunotherapy. Cytotherapy 11:206–217
108. Hernandez-Chacon JA, Li Y, Wu RC, Bernatchez C, Wang Y, Weber JS, Hwu P, Radvanyi LG (2011) Costimulation through the CD137/4-1BB pathway protects human melanoma tumor-infiltrating lymphocytes from activation-induced cell death and enhances antitumor effector function. J Immunother 34:236–250
109. Suhoski MM, Golovina TN, Aqui NA, Tai VC, Varela-Rohena A, Milone MC, Carroll RG, Riley JL, June CH (2007) Engineering artificial antigen-presenting cells to express a diverse array of co-stimulatory molecules. Mol Ther 15:981–988
110. Ahmed R, Bevan MJ, Reiner SL, Fearon DT (2009) The precursors of memory: models and controversies. Nat Rev Immunol 9:662–668
111. Sallusto F, Geginat J, Lanzavecchia A (2004) Central memory and effector memory T cell subsets: function, generation, and maintenance. Annu Rev Immunol 22:745–763
112. Merkenschlager M, Terry L, Edwards R, Beverley PC (1988) Limiting dilution analysis of proliferative responses in human lymphocyte populations defined by the monoclonal antibody UCHL1: implications for differential CD45 expression in T cell memory formation. Eur J Immunol 18:1653–1661
113. Gattinoni L, Lugli E, Ji Y, Pos Z, Paulos CM, Quigley MF, Almeida JR, Gostick E, Yu Z, Carpenito C et al (2011) A human memory T cell subset with stem cell-like properties. Nat Med 17:1290–1297
114. Powell DJ Jr, Dudley ME, Robbins PF, Rosenberg SA (2005) Transition of late-stage effector T cells to CD27+ CD28+ tumor-reactive effector memory T cells in humans after adoptive cell transfer therapy. Blood 105:241–250
115. Zhou J, Shen X, Huang J, Hodes RJ, Rosenberg SA, Robbins PF (2005) Telomere length of transferred lymphocytes correlates with in vivo persistence and tumor regression in melanoma patients receiving cell transfer therapy. J Immunol 175:7046–7052
116. Plunkett FJ, Franzese O, Finney HM, Fletcher JM, Belaramani LL, Salmon M, Dokal I, Webster D, Lawson AD, Akbar AN (2007) The loss of telomerase activity in highly differentiated CD8+CD28-CD27- T cells is associated with decreased Akt (Ser473) phosphorylation. J Immunol 178:7710–7719
117. Kim EH, Sullivan JA, Plisch EH, Tejera MM, Jatzek A, Choi KY, Suresh M (2012) Signal integration by Akt regulates CD8 T cell effector and memory differentiation. J Immunol 188:4305–4314
118. Henson SM, Franzese O, Macaulay R, Libri V, Azevedo RI, Kiani-Alikhan S, Plunkett FJ, Masters JE, Jackson S, Griffiths SJ et al (2009) KLRG1 signaling induces defective Akt (ser473) phosphorylation and proliferative dysfunction of highly differentiated CD8+ T cells. Blood 113:6619–6628

119. Appay V, Papagno L, Spina CA, Hansasuta P, King A, Jones L, Ogg GS, Little S, McMichael AJ, Richman DD et al (2002) Dynamics of T cell responses in HIV infection. J Immunol 168:3660–3666
120. Brenchley JM, Karandikar NJ, Betts MR, Ambrozak DR, Hill BJ, Crotty LE, Casazza JP, Kuruppu J, Migueles SA, Connors M et al (2003) Expression of CD57 defines replicative senescence and antigen-induced apoptotic death of CD8+ T cells. Blood 101:2711–2720
121. Chattopadhyay PK, Betts MR, Price DA, Gostick E, Horton H, Roederer M, De Rosa SC (2009) The cytolytic enzymes granyzme A, granzyme B, and perforin: expression patterns, cell distribution, and their relationship to cell maturity and bright CD57 expression. J Leukoc Biol 85:88–97
122. Huang J, Khong HT, Dudley ME, El-Gamil M, Li YF, Rosenberg SA, Robbins PF (2005) Survival, persistence, and progressive differentiation of adoptively transferred tumor-reactive T cells associated with tumor regression. J Immunother 28:258–267
123. Zhou J, Dudley ME, Rosenberg SA, Robbins PF (2005) Persistence of multiple tumor-specific T-cell clones is associated with complete tumor regression in a melanoma patient receiving adoptive cell transfer therapy. J Immunother 28:53–62
124. Fang M, Siciliano NA, Hersperger AR, Roscoe F, Hu A, Ma X, Shamsedeen AR, Eisenlohr LC, Sigal LJ (2012) Perforin-dependent CD4+ T-cell cytotoxicity contributes to control a murine poxvirus infection. Proc Natl Acad Sci U S A 109:9983–9988
125. Qui HZ, Hagymasi AT, Bandyopadhyay S, St Rose MC, Ramanarasimhaiah R, Menoret A, Mittler RS, Gordon SM, Reiner SL, Vella AT et al (2011) CD134 plus CD137 dual costimulation induces Eomesodermin in CD4 T cells to program cytotoxic Th1 differentiation. J Immunol 187:3555–3564
126. Nicol AJ, Tokuyama H, Mattarollo SR, Hagi T, Suzuki K, Yokokawa K, Nieda M (2011) Clinical evaluation of autologous gamma delta T cell-based immunotherapy for metastatic solid tumours. Br J Cancer 105:778–786
127. Schluns KS, Lefrancois L (2003) Cytokine control of memory T-cell development and survival. Nat Rev Immunol 3:269–279
128. Markley JC, Sadelain M (2010) IL-7 and IL-21 are superior to IL-2 and IL-15 in promoting human T cell-mediated rejection of systemic lymphoma in immunodeficient mice. Blood 115:3508–3519
129. Santegoets SJ, Turksma AW, Suhoski MM, Stam AG, Albelda SM, Hooijberg E, Scheper RJ, van den Eertwegh AJ, Gerritsen WR, Powell DJ Jr et al (2013) IL-21 promotes the expansion of CD27+ CD28+ tumor infiltrating lymphocytes with high cytotoxic potential and low collateral expansion of regulatory T cells. J Transl Med 11:37
130. Araki K, Youngblood B, Ahmed R (2013) Programmed cell death 1-directed immunotherapy for enhancing T-cell function. Cold Spring Harb Symp Quant Biol 78:239–247
131. Dolan DE, Gupta S (2014) PD-1 Pathway Inhibitors: changing the landscape of cancer immunotherapy. Cancer Control 21:231–237
132. Gardiner D, Lalezari J, Lawitz E, DiMicco M, Ghalib R, Reddy KR, Chang KM, Sulkowski M, Marro SO, Anderson J et al (2013) A randomized, double-blind, placebo-controlled assessment of BMS-936558, a fully human monoclonal antibody to programmed death-1 (PD-1), in patients with chronic hepatitis C virus infection. PLoS One 8, e63818
133. Peng W, Liu C, Xu C, Lou Y, Chen J, Yang Y, Yagita H, Overwijk WW, Lizee G, Radvanyi L et al (2012) PD-1 blockade enhances T-cell migration to tumors by elevating IFN-gamma inducible chemokines. Cancer Res 72:5209–5218
134. Peng W, Lizee G, Hwu P (2013) Blockade of the PD-1 pathway enhances the efficacy of adoptive cell therapy against cancer. Oncoimmunology 2, e22691
135. Radvanyi L, Pilon-Thomas S, Peng W, Sarnaik A, Mule JJ, Weber J, Hwu P (2013) Antagonist antibodies to PD-1 and B7-H1 (PD-L1) in the treatment of advanced human cancer--letter. Clin Cancer Res 19:5541
136. Brentjens RJ, Curran KJ (2012) Novel cellular therapies for leukemia: CAR-modified T cells targeted to the CD19 antigen. Hematology Am Soc Hematol Educ Program 2012: 143–151

137. Turtle CJ (2014) Chimeric antigen receptor modified T cell therapy for B cell malignancies. Int J Hematol 99:132–140
138. Khalili JS, Hwu P, Lizee G (2013) Forging a link between oncogenic signaling and immunosuppression in melanoma. Oncoimmunology 2, e22745
139. Khalili JS, Liu S, Rodriguez-Cruz TG, Whittington M, Wardell S, Liu C, Zhang M, Cooper ZA, Frederick DT, Li Y et al (2012) Oncogenic BRAF(V600E) promotes stromal cell-mediated immunosuppression via induction of interleukin-1 in melanoma. Clin Cancer Res 18:5329–5340
140. Liu C, Peng W, Xu C, Lou Y, Zhang M, Wargo JA, Chen JQ, Li HS, Watowich SS, Yang Y et al (2013) BRAF inhibition increases tumor infiltration by T cells and enhances the antitumor activity of adoptive immunotherapy in mice. Clin Cancer Res 19:393–403
141. Reis PP, Waldron L, Goswami RS, Xu W, Xuan Y, Perez-Ordonez B, Gullane P, Irish J, Jurisica I, Kamel-Reid S (2011) mRNA transcript quantification in archival samples using multiplexed, color-coded probes. BMC Biotechnol 11:46
142. Beard RE, Abate-Daga D, Rosati SF, Zheng Z, Wunderlich JR, Rosenberg SA, Morgan RA (2013) Gene expression profiling using nanostring digital RNA counting to identify potential target antigens for melanoma immunotherapy. Clin Cancer Res 19:4941–4950
143. Messina JL, Fenstermacher DA, Eschrich S, Qu X, Berglund AE, Lloyd MC, Schell MJ, Sondak VK, Weber JS, Mule JJ (2012) 12-Chemokine gene signature identifies lymph node-like structures in melanoma: potential for patient selection for immunotherapy? Sci Rep 2:765
144. Tanese K, Grimm EA, Ekmekcioglu S (2012) The role of melanoma tumor-derived nitric oxide in the tumor inflammatory microenvironment: its impact on the chemokine expression profile, including suppression of CXCL10. Int J Cancer 131:891–901
145. Ekmekcioglu S, Ellerhorst J, Smid CM, Prieto VG, Munsell M, Buzaid AC, Grimm EA (2000) Inducible nitric oxide synthase and nitrotyrosine in human metastatic melanoma tumors correlate with poor survival. Clin Cancer Res 6:4768–4775
146. Marzese DM, Scolyer RA, Huynh JL, Huang SK, Hirose H, Chong KK, Kiyohara E, Wang J, Kawas NP, Donovan NC et al (2014) Epigenome-wide DNA methylation landscape of melanoma progression to brain metastasis reveals aberrations on homeobox D cluster associated with prognosis. Hum Mol Genet 23:226–238
147. Griewank KG, Ugurel S, Schadendorf D, Paschen A (2013) New developments in biomarkers for melanoma. Curr Opin Oncol 25:145–151
148. Mann GJ, Pupo GM, Campain AE, Carter CD, Schramm SJ, Pianova S, Gerega SK, De Silva C, Lai K, Wilmott JS et al (2013) BRAF mutation, NRAS mutation, and the absence of an immune-related expressed gene profile predict poor outcome in patients with stage III melanoma. J Invest Dermatol 133:509–517
149. Gajewski TF, Louahed J, Brichard VG (2010) Gene signature in melanoma associated with clinical activity: a potential clue to unlock cancer immunotherapy. Cancer J 16:399–403
150. Ji RR, Chasalow SD, Wang L, Hamid O, Schmidt H, Cogswell J, Alaparthy S, Berman D, Jure-Kunkel M, Siemers NO et al (2012) An immune-active tumor microenvironment favors clinical response to ipilimumab. Cancer Immunol Immunother 61:1019–1031
151. Ahmadzadeh M, Felipe-Silva A, Heemskerk B, Powell DJ Jr, Wunderlich JR, Merino MJ, Rosenberg SA (2008) FOXP3 expression accurately defines the population of intratumoral regulatory T cells that selectively accumulate in metastatic melanoma lesions. Blood 112:4953–4960
152. Galon J, Pages F, Marincola FM, Angell HK, Thurin M, Lugli A, Zlobec I, Berger A, Bifulco C, Botti G et al (2012) Cancer classification using the immunoscore: a worldwide task force. J Transl Med 10:205
153. Galon J, Pages F, Marincola FM, Thurin M, Trinchieri G, Fox BA, Gajewski TF, Ascierto PA (2012) The immune score as a new possible approach for the classification of cancer. J Transl Med 10:1
154. Fridman WH, Galon J, Pages F, Tartour E, Sautes-Fridman C, Kroemer G (2011) Prognostic and predictive impact of intra- and peritumoral immune infiltrates. Cancer Res 71:5601–5605

155. Gajewski TF, Meng Y, Harlin H (2006) Immune suppression in the tumor microenvironment. J Immunother 29:233–240
156. Spranger S, Spaapen RM, Zha Y, Williams J, Meng Y, Ha TT, Gajewski TF (2013) Up-regulation of PD-L1, IDO, and T(regs) in the melanoma tumor microenvironment is driven by CD8(+) T cells. Sci Transl Med 5:200ra116
157. Pilon-Thomas S, Kuhn L, Ellwanger S, Janssen W, Royster E, Marzban S, Kudchadkar R, Zager J, Gibney G, Sondak VK et al (2012) Efficacy of adoptive cell transfer of tumor-infiltrating lymphocytes after lymphopenia induction for metastatic melanoma. J Immunother 35:615–620
158. Jin J, Sabatino M, Somerville R, Wilson JR, Dudley ME, Stroncek DF, Rosenberg SA (2012) Simplified method of the growth of human tumor infiltrating lymphocytes in gas-permeable flasks to numbers needed for patient treatment. J Immunother 35:283–292
159. Somerville RP, Devillier L, Parkhurst MR, Rosenberg SA, Dudley ME (2012) Clinical scale rapid expansion of lymphocytes for adoptive cell transfer therapy in the WAVE(R) bioreactor. J Transl Med 10:69
160. Donia M, Larsen SM, Met O, Svane IM (2014) Simplified protocol for clinical-grade tumor-infiltrating lymphocyte manufacturing with use of the Wave bioreactor. Cytotherapy 16:1117–1120
161. Dudley ME, Gross CA, Somerville RP, Hong Y, Schaub NP, Rosati SF, White DE, Nathan D, Restifo NP, Steinberg SM et al (2013) Randomized selection design trial evaluating CD8+-enriched versus unselected tumor-infiltrating lymphocytes for adoptive cell therapy for patients with melanoma. J Clin Oncol 31:2152–2159
162. Tran E, Turcotte S, Gros A, Robbins PF, Lu YC, Dudley ME, Wunderlich JR, Somerville RP, Hogan K, Hinrichs CS et al (2014) Cancer immunotherapy based on mutation-specific CD4+ T cells in a patient with epithelial cancer. Science 344:641–645
163. Ye Q, Song DG, Poussin M, Yamamoto T, Best A, Li C, Coukos G, Powell DJ Jr (2014) CD137 accurately identifies and enriches for naturally occurring tumor-reactive T cells in tumor. Clin Cancer Res 20:44–55
164. Gros A, Robbins PF, Yao X, Li YF, Turcotte S, Tran E, Wunderlich JR, Mixon A, Farid S, Dudley ME et al (2014) PD-1 identifies the patient-specific CD8(+) tumor-reactive repertoire infiltrating human tumors. J Clin Invest 124:2246–2259
165. Chacon JA, Wu RC, Sukhumalchandra P, Molldrem JJ, Sarnaik A, Pilon-Thomas S, Weber J, Hwu P, Radvanyi L (2013) Co-stimulation through 4-1BB/CD137 improves the expansion and function of CD8(+) melanoma tumor-infiltrating lymphocytes for adoptive T-cell therapy. PLoS One 8, e60031
166. Chacon JA, Pilon-Thomas S, Sarnaik AA, Radvanyi LG (2013) Continuous 4-1BB co-stimulatory signals for the optimal expansion of tumor-infiltrating lymphocytes for adoptive T-cell therapy. Oncoimmunology 2, e25581
167. Duraiswamy J, Kaluza KM, Freeman GJ, Coukos G (2013) Dual blockade of PD-1 and CTLA-4 combined with tumor vaccine effectively restores T-cell rejection function in tumors. Cancer Res 73:3591–3603
168. Ellebaek E, Iversen TZ, Junker N, Donia M, Engell-Noerregaard L, Met O, Holmich LR, Andersen RS, Hadrup SR, Andersen MH et al (2012) Adoptive cell therapy with autologous tumor infiltrating lymphocytes and low-dose Interleukin-2 in metastatic melanoma patients. J Transl Med 10:169
169. Ullenhag GJ, Sadeghi AM, Carlsson B, Ahlstrom H, Mosavi F, Wagenius G, Totterman TH (2012) Adoptive T-cell therapy for malignant melanoma patients with TILs obtained by ultrasound-guided needle biopsy. Cancer Immunol Immunother 61:725–732

Chapter 8
Harnessing Stem Cell-Like Memory T Cells for Adoptive Cell Transfer Therapy of Cancer

Enrico Lugli and Luca Gattinoni

Abstract Immunotherapies based on the adoptive transfer of naturally occurring or gene-engineered tumor-reactive T cells can result in durable complete responses in patients with metastatic cancers. Increasing findings from mouse studies and clinical trials indicate that intrinsic properties related to the differentiation state of the transferred T cells are crucial to the success of adoptive immunotherapies. There is now evidence that stem cell-like T cells with enhanced capacity for self-renewal and the ability to derive potent effector T cells might be used to improve persistence and long-term anti-tumor immunity. Here, we describe the molecular, metabolic and cellular aspects of T cell differentiation and their relevance to cancer immunotherapy. We also discuss current efforts and new approaches that might potentiate T cell-based immunotherapies through the modulation of T cell fate and differentiation.

Keywords Adoptive T cell therapy • T celldifferentiation • T memory stem cells • Transcription factors • Immune metabolism • Gene therapy • Reprogramming • Induced pluripotent stem cells • Small molecules • Homeostatic cytokines

Introduction

Adoptive T cell-based therapies have emerged as a potent and highly effective treatment for patients with advanced solid cancer and hematologic malignancies [1–3]. These therapeutic modalities are based on the ex vivo expansion and re-infusion of autologous or allogeneic tumor-specific T cells to patients. Early efforts to target malignancy focused on the use of tumor-infiltrating lymphocytes (TIL) [4], bulk

E. Lugli, Ph.D. (✉)
Principal Investigator, Laboratory of Translational Immunology, Humanitas Clinical and Research Center, Via Alessandro Manzoni 113, Rozzano, Milan, Italy
e-mail: enrico.lugli@humanitasresearch.it

L. Gattinoni, M.D. (✉)
Experimental transplantation and Immunology Branch, Center for Cancer Research, National Cancer Institute, 10 Center Drive, Bethesda, MD, USA
e-mail: gattinol@mail.nih.gov

P.A. Ascierto et al. (eds.), *Developments in T Cell Based Cancer Immunotherapies*, Cancer Drug Discovery and Development,
DOI 10.1007/978-3-319-21167-1_8

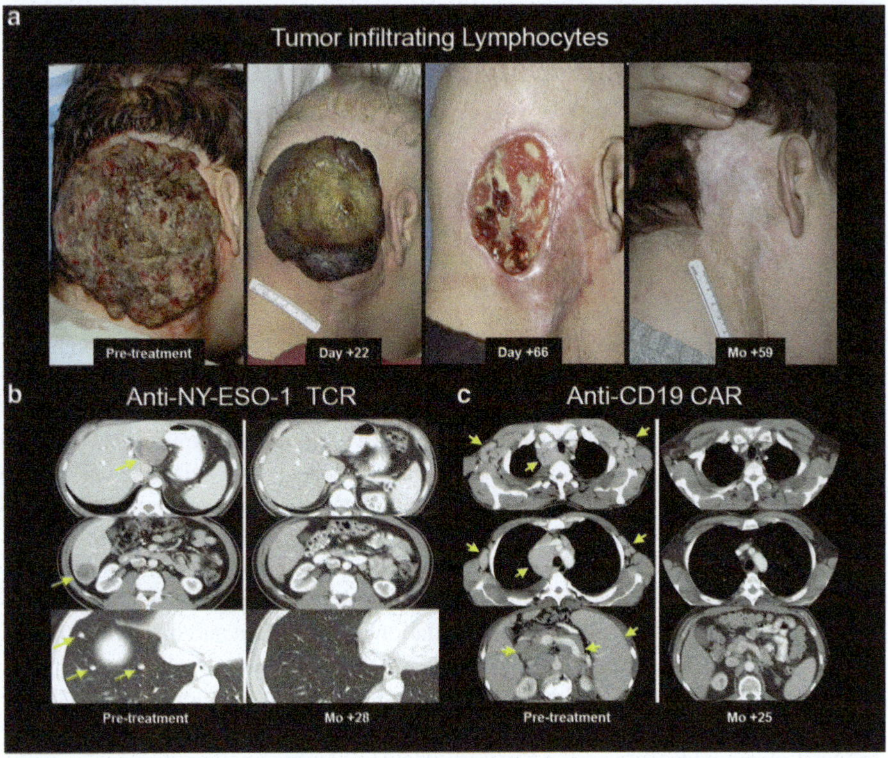

Fig. 8.1 Objective clinical regressions in patients with metastatic tumor treated with cell transfer therapy. (**A**) Regression of a large fungating scalp mass in a melanoma patient treated with ex vivo expanded tumor-infiltrating lymphocyte. (**B**) Regression of multiple liver and lung metastases in a melanoma patient treated with peripheral blood lymphocytes genetically engineered to express a NY-ESO-1-specific T cell receptor (TCR). (**C**) Regression of splenomegaly and multiple adenopathies in a B-cell lymphoma patient treated with peripheral blood lymphocytes genetically engineered to express an anti-CD19 chimeric antigen receptor (CAR). All patients were treated in the Surgery Branch of the National Cancer Institute, USA

lymphocyte populations in vitro-sensitized against tumor antigens [5] and tumor antigen-specific clones [6–8]. Recent advances in gene transfer technology have permitted to convey de novo cancer reactivity to T cells through genetic engineering of tumor-reactive T cell receptors (TCR) [9–11] or chimeric antigen receptors (CAR), which consist of a single-chain variable fragment of a tumor-specific antibody linked to trans-membrane and cytoplasmic domains of T cell signaling molecules [12–17]. Transfer of naturally occurring or genetically engineered tumor-reactive T cells has resulted in dramatic and possibly curative responses in some patients (Fig. 8.1). Increasing experimental and clinical evidence indicate that the differentiation state, the self-renewal capacity and the ability to derive large numbers of potent effectors critically influence the ability of tumor-reactive T cells to mediate effective anti-tumor immune responses. Here, we describe the molecular, metabolic and cellular aspects of T cell differentiation and the impact they have in anti-cancer immunotherapy. Finally, we highlight current efforts and promising strategies that might potentiate adoptive immunotherapies through the modulation of T cell fate and differentiation.

T Cell Differentiation

Peripheral T lymphocytes are mature cells of the adaptive immune system but, differently to other committed cells of the body, they display a tremendous heterogeneity and high degree of plasticity. Mature T cells are released from the thymus into the periphery and harbor a given specificity that is encoded at TCR level. Following cognate antigen (Ag) recognition, naïve T (T_N) cells clonally expand into effectors, the vast majority of which migrate to peripheral tissues and inflamed sites to remove the infected targets [18]. As a consequence of infection clearance, ~90–95 % of activated cells dies while a small pool of T cells ultimately develops into long-lived memory cells capable to persist in the long term in the putative absence of Ag [18]. Analogous to other tissues in which terminally differentiated cells are replaced by the progeny of somatic stem cells [19], survival and maintenance of memory T cells is thought to occur in a stem cell-like fashion, where less differentiated cells give rise to more committed progeny.

The Diversity of T Cell Subsets

While naïve T cells constitute a fairly homogenous population, memory T cells are highly heterogeneous in terms of phenotypic and functional composition [20]. Seminal studies in the late 1990s segregated T cells into central memory T (T_{CM}) and effector memory T (T_{EM}) cell subsets on the basis of migratory capacity (secondary lymphoid vs. peripheral tissues, respectively) and immediacy of effector functions/killing activity ($T_{EM}>T_{CM}$) [21]. CD27, a member of the tumor necrosis factor receptor superfamily, and the lymphoid homing molecules C-C chemokine receptor 7 (CCR7) and L-selectin (CD62L) were initially used together with CD45RA, the long isoform the CD45 protein, to define heterogeneity in the human memory T cell compartment. Alternatively, CD45RO, the CD45 short isoform, can replace CD45RA as the expression of the two molecules is generally mutually exclusive on the cell surface. Single cell analysis of effector functions revealed that $CD45RA^{-}CD8^{+}$ memory T cells expressing $CD27^{+}$ secrete both IFN-γ and IL-2 but lack immediate killing capacity, while $CD45RA^{+}CD27^{-}$ cells produce IFN-γ and TNF but lack IL-2 production and simultaneously display immediate cytotoxic activity ex vivo [22]. A similar cytokine production profile is shared with $CD4^{+}$ T_{CM} and T_{EM} cells [23]. An additional subset of terminally differentiated cells, named terminal effectors (T_{TE}) is abundant in the $CD8^{+}$ but rare in the $CD4^{+}$ T cell compartment. These cells

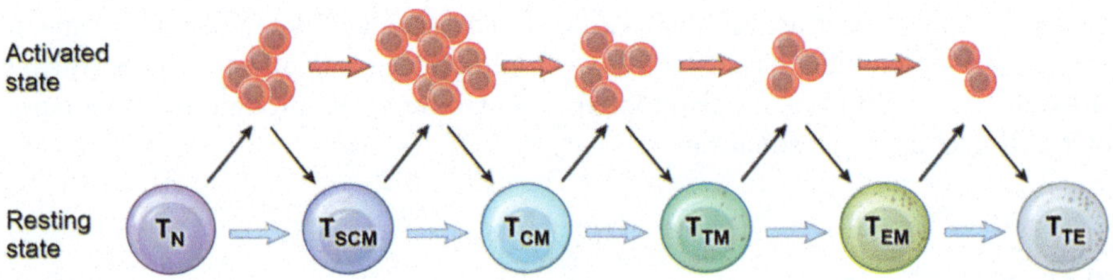

	T_N	T_{SCM}	T_{CM}	T_{TM}	T_{EM}	T_{TE}
CD45RA	+	+	–	–	–	+
CD45RO	–	–	+	+	+	–
CCR7	+	+	+	–	–	–
CD28	+	+	+	+	–	–
CD27	+	+	+	+	–	–
IL-7Rα	+	+	+	+/–	–	–
CD95	–	+	+	+	+	+
CD57	–	–	–	–	+/–	+

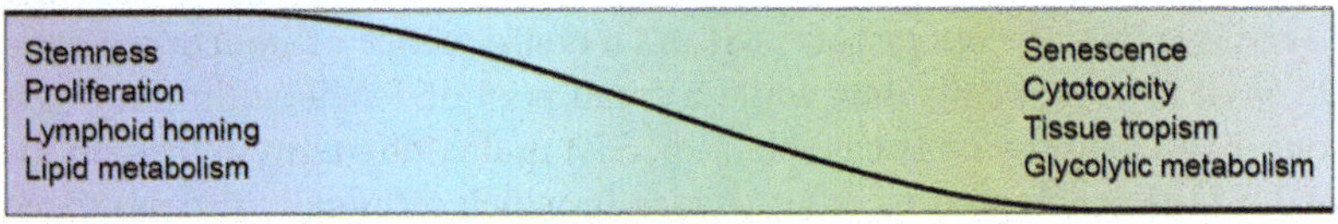

Fig. 8.2 Heterogeneity and functionality of peripheral T cell subsets in humans. The combinatorial expression of multiple markers on the cell surface, as determined by polychromatic flow cytometry, allows to identify up to six subpopulations in the peripheral blood and tissues of healthy individuals. With peripheral maturation, T cells progressively lose or acquire specific functional capacities, as shown at the bottom of the figure. Following antigen recognition, a given T cell is activated (in *red*) and undergoes clonal expansion. The extent of proliferation is dependent on the initial differentiation status. Among peripheral T cell subsets, T_{SCM} cells retain the greatest proliferative capacity in vivo. When infected/tumor cells are removed, some activated T cells escape clonal deletion and generate long-lived memory T cells that can divide by homeostatic proliferation and generate more differentiated progeny. In the model proposed here, duration of antigenic stimulation dictates the stage of differentiation of activated cells returning to quiescence

re-express CD45RA [22], lack CCR7 [23] and costimulatory molecules [22] and bear high levels of the carbohydrate epitope HNK-1, otherwise known as CD57 [24]. T_{TE} cells stain positive for cytolytic molecules and display immediate effector functions [25] but are defective in proliferative and survival capacities [24].

In vivo analysis of memory T cell dynamics and tropism in nonhuman primates allowed the identification of the so-called transitional memory or T_{TM} cells that were proposed to be intermediate between T_{CM} and T_{EM} cells [26]. Specifically, these cells have down-regulated CCR7 and CD62L but retain the expression of the costimulatory molecules CD28 and, in humans, CD27 [26, 20]. T_{TM} cells are highly responsive to IL-15 treatment in vivo [26, 27] and migrate to effector sites following stimulation. The phenotypic combinations generally used in our laboratories to define human memory T cell subsets are indicated in Fig. 8.2.

Because of the phenotypic differences between human and mouse T cells, murine memory T cell subsets are classically defined on the basis of CD44 and CD62L.

Naïve T cells express CD62L but lack CD44, which is upregulated by CD62L$^+$ T$_{CM}$ cells and CD62L$^-$ T$_{EM}$ cells. Additional markers are available to classify memory T cells in mice, including CCR7, CD27 and a glycoform of the CD43 molecule. Both CD27$^+$CD43$^+$ and CD27$^+$CD43$^-$ memory T cells contain conventional T$_{CM}$ and T$_{EM}$ cells, while CD27$^-$CD43$^-$ T cells are classified as terminal effectors. These cells express high levels of the Killer cell lectin-like receptor subfamily G member 1 (KLRG-1), and low levels of the Interleukin-7 receptor α (IL-7Rα) and IL-2Rβ chains, otherwise known as CD127 and CD122, respectively [28].

Sallusto and Lanzavecchia for the first time proposed a precursor-progeny relationship between T$_{CM}$ and T$_{EM}$ cells [23] on the basis of the evidence that T$_{CM}$ were capable to derive T$_{EM}$ cells in vitro, while the opposite was not observed [29, 30]. Several studies in mice, nonhuman primates and humans later demonstrated that T$_{CM}$ cells serve as early-differentiated progenitors capable of self-renewing and generating more-differentiated progeny [21, 31, 32]. T$_{CM}$ cells were thus thought to act as a reservoir of memory T cells, capable to continuously regenerate the memory T cell compartment in physiology and following injury (e.g., lymphopenia) in a stem cell-like manner [33, 31, 34].

Identification of T Memory Stem Cells

In 2005, the Emerson's group reported that a novel memory T cell population, characterized by a largely naïve-like phenotype but expressing the memory markers IL-2Rβ and the chemokine C-X-C motif receptor 3 (CXCR3), was responsible of maintaining graft-versus-host disease upon serial transplantations in mice [35]. Unexpectedly, classically-defined T$_{CM}$ cells failed to do so when adoptively-transferred as purified fractions. These cells were termed T memory stem cells (T$_{SCM}$) as they could differentiate into T$_{CM}$, T$_{EM}$ and T$_{TE}$ cells while maintaining their own pool size through self-renewal. T$_{SCM}$ cells can be successfully generated from naïve precursors by activating the Wnt signaling pathway using either a physiological Wnt ligand, Wnt3A, or inhibitors of glycogen synthase kinase-3β (GSK-3β) [36]. The generated T$_{SCM}$ cells maintained the undifferentiated CD44$^-$CD62L$^+$ naïve-like phenotype but acquired several memory attributes, including the capacity to rapidly produce effector cytokines, persist in MHC class I-deficient hosts and reconstitute the diversity of T cell subsets upon serial transplantations [36]. Recently, we have described a T$_{SCM}$ cell population in humans [37]. Similar to mouse cells, human T$_{SCM}$ cells display a largely naïve-like phenotype together with few memory markers such as CD95 and IL-2Rβ (Fig. 8.2). Moreover, these cells exhibit enhanced stem cell-like properties and superior reconstitution capacity in immunodeficient hosts compared to T$_{CM}$ cells [37]. T$_{SCM}$ are precursors of T$_{CM}$ cells as regards to peripheral differentiation as revealed by their phenotypic and gene expression properties. Despite displaying a transcriptional signature characteristic of memory cells, T$_{SCM}$ cells retain a core of genes expressed by T$_N$ cells [37] and share the recirculation patterns and distribution of T$_N$ cells in vivo. Indeed, they show relative abundance in lymphoid tissues compared to the spleen and bone marrow and are virtually absent from mucosal surfaces [38]. To date, murine T_{SCM} cells have not been definitively described for pathogen-specific T cells. Conversely, nonhuman primates bear virus-specific $CD8^+$ T_{SCM} cells following simian immunodeficiency virus (SIV) infection. SIV-specific T_{SCM} cells are generated early during acute infection [38], suggesting that a specific gene expression program underlies their differentiation. T_{SCM} cells are subsequently maintained in the long term even under chronic stimulation by the cognate antigen, possibly because of the selective overexpression of transcripts regulating self-renewal (*LEF1*) and mediating protection from apoptosis (*MCL1* and *BCL2*) [38]. The superior abilities of T_{SCM} cells to self-renew, resist apoptosis and survive for long periods of time have been corroborated by a recent study in HIV patients showing that T_{SCM} cells make increasing contributions to the total viral $CD4^+$ T cell reservoir over time [39]. Finally, the importance of T_{SCM} cells in the maintenance of the immune homeostasis is suggested by new findings in nonhuman primates reveling a perturbation of the T_{SCM} cell compartment during pathogenic but not nonpathogenic SIV [40].

Environmental and Cell Intrinsic Cues Regulating T Cell Differentiation

Transcriptional Control of T Cell Differentiation

Despite the human memory T cell compartment has been extensively characterized in the past years at the phenotypic and functional level, the molecular determinants leading to the formation of long-lived memory T cells have been mostly defined using mouse models. The study of the transcriptional regulators of memory T cell differentiation exploded after the identification of a population of cells capable to survive the effector phase of the immune response and enter into the memory pool and it has been extensively reviewed elsewhere [41, 42]. At least two independent reports identified heterogeneity in the effector T cell pool at the peak of the immune response (i.e., at day 7 post infection in mouse models) on the basis of IL-7Rα and KLRG-1 [43] or CD62L [44] expression. Adoptive transfer of discrete populations indicated that IL-$7R\alpha^+$ KLRG-1^- cells are activated T cells capable to persist and further differentiate into long-lived memory T cells [43]. Conversely, IL-$7R\alpha^-$KLRG-1^+ T cells are potent short-lived effectors, able to migrate to inflamed sites and remove infected targets. Gene expression profiling of these two T cell populations helped revealing the role of specific transcription factors and molecular regulators in driving memory vs. effector differentiation [45, 46].

Two T-box transcription factors, T-bet (encoded by *Tbx21*) and Eomesodermin (Eomes, encoded by *Eomes*) play a pivotal role in this regard. Their expression seems to be reciprocal with progressive memory differentiation in the mouse, as *Eomes* is highly expressed in T_{CM} but not in T_{EM}, while the opposite pattern is observed for T-bet. Conversely, *EOMES* mRNA is low in human naïve T cells and is progressively upregulated in more differentiated memory subsets [37], highlighting potentially different mechanisms at the basis of memory T cell differentiation in the two species. Both proteins are involved in effector differentiation following antigen encounter by naïve T cells, as they are required for the optimal acquisition of killing activities [47]. Perforin, granzyme B and IFN-γ as well as CXCR3, which directs effector cells towards inflamed sites, are partially under the control of these transcription factors [48, 49]. *Tbx21*$^{-/-}$ mice fail to develop short-lived effectors in response to antigen stimulation but are enriched in memory precursors [46]. Conversely, T cells lacking Eomes generate high levels of memory precursors, which, however, fail to differentiate into long-lived memory T cells [50]. This is, at least in part, due to their inability to respond to IL-15 because of the lack of IL-2Rβ chain on the surface of *Eomes*$^{-/-}$ and *Tbx21*$^{-/-}$ T cells [47]. IL-15-dependent survival and homeostatic proliferation of memory T cells and their localization to the bone marrow niche is consequently abrogated [50].

Similarly to Tbet, the transcription factor Blimp-1 encoded by the *Prdm1* gene regulates effector differentiation from naïve precursors, and ensures effector recall responses from mature memory T cells [51, 52]. Blimp-1-deficient T cells are highly enriched in memory precursors following infection with different viruses, and more rapidly develop into IL-2-secreting T_{CM} cells than wild-type mice [51]. Consistent with their defective effector differentiation and accelerated memory formation, Ag-specific *Prdm1*$^{-/-}$ T cells displayed decreased levels of T-bet and increased levels of Eomes [52]. Blimp-1 limits the ability of short-lived effectors to enter in the memory pool by repressing the inhibitor of DNA binding 3 (*Id3*), an important regulator of genome stability [53]. Accordingly, Id3^{high} T cells at the peak of the immune response identified CD8^{+} T cells capable to develop into long-lived memory T cells [54]. In turn, Bach2, originally identified in B cells as a regulator of class switch recombination and somatic hypermutation of immunoglobulin genes, represses Blimp-1 expression in both B cells and CD8^{+} T cells, induces the expression of Id3 and enhances the formation of memory T cells by increasing the frequency of CD62L^{+}KLRG-1^{-} T cells [55]. In CD4^{+} cells, the T cell-specific transcription factor Menin binds to the Bach2 locus and ensures Bach2 expression through histone acetylation [56]. Menin overexpression inhibits senescence in CD4^{+} T cells and restricts the acquisition of the senescence-associated secretory phenotype, characterized by the overexpression of pro-inflammatory cytokines and matrix remodeling factors [56]. The role of the Menin-Bach2 axis in the function of memory T cells during a recall response is, at present, still to be defined.

Members of Forkhead box O (FoxO) protein family are also emerging as key regulators of memory T cell differentiation and homeostasis. In the unphosphorylated form, Foxo proteins are found in the nucleus where they regulate gene expression by binding regulatory DNA motifs [57]. FoxO proteins are mostly involved in the regulation of genes involved in cell cycle and apoptosis, and can either induce or block these specific functions. Upon phosphorylation, the FoxO proteins are retained in the cytoplasm and are subsequently inactivated through proteasomal-mediated degradation [57]. Growth factor, hormone and cytokine stimulations activate the AKT pathway, which in turn mediate FoxO proteins inactivation. The best characterized members of the FoxO family, as regards to memory T cell biology, are FoxO3a and Foxo1. The former was initially described as a regulator of human CD4^{+} T cell survival. T_{CM} cells harbor slightly lower levels of the native protein compared to T_{EM} cells but show a two- to five-fold higher levels of the phosphorylated form on multiple residues *ex vivo*, thus resulting in increased in vitro resistance to spontaneous and Fas ligand (FasL also known as CD95L)-induced apoptosis [58]. T_{CM} cells also show reduced levels of the pro-apoptotic protein Bcl2-Like 11 (also known as Bim), which is as direct FoxO3a transcriptional target [58]. Accordingly, FoxO3a-deficient mice show reduced levels of Bim and Bcl2 binding component 3 (also known as PUMA) pro-apoptotic proteins, generate higher numbers of memory precursors following cognate antigen activation and preferentially persist in the long-term [59].

Differently, FoxO1 was found to promote memory T cell persistence through the induction of genes promoting cell survival (*BCL2*, *Il7r*, *Tcf7*) and maintenance of the T_{CM} cell status (*Sell*, encoding for CD62L, and *Ccr7*) [60]. ChIP sequencing analysis demonstrated that *Tcf7* and *Ccr7* are direct transcriptional targets of FoxO1 [60]. Deletion of FoxO1 from activated CD8^{+}T cells did not affect effector differentiation but abrogated the formation of memory T cells [60, 61]. Moreover, the few T cells transitioning to the memory phase showed virtually no capacity to respond to recall antigenic stimulation.

Therefore, two members of the same protein family that are inactivated downstream of the Akt pathway by similar mechanisms regulate memory T cell differentiation by promoting the transcription of genes with opposite functions. It is possible to speculate that FoxO1 and FoxO3a are repressed by different concentrations of Akt stimulators or with different kinetics, thus deciding whether a T cell transits to the memory phase or dies during the effector phase. In addition, different concentrations of the two proteins at the single cell level might increase the heterogeneity of the memory precursors population even further.

Metabolic Regulation of T Cell Differentiation

A single naïve T cell is able to generate thousands of daughter cells by dividing every 4–6 h in response to antigen recognition [62]. It is becoming increasingly evident that such a fast dynamics requires profound changes not only at the transcriptional level but also at the metabolic level [63]. In the absence of antigen, naïve T cells are quiescent, are characterized by a very high nuclear:cytoplasmic ratio (indicative of little protein synthesis), divide only rarely in response to homeostatic proliferation, mainly mediated by IL-7, and have low energetic demand [63]. Memory T cells are also quiescent in the absence of antigenic stimulation but display greater mitochondrial mass, which provides a bioenergetic advantage for supporting rapid recall responses after antigen re-exposure [64, 65]. Quiescent naïve and memory T cells rely almost completely on energy derived from mitochondrial oxidative phosphorylation (OXPHOS) and fatty acid β-oxidation (FAO), however following antigen-induced activation they switch to a glycolytic metabolism even in oxygen-replete microenvironments [63, 66]. It remains unclear why T cells adopt a less efficient pathway for ATP generation under conditions of high-energy demand. For years, it has been proposed that this phenomenon, also known as the Warburg effect, was necessary to generate precursors of deoxyribonucleotides that are subsequently used for DNA replication [67]. However, recent findings indicate that glycolysis is critical for effector differentiation as it is required for the post-transcriptional regulation of specific effector function such as IFN-γ production [68]. Moreover, T cells activated in limiting concentrations of glucose failed to upregulate killing molecules, such as perforin and granzymes [69].

These observations clearly indicate that distinct T cell subsets exhibit unique metabolic programs. Whether these metabolic characteristics reflect functional changes orchestrated by diverse transcriptional programs or rather instructively dictate T cell fate decisions has just begun to be addressed. The first evidence that memory T cell formation is regulated at the metabolic level came from the analysis of mice lacking tumor necrosis factor (TNF) receptor-associated factor 6 (TRAF6) [70]. Despite effector differentiation was maintained in *Traf6*$^{-/-}$ mice, the generation of memory T cells was abrogated almost completely. Gene expression studies revealed that *Traf6*$^{-/-}$ T cells were incapable to induce transcripts involved in FAO, and displayed a reduced 5′ adenosine monophosphate-activated protein kinase (AMPK), a master regulator of fatty acid metabolism [70]. Accordingly, the overexpression of carnitine palmitoyltransferase 1a (*Cpt1a*), the rate-limiting enzyme of FAO, was sufficient to augment CD8^{+} T cell memory and hence potentiate recall immune responses [71]. These findings clearly demonstrate that FAO can influence the establishment of immunological memory, indicating that changes in metabolism play a direct role in regulating T cell differentiation. Recently, we have demonstrated that also glycolysis can directly influence the generation of memory and effector T cells [72]. Activated T cells displaying high glycolytic activity tended to be short-lived, while cells with low glycolytic metabolism established memory. Moreover, enforcing glycolysis by overexpression of the glycolytic enzyme *Pgam1* severely impaired the ability of CD8^{+} T cells to persist in the long term whereas inhibition of glycolytic flux using 2-deoxyglucose (2-DG) increased CD8^{+} T cell memory formation by preserving the expression of *Tcf7*, *Lef1* and *Bcl6* and repressing the upregulation of *Prdm1* and effector-associated genes.

AKT, also known as protein kinase B, has been shown to integrate cell growth signals with glycolytic metabolism in a variety of cellular systems. Increased AKT activity in effector T cells leads to the activation of the mammalian target of rapamycin (mTOR) which favors cell growth, protein synthesis and proliferation [73]. In mammalian cells, mTOR is found in two different multiprotein complexes, mTORC1 and mTORC2, that are activated by a plethora of extrinsic and intrinsic signals in T cells, including antigen, cytokines, glucose and amino acids, among others. For more detailed information on mTORC1 and mTORC2, excellent reviews were recently published [74, 75]. The increased function of mTOR has been linked to a number of molecular events involved in effector differentiation. TCR-dependent immune activation leading to the down-regulation of CCR7 and CD62L depends on increased PI3K-mTOR function, which subsequently controls migratory capacity in vitro and in vivo [166]. Furthermore, excessive mTOR stimulation drives T cells towards a terminally differentiated effector state whereas mTOR inhibition by the immunosuppressive drug rapamycin resulted in increased numbers of memory T cells [76]. These results appear counterintuitive as rapamycin is widely used in solid organ and hematopoietic stem cell transplantations to inhibit allogeneic T cell immune responses [73]. Despite the exact mechanisms at the basis of mTOR inhibition by rapamycin and the outcome of the T cell response are still to be defined, it is possible to hypothesize that the dose, the timing and the duration of administration are important variables in this regard.

In summary, T cell fate is tightly regulated at the transcriptional, signaling and metabolic levels. What is clear is that these different aspects are not independent but are closely interconnected. Most importantly, they can be modulated by using small molecules that are approved for clinical use [77]. The implications of memory T cell differentiation in the regulation of anti-tumor immunity at the preclinical and clinical levels are described below.

T Cell Differentiation and Adoptive Immunotherapy Efficacy

Lessons from Mouse Models

It had long been unclear whether the differentiation state represented a crucial determinant of the ability of tumor-reactive T cells to mediate anti-tumor responses upon adoptive transfer. Because of the vast heterogeneity of T cell preparations employed in adoptive immunotherapy studies it was impossible to precisely separate the therapeutic contribution of defined T cell subsets from the impact of distinct TCRs before the advent of TCR transgenic mice. Since the final goal of adoptive immunotherapy is to generate T cells capable of patrolling the body in search of cancer cells to destroy, it was initially assumed that T_{TE} and T_{EM} cells were the ideal T cells to transfer, as they possess a propensity to migrate into peripheral tissues and display immediate cytotoxic functions upon antigen encounter. Accordingly, the potency of T cell products was exclusively determined by assessing the ability of tumor-reactive T cells to release IFN-γ and kill tumor cells upon in vitro co-culture [78]. It was somehow surprising to realize that these two subsets, on the contrary, were poorly capable of destroying tumors upon adoptive transfer compared to less differentiated T cells. First evidence came from two sets of experiments conducted in the pmel-1 model of adoptive immunotherapy that employs gp100-specific $CD8^+$ T cells derived from the TCR transgenic mouse pmel-1 to target B16 melanoma [79]. Terminally differentiated KLRG-1^+ T_{TE} cells generated from reiterative stimulations of pmel-1 cells with cognate antigen and IL-2 were found to be 100-fold less effective in vivo on a per-cell basis than T cells at an early stage of differentiation [80]. Parallel experiments evaluating the antitumor efficacy of tumor-specific $CD8^+$ memory subsets revealed that less differentiated T_{CM} cells were capable of inducing durable complete responses while mice receiving T_{EM} cells ultimately succumbed to unrestrained tumor growth [81]. The inadequacy of $CD62L^-$ T cell subsets to mediate profound immune responses following adoptive transfer has been also documented by several other groups in diverse settings including models of tumor treatment [82–84], viral protection [85, 86] and allogeneic hematopoietic stem cell transplantation [87, 88].

The paradoxical inability of T_{TE} and T_{EM} at triggering tumor regression upon adoptive transfer finds its roots in several biological hurdles that are integral components of the effector differentiation program. For instance, the inefficient trafficking to peripheral lymphoid tissues due to the loss of CD62L and CCR7 expression can disrupt the intimate interactions with dendritic cells that are fundamental for the induction of productive T cell responses [80]. Indeed, anti-tumor responses were virtually abrogated in hosts devoid of secondary lymph nodes and with a disrupted splenic structure [81]. Moreover, *Sell*$^{-/-}$ tumor-specific $CD8^+$ T cells were impaired in their ability to inhibit tumor growth compared with wild-type T cells [80, 81]. A profound reshaping of the co-stimulatory and inhibitory receptor landscape also accompanies the differentiation process. Down-regulation of CD28 and CD27 expression in T_{TE} and T_{EM} cells can limit co-stimulatory signals resulting in decreased cell proliferation and long-term survival [89–92]. This dysfunctionality can also be aggravated by the concomitant overexpression of KLRG-1 and multiple inhibitory receptors such as PD-1, LAG-3 and 2B4, which have detrimental effects on cell growth and function [93, 94]. T_{TE} and T_{EM} cells might also not receive sufficient pro-survival and activating signals from common γ chain (γ_C) cytokines. As T cells progressively differentiate into T_{TE} cells, they lose the ability to utilize IL-2 in autocrine fashion [95, 80] and to respond to IL-7 cues due to the down-regulation of IL-7rα expression [43, 44]. Finally, gradual telomere erosion [96, 23, 97] and up-regulation of pro-apoptotic molecules, including BID (B-cell lymphoma 2 (BCL-2)-homology domain 3 (BH3)-interacting-domain death agonist) and BAD (BCL-2-antagonist of cell death) [80, 81] might ultimately result in T_{TE} replicative senescence and cell death. Altogether, these phenotypic and functional changes characterizing the differentiation program severely impair the ability of T_{TE} and T_{EM} cells to engraft, expand and persist long-term following adoptive transfer into tumor-bearing hosts [80, 81].

It is now clear that the proliferative potential and survival capacity are key attributes to seek in tumor-reactive T cells for adoptive transfer. Among T cell memory subsets, T_{SCM} cells possess a robust proliferative capacity and a superior ability to persist in the long-term, which make them a desirable cell population to employ in adoptive immunotherapy [98]. When tested in the pmel-1 model, minuscule numbers of T_{SCM} cells mediated dramatic tumor regression of large established B16 melanoma [36]. Paralleling their engraftment and proliferative potentials, the ability of memory T cells to mediate tumor regression progressively decreased from T_{SCM} cells to T_{CM} cells and T_{EM} cells [36]. These findings were corroborated in subsequent experiments using human T cell memory subsets genetically engineered to express an anti-mesothelin CAR to treat human mesothelioma xenografts in immunodeficient mice [37]. In conclusion, findings made in mice have established an inverse relationship between T cell differentiation status and the relative capacities of transferred T cells to engraft, proliferate, and mediate antitumor immunity. These data strongly support the use of the less differentiated CD62L^{+} subsets and particularly T_{SCM} cells over the CD62L^{-} T_{EM} and T_{TE} cells for adoptive immunotherapies.

Insights from the Clinic

The question of which T cell subset is more effective for adoptive immunotherapy becomes murkier when considering the available clinical data. Clinical trials employing well-defined tumor-reactive T cell subsets are still lacking due to the technical complexity associated with the isolation procedures, however some conclusions can be drawn from key observations and retrospective analyses. Consistent to what was observed in mouse studies [80, 84], tumor-specific CD8^{+} T cell clones that were generated and expanded ex vivo through multiple stimulations in the presence of IL-2, a cytokine that promotes terminal differentiation [99], failed to persist after infusion and did not meditate clinically meaningful tumor regressions [6–8]. Conversely, T cell persistence has been highly correlated with tumor responses across multiple clinical trials [100–102, 15] and has been linked to intrinsic T cell properties that are reflective of their differentiation state and replicative history. Early studies revealed that a short duration of tumor infiltrating lymphocyte (TIL) culture or a relatively rapid doubling time were associated with clinical responses [103, 104]. Additional parameters such as the length of telomeres [105, 100], the expression of CD27 [106, 100] and CD28 [105] and the frequency of T_{CM} cells in the infusion product [107] have also been correlated with tumor responses in patients with cancer. However, recent TIL analyses from a limited cohort of patients failed to observe a correlation with telomere length and tumor responses [108]. Furthermore, in this set of patients, objective responses were associated with the infusion of CD45RA^{-}CD62L^{-}CD27^{-} T_{TE} [108]. These discrepancies might be related to the prevalence of tumor-specific T cells within a given T cell subset. For instance, T_{TE} might have been relatively enriched for tumor-reactive T cells or highly avid TCR clonotypes. Taken together, the majority of data in humans is consistent with the notion that less-differentiated T cells confer superior antitumor efficacy relative to T_{EM} and T_{TE} cells.

Potentiating Adoptive T Cell Therapies by Modulating T Cell Differentiation

Restraining T Cell Differentiation

Current approaches employed to generate T cells for adoptive transfer often rely on variations of a protocol established more than 20 years ago [109, 110], before the implication of T cell differentiation on in vivo anti-tumor efficacy was completely

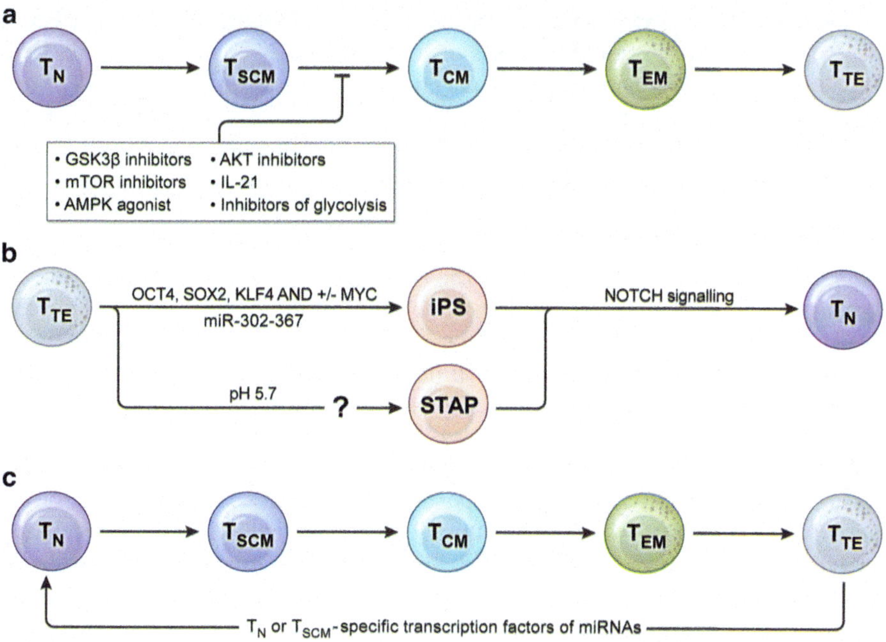

Fig. 8.3 Programming and reprogramming T cell fates for therapeutic use. (**A**) Restraining T cell differentiation. Differentiation of activated naive T (T_N) cells can be withheld by small molecules targeting key metabolic and developmental pathways or γ_C cytokines alternative to interleukin-2 (IL-2), such as IL-21. (**B**) Two step reprogramming of terminally differentiated effector T (T_{TE}) cells through an induced pluripotent stem (iPS) cell intermediate or a possible stimulus-triggered acquisition of pluripotency (STAP) cell intermediate. T_{TE} cells are reprogrammed into iPS cells by enforced expression of OCT4, Kruppel-like factor 4 (KLF4) and sex determining region Y (SRY) BOX 2 (SOX2) with or without MYC or by ectopic expression of the microRNA (miRNA) cluster 302–367. T_{TE} cells might also be converted into STAP cells by exposure to strong external stimuli such as a transient low-pH stressor. iPS and STAP cells can be then re-differentiated into T_N cells through the induction of NOTCH signaling. (**C**) Direct reprogramming of T_{TE} into T_N or memory stem (T_{SCM}) cells by enforced expression of T_N or T_{SCM}-associated transcription factors or miRNAs. T_{CM}, central memory; T_{EM}, effector memory; GSK-3β, glycogen synthase 3β

appreciated [111]. This strategy consist of potent activating stimuli, such as anti-CD3 antibodies, high concentrations of IL-2 and allogeneic feeders, which result in the generation of large numbers of tumor-reactive T cells but inevitably drive cells towards terminal differentiation and senescence.

To limit the negative influence of ex vivo expansion on T cell differentiation, new methods have been investigated (Fig. 8.3A). Early studies focused on the use of γ_C cytokines alternative to IL-2, such as IL-15 and IL-21. IL-15 was found to sustain T cell expansion without the robust pro-differentiating activity that is typical of IL-2. Contrary to IL-2, which generates T_{TE} and T_{EM} cells, stimulation of T cells in the presence of IL-15 promoted the formation of T cells with the phenotypic, functional and metabolic qualities found in naturally occurring T_{CM} cells [81, 80, 71, 112]. IL-15-generated tumor-reactive T cells exhibited enhanced anti-tumor responses compared to those grown in the presence of IL-2 [113]. Additionally, human T cells activated in the presence of low doses of IL-15 and IL-7 generated T_{SCM}-like cells capable of expanding and mediating GVHD on serial transplantation [114]. More recently, several investigators have evaluated the activity of IL-21 on the expansion and differentiation of tumor-specific $CD8^+$ T cells. In both mouse and human studies, IL-21 profoundly repressed T cell differentiation as manifested by the generation of T cells lacking the expression of conventional memory cell markers [115, 116]. T cells activated in IL-21 maintained a naive-like phenotype and the ability to secrete high amounts of IL-2 [116, 115, 117]. Although a comprehensive phenotypic characterization of these cells was not done in these studies, it is likely that IL-21-generated cells might comprise T_{SCM}-like cells.

In the past decade there has been an increasing understanding of the signaling pathways and transcriptional circuitry that regulate memory and effector T cell differentiation [42, 41]. Many of these pathways can now be targeted by small molecules, which are already approved, or under clinical evaluation for other indications. This raises the exciting possibility of repurposing these drugs to modulate T cell differentiation to potentiate T cell therapeutic fitness [77]. As discussed above, mTOR has emerged as a key modulator of $CD8^+$ T cell fate commitment and its modulation by rapamycin as well as by temsirolimus, a rapamycin analogue that is approved for treating advanced renal cell carcinoma, were shown to enhance the formation of $CD8^+$ memory T cells and augment their anti-tumor efficacy [118, 119].

Analogous results can also be obtained by targeting molecules upstream from mTOR such as AMPK and AKT. Metformin, an AMPK agonist used for the treatment of type 2 diabetes [120], has been shown to enhance T cell survival, memory responses and anti-tumor treatment [70]. Pharmacological blockade of AKT limited the acquisition of effector molecules and function while preserving a T_{CM}-like phenotype and migratory capacity [69]. Although the AKT inhibitor employed in this study is not available for use in humans, several other AKT inhibitors are currently under clinical evaluation for the treatment of solid tumors and hematologic malignancies [77]. Additionally, saracatinib, an Src family inhibitor undergoing clinical investigation for the treatment of cancer and Alzheimer's disease, was found to enhance the generation of T_{CM} cells in responses to vaccination and confer superior protection against tumor challenge through an unresolved signaling pathway regulating the AKT–mTOR pathway [121].

Another important pathway that has recently been implicated in the regulation of T cell differentiation and memory formation is the WNT–β-catenin signaling pathway [122, 123]. GSK3-β inhibitors that are under clinical evaluation for Alzheimer's disease and other neurodegenerative diseases [124] could be use to induce downstream signals of the WNT–β-catenin pathway to generate T_{SCM}-like cells capable of triggering potent anti-tumor immune responses [37, 36].

More recently, direct targeting of metabolic rate-limiting enzymes has been shown to be an effective strategy to restrain differentiation and enhance T cell memory and anti-tumor function [71, 72]. Inhibition of glycolytic flux by 2-DG, a hexokinase inhibitor currently under evaluation in clinical trials because of its direct negative impact on glycolytic tumor cells [125], limited T cell differentiation, resulting in the generation of T cells with improved anti-tumor efficacy [72]. Taken together, these studies underscore the ever-increasing number of reagents available that can be immediately integrated in the next generation protocols for the production of tumor-specific T cells for adoptive immunotherapy.

Reprogramming Terminally Differentiated T Cells

Restraining T cell differentiation during ex vivo expansion can be an effective strategy for therapies based on the adoptive transfer of T cells genetically engineered with a tumor-reactive TCR or CAR as a large fraction of peripheral blood lymphocytes comprises T_N and T_{CM} cells [37, 126]. However, this approach become less relevant for therapies relying on naturally occurring tumor-specific T cells, which are often found in a state of terminal differentiation and exhaustion due to chronic antigen stimulation in the tumor-bearing host [127–130]. Recent advances of regenerative medicine demonstrating successful reprogramming of mature cell lineages into induced pluripotent stem (iPS) cells have opened the exciting possibility of rejuvenating exhausted and senescent T cells [131]. Since Yamanaka's seminal discovery, mature T lymphocytes have been reprogrammed into iPS cells by enforcing expression of the transcription factors *OCT4*, *SOX2*, *KLF4* and *MYC* [132–134]. Importantly, T cell-derived iPS cells retain the rearranged variable (V), diversity (D) and joining regions (J) of the TCR chains, indicating that iPS cells generated from tumor-specific T cells could maintain their anti-tumor reactivity. Increasing understanding of the signaling required for T cell development during thymopoiesis has led to the development of ex vivo protocols that support the generation of T cells from stem cell precursors including iPS cells, providing a feasible methodology for re-differentiating T cell-derived iPS cells [135–139]. Recently, Vizcardo et al. [140] and Nishimura et al. [141] have put this two-step reprogramming concept into practice (Fig. 8.3B). These groups obtained iPS cells from T cell clones specific for the melanoma-associated antigen MART-1 or the HIV-1 protein Nef, respectively, and re-differentiated them into mature T cells by coculture with OP9 feeder cells overexpressing the Notch ligand Delta like-1 (DLL1). Although reprogrammed T cells maintained the original TCR rearrangement and the ability to mediate specific effector functions, it remains unclear whether these cells could be truly considered rejuvenated. These cells exhibited elongated telomeres, indicating an increased proliferative potential, however they displayed phenotypic traits of T_{EM} cells rather than naïve or early memory subsets. A recent paper indicated that reprogrammed T cells from iPS cells acquired the phenotype and functional characteristics of innate-like γδ T cells [142]. When redirected with a CD19-specific CAR, they were able to mediate potent tumor regression in a human lymphoma xenograft tumor model.

Although feasible, the two-step reprogramming approach is currently inefficient both in terms of the frequency of cells successfully reprogrammed and the duration necessary to achieve full reprogramming. Recently, Obokata and colleagues have reported that pluripotency could be induced faster and more efficiently by exposing lymphocytes to strong external stimuli such as a transient low-pH stressor [143]. Since then, misconduct proceedings have surfaced calling into question the validity of the results [144]. If confirmed and reproduced in adult human cells, however, Obokata's findings could make the two-step reprogramming method more practical (Fig. 8.3B).

An alternative approach to overcome the inefficiencies of two-step reprogramming might be the direct reprogram of terminally differentiated T cells into T_N and T_{SCM} cells. An increasing number of reports have revealed that direct reprogramming can be used to differentiate diverse mature cell types into alternative differentiated lineages such as neurons [145, 146], hepatocytes [147], cardiomyocytes [148], blood progenitors [149], and pancreatic β cells [150, 151] by ectopic expression of cell-specific transcription factors. Adapting this strategy, enforced expression of transcription factors or miRNAs essential to T_N and T_{SCM} identity might result in the intra-lineage reprogramming of T_{TE} cells into less-differentiated T cells (Fig. 8.3C).

Current Clinical Efforts and Future Directions

The realization that T cell differentiation and in vivo anti-tumor effectiveness are inversely correlated has recently prompted a series of new clinical trials designed to test the efficacy of less-differentiated T cell populations (Table 8.1). Because prolonged TIL cultures can drive cells towards terminal differentiation, changes were made to standard TIL protocols to shorten the period of ex vivo expansion [152–154]. These minimally cultured TILs were called young TILs, as they possessed characteristic of less-differentiated T cells including longer telomeres and higher expression of CD27 and CD28 compared to TIL cultured with a conventional protocol. Early trials using young TIL preparations have demonstrated anti-tumor efficacies comparable to standard TILs with objective responses of 28–58 % [155, 156, 153, 157, 104]. It is unclear the reason why young TILs did not induce increased anti-tumor responses compared to the historical experience using standard TILs. However, it is becoming evident that "young TILs" is a misnomer as the rapid expansion protocol employed to expand TILs prior to infusion virtually nullify the benefit of initial short TIL cultures resulting in low frequencies of less-differentiated T cells [152]. Furthermore, in a recent young TIL trial characterized by an unusually low response rate, infused TILs, rather than being minimally cultured, were grown for period of time comparable to standard TILs [156].

Alternative γ_C cytokines have been used in recent clinical protocols for the generation of tumor-reactive T cells for adoptive immunotherapy. Autologous MART1-specific $CD8^+$ T cells were generated in vitro using artificial antigen presenting cells in the presence of a combination of IL-2 and IL-15 [158]. This approach resulted in the generation of a mixture of T_{CM} and T_{EM} memory cells capable of engrafting and persist in patients for prolonged periods in the absence of previous lymphodepletion conditioning or cytokine support. Notably, these cells trafficked to the tumor and

Table 8.1 Adoptive immunotherapy trials designed to administer less differentiated T cells

Target	Cancer	Cell product	ID	Center	Status
Undefined	Melanoma	Young tumor infiltrating lymphocytes	NCT00513604	NCI	Completed (Dudley et al. [156])
Undefined	Melanoma	Young tumor infiltrating lymphocytes	NCT00287131	Sheba Medical Center	Completed (Besser et al. [104])
Undefined	GI cancer	Young tumor infiltrating lymphocytes	NCT01174121	NCI	Recruiting
E6; E7	HPV-associated cancers	Young tumor infiltrating lymphocytes	NCT01585428	NCI	Recruiting
MART-1	Melanoma	IL-15/IL-2 modulated, MART-1 specific $CD8^+$ T cells	NCT00512889	Dana-Farber Cancer Institute	Completed (Butler et al. [158])
WT-1	Melanoma	IL-21 modulated, WT-1 specific $CD8^+$ T cell clones	NCT00052520	Fred Hutchinson Cancer Research	Completed (Chapuis et al. [159])
MART-1	Melanoma	IL-21 modulated, MART-1 specific $CD8^+$ T cell clones	NCT01106235	Fred Hutchinson Cancer Research	Completed
gp100	Melanoma	gp100-specific $CD8^+$ T cell clones derived from high *IL-2*: *IFNG* index precursors	NCT00665470	NCI	Completed (Wang et al. [162])
CD19	$CD19^+$ B cell malignancies	CD19-CAR specific $CD8^+$ T-cells derived from virus-specific T_{CM}	NCT01475058	Fred Hutchinson Cancer Research	Recruiting
CD19	B-Lineage NHL	CD19-CAR $CD8^+$ T cells derived from T_{CM} enriched cells	NCT01318317	City of Hope Medical Center	Active, non-recruiting
CD19	B-Lineage NHL	CD19-CAR T cells derived from T_{CM} enriched cells	NCT01815749	City of Hope Medical Center	Recruiting
NY-ESO-1	Melanoma	NY-ESO-1 TCR T cells derived from $CD62L^+$ cells	NCT02062359	NCI	Not yet recruiting

GI gastrointestinal, *HPV* human papilloma virus, *IL* interleukin, *IFNG* interferon-γ, *NHLnon-Hodgkin*'s lymphoma, T_{CM} central memory T cells, *NCI* National Cancer Institute mediated biological and clinical responses. IL-21 has also been used to limit terminal differentiation of WT1-specific donor-derived $CD8^+$ T cell clones [159]. WT1-specific clonal populations generated with exposure to IL-21 displayed higher CD27, CD28, or IL-7Rα compared to clones generated in the absence of this cytokine. Consistent with previous studies [6–8], clones generated without IL-21 failed to persist longer than two weeks in vivo. Remarkably, WT-1-specific clones generated in the presence of IL-21 survived long-term after infusion, establishing immunological memory. Most importantly, IL-21 generated clones exhibited direct evidence of anti-leukemic activity.

Another promising strategy currently under clinical evaluation is the use of cell products derived from the expansion of T_{CM} cells. These studies stem from preclinical evidence in mice and nonhuman primates indicating that progenies of isolated T_{CM} cells have enhanced capacity to persist and form long-lived memory cells following adoptive transfer compared to T_{EM}-derived cells [160, 161]. Since reagents necessary to isolate T_{CM} in a high-throughput manner were not initially available, Wang and colleagues took advantage of the well-known ability of T_{CM} cells to produce greater amount of IL-2 than their T_{EM} counterparts to develop a PCR-based assay for early detection of T_{CM} clones [162]. This strategy enabled the isolation, expansion, and transfer of rare human melanoma-specific $CD8^+$ T_{CM} cells. TCM-derived T cell clones engrafted and persisted at high frequencies in 4 out of 5 patients, 1 month after the transfer, and were associated with some minor and mixed tumor regression [162]. More recently, the development of GMP-compliant beads for T_{CM} cell isolation [163] has led to the initiation of a series of trials employing CD19-CAR engineered $CD8^+$ T cells derived from T_{CM} cells for the treatment of B cell malignancies (Table 8.1). Recently, two studies have suggested that T cells derived from naïve rather than T_{CM} cells might allow superior efficacy upon adoptive transfer [164, 126]. T_N-derived cells exhibited greater proliferative potential and mediated enhanced anti-tumor function compared to T_{CM} cells in a murine tumor model [164]. Moreover, in human studies, T_N-derived cells displayed higher expression of CD27 and longer telomeres compared to cells derived from conventional memory subsets, indicating that the T_N-derived progeny possess traits that correlate with tumor responses in clinical trials [126]. Thus, a new study employing tumor-specific T cells derived from $CD62L^+$ precursors, which comprise naïve, T_{SCM} and T_{CM} cell has recently been planned (Table 8.1).

In summary, several new studies investigating the safety and efficacy of less-differentiated cells have been initiated or planned. Preliminary results are starting to reveal significant improvements in terms of T cell persistence, which hopefully, will translate into increased tumor response rates. New clinical-grade protocols for T_{SCM} cell generation are also under development, paving the way for a rapid translation of T_{SCM} into future clinical trials [165, 114].

Acknowledgments This work was supported by the Intramural Research Programs of the US National Institutes of Health, National Cancer Institute and by grants from the Associazione Italiana per la Ricerca sul Cancro (MFAG10607), Fondazione Cariplo (2012/0683), Italian Ministry of Health (Bando Giovani Ricercatori GR-2011-02347324) and the European Community

(Marie Curie Career Integration Grant 322093) to E.L. The authors would like to thank Steven Rosenberg for providing images of the patients treated in the Surgery Branch, and Yun Ji and Alessandra Roberto for critical discussion. E.L. is an International Society for the Advancement of Cytometry (ISAC) scholar.

References

1. Maus MV, Fraietta JA, Levine BL et al (2014) Adoptive immunotherapy for cancer or viruses. Annu Rev Immunol 32:189–225. doi:10.1146/annurev-immunol-032713-120136
2. Restifo NP, Dudley ME, Rosenberg SA (2012) Adoptive immunotherapy for cancer: harnessing the T cell response. Nat Rev Immunol 12:269–281. doi:10.1038/nri3191
3. Bollard CM, Rooney CM, Heslop HE (2012) T-cell therapy in the treatment of post-transplant lymphoproliferative disease. Nat Rev Clin Oncol 9:510–519. doi:10.1038/nrclinonc.2012.111
4. Dudley ME, Wunderlich JR, Robbins PF et al (2002) Cancer regression and autoimmunity in patients after clonal repopulation with antitumor lymphocytes. Science 298:850–854. doi:10.1126/science.10765141076514
5. Mackensen A, Meidenbauer N, Vogl S et al (2006) Phase I study of adoptive T-cell therapy using antigen-specific CD8+ T cells for the treatment of patients with metastatic melanoma. J Clin Oncol 24:5060–5069. doi:10.1200/JCO.2006.07.1100
6. Dudley ME, Wunderlich J, Nishimura MI et al (2001) Adoptive transfer of cloned melanoma-reactive T lymphocytes for the treatment of patients with metastatic melanoma. J Immunother 24:363–373
7. Dudley ME, Wunderlich JR, Yang JC et al (2002) A phase I study of nonmyeloablative chemotherapy and adoptive transfer of autologous tumor antigen-specific T lymphocytes in patients with metastatic melanoma. J Immunother 25:243–251
8. Yee C, Thompson JA, Byrd D et al (2002) Adoptive T cell therapy using antigen-specific CD8+ T cell clones for the treatment of patients with metastatic melanoma: in vivo persistence, migration, and antitumor effect of transferred T cells. Proc Natl Acad Sci U S A 99:16168–16173. doi:10.1073/pnas.242600099
9. Morgan RA, Dudley ME, Wunderlich JR et al (2006) Cancer regression in patients after transfer of genetically engineered lymphocytes. Science 314:126–129. doi:10.1126/science.1129003
10. Johnson LA, Morgan RA, Dudley ME et al (2009) Gene therapy with human and mouse T-cell receptors mediates cancer regression and targets normal tissues expressing cognate antigen. Blood 114:535–546. doi:10.1182/blood-2009-03-211714
11. Robbins PF, Morgan RA, Feldman SA et al (2011) Tumor regression in patients with metastatic synovial cell sarcoma and melanoma using genetically engineered lymphocytes reactive with NY-ESO-1. J Clin Oncol 29:917–924. doi:10.1200/JCO.2010.32.2537
12. Brentjens RJ, Davila ML, Riviere I et al (2013) CD19-Targeted T cells rapidly induce molecular remissions in adults with chemotherapy-refractory acute lymphoblastic leukemia. Sci Transl Med 5:177ra138. doi:10.1126/scitranslmed.3005930
13. Grupp SA, Kalos M, Barrett D et al (2013) Chimeric antigen receptor-modified T cells for acute lymphoid leukemia. N Engl J Med 368:1509–1518. doi:10.1056/NEJMoa1215134
14. Porter DL, Levine BL, Kalos M et al (2011) Chimeric antigen receptor-modified T cells in chronic lymphoid leukemia. N Engl J Med 365:725–733. doi:10.1056/NEJMoa1103849
15. Pule MA, Savoldo B, Myers GD et al (2008) Virus-specific T cells engineered to coexpress tumor-specific receptors: persistence and antitumor activity in individuals with neuroblastoma. Nat Med 14:1264–1270. doi:10.1038/nm.1882
16. Kochenderfer JN, Dudley ME, Carpenter RO et al (2013) Donor-derived CD19-targeted T cells cause regression of malignancy persisting after allogeneic hematopoietic stem cell transplantation. Blood 122:4129–4139. doi:10.1182/blood-2013-08-519413

17. Kochenderfer JN, Wilson WH, Janik JE et al (2010) Eradication of B-lineage cells and regression of lymphoma in a patient treated with autologous T cells genetically engineered to recognize CD19. Blood 116:4099–4102. doi:10.1182/blood-2010-04-281931
18. Zhang N, Bevan MJ (2011) CD8(+) T cells: foot soldiers of the immune system. Immunity 35:161–168. doi:10.1016/j.immuni.2011.07.010
19. Biteau B, Hochmuth CE, Jasper H (2011) Maintaining tissue homeostasis: dynamic control of somatic stem cell activity. Cell Stem Cell 9:402–411. doi:10.1016/j.stem.2011.10.004
20. Mahnke YD, Brodie TM, Sallusto F et al (2013) The who's who of T-cell differentiation: human memory T-cell subsets. Eur J Immunol 43:2797–2809. doi:10.1002/eji.201343751
21. Sallusto F, Geginat J, Lanzavecchia A (2004) Central memory and effector memory T cell subsets: function, generation, and maintenance. Annu Rev Immunol 22:745–763. doi:10.1146/annurev.immunol.22.012703.104702
22. Hamann D, Baars PA, Rep MH et al (1997) Phenotypic and functional separation of memory and effector human CD8+ T cells. J Exp Med 186:1407–1418
23. Sallusto F, Lenig D, Forster R et al (1999) Two subsets of memory T lymphocytes with distinct homing potentials and effector functions. Nature 401:708–712. doi:10.1038/44385
24. Brenchley JM, Karandikar NJ, Betts MR et al (2003) Expression of CD57 defines replicative senescence and antigen-induced apoptotic death of CD8+ T cells. Blood 101:2711–2720. doi:10.1182/blood-2002-07-2103
25. Chattopadhyay PK, Betts MR, Price DA et al (2009) The cytolytic enzymes granyzme A, granzyme B, and perforin: expression patterns, cell distribution, and their relationship to cell maturity and bright CD57 expression. J Leukoc Biol 85:88–97. doi:10.1189/jlb.0208107
26. Picker LJ, Reed-Inderbitzin EF, Hagen SI et al (2006) IL-15 induces CD4 effector memory T cell production and tissue emigration in nonhuman primates. J Clin Invest 116:1514–1524. doi:10.1172/JCI27564
27. Lugli E, Goldman CK, Perera LP et al (2010) Transient and persistent effects of IL15 on lymphocyte homeostasis in nonhuman primates. Blood 116:3238–3248. doi:10.1182/blood-2010-03-275438
28. Olson JA, McDonald-Hyman C, Jameson SC et al (2013) Effector-like CD8(+) T cells in the memory population mediate potent protective immunity. Immunity 38:1250–1260. doi:10.1016/j.immuni.2013.05.009
29. Geginat J, Sallusto F, Lanzavecchia A (2003) Cytokine-driven proliferation and differentiation of human naive, central memory and effector memory CD4+ T cells. Pathol Biol (Paris) 51:64–66
30. Geginat J, Lanzavecchia A, Sallusto F (2003) Proliferation and differentiation potential of human CD8+ memory T-cell subsets in response to antigen or homeostatic cytokines. Blood 101:4260–4266. doi:10.1182/blood-2002-11-3577
31. Stemberger C, Neuenhahn M, Gebhardt FE et al (2009) Stem cell-like plasticity of naive and distinct memory CD8+ T cell subsets. Semin Immunol 21:62–68. doi:10.1016/j.smim.2009.02.004. pii: S1044-5323(09)00009-8
32. Mueller SN, Gebhardt T, Carbone FR et al (2013) Memory T cell subsets, migration patterns, and tissue residence. Annu Rev Immunol 31:137–161. doi:10.1146/annurev-immunol-032712-095954
33. Lanzavecchia A, Sallusto F (2002) Progressive differentiation and selection of the fittest in the immune response. Nat Rev Immunol 2:982–987. doi:10.1038/nri959
34. Fearon DT, Manders P, Wagner SD (2001) Arrested differentiation, the self-renewing memory lymphocyte, and vaccination. Science 293:248–250. doi:10.1126/science.1062589
35. Zhang Y, Joe G, Hexner E et al (2005) Host-reactive CD8+ memory stem cells in graft-versus-host disease. Nat Med 11:1299–1305. doi:10.1038/nm1326
36. Gattinoni L, Zhong XS, Palmer DC et al (2009) Wnt signaling arrests effector T cell differentiation and generates CD8+ memory stem cells. Nat Med 15:808–813. doi:10.1038/nm.1982
37. Gattinoni L, Lugli E, Ji Y et al (2011) A human memory T cell subset with stem cell-like properties. Nat Med 17:1290–1297. doi:10.1038/nm.2446

38. Lugli E, Dominguez MH, Gattinoni L et al (2013) Superior T memory stem cell persistence supports long-lived T cell memory. J Clin Invest 123:594–599. doi:10.1172/JCI66327
39. Buzon MJ, Sun H, Li C et al (2014) HIV-1 persistence in CD4(+) T cells with stem cell-like properties. Nat Med 20:139–142. doi:10.1038/nm.3445
40. Cartwright EK, McGary CS, Cervasi B et al (2014) Divergent CD4+ T memory stem cell dynamics in pathogenic and nonpathogenic simian immunodeficiency virus infections. J Immunol 192:4666–4673. doi:10.4049/jimmunol.1303193
41. Kaech SM, Cui W (2012) Transcriptional control of effector and memory CD8+ T cell differentiation. Nat Rev Immunol 12:749–761. doi:10.1038/nri3307
42. Thaventhiran JE, Fearon DT, Gattinoni L (2013) Transcriptional regulation of effector and memory CD8+ T cell fates. Curr Opin Immunol 25:321–328. doi:10.1016/j.coi.2013.05.010
43. Kaech SM, Tan JT, Wherry EJ et al (2003) Selective expression of the interleukin 7 receptor identifies effector CD8 T cells that give rise to long-lived memory cells. Nat Immunol 4:1191–1198. doi:10.1038/ni1009
44. Huster KM, Busch V, Schiemann M et al (2004) Selective expression of IL-7 receptor on memory T cells identifies early CD40L-dependent generation of distinct CD8+ memory T cell subsets. Proc Natl Acad Sci U S A 101:5610–5615. doi:10.1073/pnas.0308054101
45. Sarkar S, Kalia V, Haining WN et al (2008) Functional and genomic profiling of effector CD8 T cell subsets with distinct memory fates. J Exp Med 205:625–640. doi:10.1084/jem.20071641
46. Joshi NS, Cui W, Chandele A et al (2007) Inflammation directs memory precursor and short-lived effector CD8(+) T cell fates via the graded expression of T-bet transcription factor. Immunity 27:281–295. doi:10.1016/j.immuni.2007.07.010
47. Intlekofer AM, Takemoto N, Wherry EJ et al (2005) Effector and memory CD8+ T cell fate coupled by T-bet and eomesodermin. Nat Immunol 6:1236–1244. doi:10.1038/ni1268
48. Pearce EL, Mullen AC, Martins GA et al (2003) Control of effector CD8+ T cell function by the transcription factor eomesodermin. Science 302:1041–1043. doi:10.1126/science.1090148
49. Cruz-Guilloty F, Pipkin ME, Djuretic IM et al (2009) Runx3 and T-box proteins cooperate to establish the transcriptional program of effector CTLs. J Exp Med 206:51–59. doi:10.1084/jem.20081242
50. Banerjee A, Gordon SM, Intlekofer AM et al (2010) Cutting edge: the transcription factor eomesodermin enables CD8+ T cells to compete for the memory cell niche. J Immunol 185:4988–4992. doi:10.4049/jimmunol.1002042
51. Rutishauser RL, Martins GA, Kalachikov S et al (2009) Transcriptional repressor Blimp-1 promotes CD8(+) T cell terminal differentiation and represses the acquisition of central memory T cell properties. Immunity 31:296–308. doi:10.1016/j.immuni.2009.05.014
52. Kallies A, Xin A, Belz GT et al (2009) Blimp-1 transcription factor is required for the differentiation of effector CD8(+) T cells and memory responses. Immunity 31:283–295. doi:10.1016/j.immuni.2009.06.021
53. Ji Y, Pos Z, Rao M et al (2011) Repression of the DNA-binding inhibitor Id3 by Blimp-1 limits the formation of memory CD8+ T cells. Nat Immunol 12:1230–1237. doi:10.1038/ni.2153
54. Yang CY, Best JA, Knell J et al (2011) The transcriptional regulators Id2 and Id3 control the formation of distinct memory CD8+ T cell subsets. Nat Immunol 12:1221–1229. doi:10.1038/ni.2158
55. Hu G, Chen J (2013) A genome-wide regulatory network identifies key transcription factors for memory CD8(+) T-cell development. Nat Commun 4:2830. doi:10.1038/ncomms3830
56. Kuwahara M, Suzuki J, Tofukuji S et al (2014) The Menin-Bach2 axis is critical for regulating CD4 T-cell senescence and cytokine homeostasis. Nat Commun 5:3555. doi:10.1038/ncomms4555
57. Hedrick SM, Hess Michelini R, Doedens AL et al (2012) FOXO transcription factors throughout T cell biology. Nat Rev Immunol 12:649–661. doi:10.1038/nri3278
58. Riou C, Yassine-Diab B, Van grevenynghe J et al (2007) Convergence of TCR and cytokine signaling leads to FOXO3a phosphorylation and drives the survival of CD4+ central memory T cells. J Exp Med 204:79–91. doi:10.1084/jem.20061681

59. Tzelepis F, Joseph J, Haddad EK et al (2013) Intrinsic role of FoxO3a in the development of CD8+ T cell memory. J Immunol 190:1066–1075. doi:10.4049/jimmunol.1200639
60. Kim MV, Ouyang W, Liao W et al (2013) The transcription factor Foxo1 controls central-memory CD8+ T cell responses to infection. Immunity 39:286–297. doi:10.1016/j.immuni.2013.07.013
61. Hess Michelini R, Doedens AL, Goldrath AW et al (2013) Differentiation of CD8 memory T cells depends on Foxo1. J Exp Med 210:1189–1200. doi:10.1084/jem.20130392
62. Badovinac VP, Haring JS, Harty JT (2007) Initial T cell receptor transgenic cell precursor frequency dictates critical aspects of the CD8(+) T cell response to infection. Immunity 26:827–841. doi:10.1016/j.immuni.2007.04.013. pii: S1074-7613(07)00286-5
63. Pearce EL, Poffenberger MC, Chang CH et al (2013) Fueling immunity: insights into metabolism and lymphocyte function. Science 342:1242454. doi:10.1126/science.1242454
64. van der Windt GJ, O'Sullivan D, Everts B et al (2013) CD8 memory T cells have a bioenergetic advantage that underlies their rapid recall ability. Proc Natl Acad Sci U S A 110:14336–14341. doi:10.1073/pnas.1221740110
65. Gubser PM, Bantug GR, Razik L et al (2013) Rapid effector function of memory CD8+ T cells requires an immediate-early glycolytic switch. Nat Immunol 14:1064–1072. doi:10.1038/ni.2687
66. MacIver NJ, Michalek RD, Rathmell JC (2013) Metabolic regulation of T lymphocytes. Annu Rev Immunol 31:259–283. doi:10.1146/annurev-immunol-032712-095956
67. Vander Heiden MG, Cantley LC, Thompson CB (2009) Understanding the Warburg effect: the metabolic requirements of cell proliferation. Science 324:1029–1033. doi:10.1126/science.1160809
68. Chang CH, Curtis JD, Maggi LB Jr et al (2013) Posttranscriptional control of T cell effector function by aerobic glycolysis. Cell 153:1239–1251. doi:10.1016/j.cell.2013.05.016
69. Macintyre AN, Finlay D, Preston G et al (2011) Protein kinase B controls transcriptional programs that direct cytotoxic T cell fate but is dispensable for T cell metabolism. Immunity 34:224–236. doi:10.1016/j.immuni.2011.01.012
70. Pearce EL, Walsh MC, Cejas PJ et al (2009) Enhancing CD8 T-cell memory by modulating fatty acid metabolism. Nature 460:103–107. doi:10.1038/nature08097
71. van der Windt GJ, Everts B, Chang CH et al (2012) Mitochondrial respiratory capacity is a critical regulator of CD8+ T cell memory development. Immunity 36:68–78. doi:10.1016/j.immuni.2011.12.007
72. Sukumar M, Liu J, Ji Y et al (2013) Inhibiting glycolytic metabolism enhances CD8+ T cell memory and antitumor function. J Clin Invest 123:4479–4488. doi:10.1172/JCI69589
73. Yang K, Chi H (2012) mTOR and metabolic pathways in T cell quiescence and functional activation. Semin Immunol 24:421–428. doi:10.1016/j.smim.2012.12.004
74. Laplante M, Sabatini DM (2012) mTOR signaling in growth control and disease. Cell 149:274–293. doi:10.1016/j.cell.2012.03.017
75. Chi H (2012) Regulation and function of mTOR signalling in T cell fate decisions. Nat Rev Immunol 12:325–338. doi:10.1038/nri3198
76. Araki K, Turner AP, Shaffer VO et al (2009) mTOR regulates memory CD8 T-cell differentiation. Nature 460:108–112. doi:10.1038/nature08155
77. Gattinoni L, Klebanoff CA, Restifo NP (2009) Pharmacologic induction of CD8+ T cell memory: better living through chemistry. Sci Transl Med 1:11ps12. doi:10.1126/scitranslmed.3000302
78. Dudley ME, Rosenberg SA (2003) Adoptive-cell-transfer therapy for the treatment of patients with cancer. Nat Rev Cancer 3:666–675. doi:10.1038/nrc1167
79. Overwijk WW, Theoret MR, Finkelstein SE et al (2003) Tumor regression and autoimmunity after reversal of a functionally tolerant state of self-reactive CD8+ T cells. J Exp Med 198:569–580. doi:10.1084/jem.20030590
80. Gattinoni L, Klebanoff CA, Palmer DC et al (2005) Acquisition of full effector function in vitro paradoxically impairs the in vivo antitumor efficacy of adoptively transferred CD8+ T cells. J Clin Invest 115:1616–1626. doi:10.1172/JCI24480

81. Klebanoff CA, Gattinoni L, Torabi-Parizi P et al (2005) Central memory self/tumor-reactive CD8+ T cells confer superior antitumor immunity compared with effector memory T cells. Proc Natl Acad Sci U S A 102:9571–9576. doi:10.1073/pnas.0503726102
82. Stark FC, Sad S, Krishnan L (2009) Intracellular bacterial vectors that induce CD8(+) T cells with similar cytolytic abilities but disparate memory phenotypes provide contrasting tumor protection. Cancer Res 69:4327–4334. doi:10.1158/0008-5472.CAN-08-3160
83. Le HK, Graham L, Miller CH et al (2009) Incubation of antigen-sensitized T lymphocytes activated with bryostatin 1 + ionomycin in IL-7 + IL-15 increases yield of cells capable of inducing regression of melanoma metastases compared to culture in IL-2. Cancer Immunol Immunother 58:1565–1576. doi:10.1007/s00262-009-0666-y
84. Sussman JJ, Parihar R, Winstead K et al (2004) Prolonged culture of vaccine-primed lymphocytes results in decreased antitumor killing and change in cytokine secretion. Cancer Res 64:9124–9130. doi:10.1158/0008-5472.CAN-03-0376
85. Wherry EJ, Teichgraber V, Becker TC et al (2003) Lineage relationship and protective immunity of memory CD8 T cell subsets. Nat Immunol 4:225–234. doi:10.1038/ni889
86. Nolz JC, Harty JT (2011) Protective capacity of memory CD8+ T cells is dictated by antigen exposure history and nature of the infection. Immunity 34:781–793. doi:10.1016/j.immuni.2011.03.020
87. Chen BJ, Cui X, Sempowski GD et al (2004) Transfer of allogeneic CD62L- memory T cells without graft-versus-host disease. Blood 103:1534–1541. doi:10.1182/blood-2003-08-2987
88. Bondanza A, Valtolina V, Magnani Z et al (2006) Suicide gene therapy of graft-versus-host disease induced by central memory human T lymphocytes. Blood 107:1828–1836. doi:10.1182/blood-2005-09-3716
89. Acuto O, Michel F (2003) CD28-mediated co-stimulation: a quantitative support for TCR signalling. Nat Rev Immunol 3:939–951. doi:10.1038/nri1248
90. Topp MS, Riddell SR, Akatsuka Y et al (2003) Restoration of CD28 expression in CD28– CD8+ memory effector T cells reconstitutes antigen-induced IL-2 production. J Exp Med 198:947–955. doi:10.1084/jem.20021288
91. Hendriks J, Gravestein LA, Tesselaar K et al (2000) CD27 is required for generation and long-term maintenance of T cell immunity. Nat Immunol 1:433–440. doi:10.1038/80877
92. Hendriks J, Xiao Y, Borst J (2003) CD27 promotes survival of activated T cells and complements CD28 in generation and establishment of the effector T cell pool. J Exp Med 198:1369–1380. doi:10.1084/jem.20030916
93. Blackburn SD, Shin H, Haining WN et al (2009) Coregulation of CD8+ T cell exhaustion by multiple inhibitory receptors during chronic viral infection. Nat Immunol 10:29–37. doi:10.1038/ni.1679
94. Henson SM, Akbar AN (2009) KLRG1--more than a marker for T cell senescence. Age (Dordr) 31:285–291. doi:10.1007/s11357-009-9100-9
95. Pantaleo G, Harari A (2006) Functional signatures in antiviral T-cell immunity for monitoring virus-associated diseases. Nat Rev Immunol 6:417–423. doi:10.1038/nri1840
96. Hodes RJ, Hathcock KS, Weng NP (2002) Telomeres in T and B cells. Nat Rev Immunol 2:699–706. doi:10.1038/nri890
97. Romero P, Zippelius A, Kurth I et al (2007) Four functionally distinct populations of human effector-memory CD8+ T lymphocytes. J Immunol 178:4112–4119
98. Gattinoni L, Klebanoff CA, Restifo NP (2012) Paths to stemness: building the ultimate antitumour T cell. Nat Rev Cancer 12:671–684. doi:10.1038/nrc3322
99. Kalia V, Sarkar S, Subramaniam S et al (2010) Prolonged interleukin-2Ralpha expression on virus-specific CD8+ T cells favors terminal-effector differentiation in vivo. Immunity 32:91–103. doi:10.1016/j.immuni.2009.11.010
100. Rosenberg SA, Yang JC, Sherry RM et al (2011) Durable complete responses in heavily pretreated patients with metastatic melanoma using T-cell transfer immunotherapy. Clin Cancer Res 17:4550–4557. doi:10.1158/1078-0432.CCR-11-0116

101. Robbins PF, Dudley ME, Wunderlich J et al (2004) Cutting edge: persistence of transferred lymphocyte clonotypes correlates with cancer regression in patients receiving cell transfer therapy. J Immunol 173:7125–7130
102. Kalos M, Levine BL, Porter DL et al (2011) T cells with chimeric antigen receptors have potent antitumor effects and can establish memory in patients with advanced leukemia. Sci Transl Med 3:95ra73. doi:10.1126/scitranslmed.3002842
103. Schwartzentruber DJ, Hom SS, Dadmarz R et al (1994) In vitro predictors of therapeutic response in melanoma patients receiving tumor-infiltrating lymphocytes and interleukin-2. J Clin Oncol 12:1475–1483
104. Besser MJ, Shapira-Frommer R, Treves AJ et al (2010) Clinical responses in a phase II study using adoptive transfer of short-term cultured tumor infiltration lymphocytes in metastatic melanoma patients. Clin Cancer Res 16:2646–2655. doi:10.1158/1078-0432.CCR-10-0041
105. Zhou J, Shen X, Huang J et al (2005) Telomere length of transferred lymphocytes correlates with in vivo persistence and tumor regression in melanoma patients receiving cell transfer therapy. J Immunol 175:7046–7052
106. Huang J, Kerstann KW, Ahmadzadeh M et al (2006) Modulation by IL-2 of CD70 and CD27 expression on CD8+ T cells: importance for the therapeutic effectiveness of cell transfer immunotherapy. J Immunol 176:7726–7735
107. Louis CU, Savoldo B, Dotti G et al (2011) Antitumor activity and long-term fate of chimeric antigen receptor-positive T cells in patients with neuroblastoma. Blood 118:6050–6056. doi:10.1182/blood-2011-05-354449
108. Radvanyi LG, Bernatchez C, Zhang M et al (2012) Specific lymphocyte subsets predict response to adoptive cell therapy using expanded autologous tumor-infiltrating lymphocytes in metastatic melanoma patients. Clin Cancer Res 18:6758–6770. doi:10.1158/1078-0432.CCR-12-1177
109. Topalian SL, Muul LM, Solomon D et al (1987) Expansion of human tumor infiltrating lymphocytes for use in immunotherapy trials. J Immunol Methods 102:127–141
110. Riddell SR, Watanabe KS, Goodrich JM et al (1992) Restoration of viral immunity in immunodeficient humans by the adoptive transfer of T cell clones. Science 257:238–241
111. Gattinoni L, Powell DJ Jr, Rosenberg SA et al (2006) Adoptive immunotherapy for cancer: building on success. Nat Rev Immunol 6:383–393. doi:10.1038/nri1842
112. Manjunath N, Shankar P, Wan J et al (2001) Effector differentiation is not prerequisite for generation of memory cytotoxic T lymphocytes. J Clin Invest 108:871–878. doi:10.1172/JCI13296
113. Klebanoff CA, Finkelstein SE, Surman DR et al (2004) IL-15 enhances the in vivo antitumor activity of tumor-reactive CD8+ T cells. Proc Natl Acad Sci U S A 101:1969–1974. doi:10.1073/pnas.0307298101
114. Cieri N, Camisa B, Cocchiarella F et al (2013) IL-7 and IL-15 instruct the generation of human memory stem T cells from naive precursors. Blood 121:573–584. doi:10.1182/blood-2012-05-431718
115. Hinrichs CS, Spolski R, Paulos CM et al (2008) IL-2 and IL-21 confer opposing differentiation programs to CD8+ T cells for adoptive immunotherapy. Blood 111:5326–5333. doi:10.1182/blood-2007-09-113050
116. Albrecht J, Frey M, Teschner D et al (2011) IL-21-treated naive CD45RA+ CD8+ T cells represent a reliable source for producing leukemia-reactive cytotoxic T lymphocytes with high proliferative potential and early differentiation phenotype. Cancer Immunol Immunother 60:235–248. doi:10.1007/s00262-010-0936-8
117. Li Y, Bleakley M, Yee C (2005) IL-21 influences the frequency, phenotype, and affinity of the antigen-specific CD8 T cell response. J Immunol 175:2261–2269
118. Rao RR, Li Q, Odunsi K et al (2010) The mTOR kinase determines effector versus memory CD8+ T cell fate by regulating the expression of transcription factors T-bet and Eomesodermin. Immunity 32:67–78. doi:10.1016/j.immuni.2009.10.010
119. Wang Y, Wang XY, Subjeck JR et al (2011) Temsirolimus, an mTOR inhibitor, enhances anti-tumour effects of heat shock protein cancer vaccines. Br J Cancer 104:643–652. doi:10.1038/bjc.2011.15

120. Pernicova I, Korbonits M (2014) Metformin--mode of action and clinical implications for diabetes and cancer. Nat Rev Endocrinol 10:143–156. doi:10.1038/nrendo.2013.256
121. Takai S, Sabzevari H, Farsaci B et al (2012) Distinct effects of saracatinib on memory CD8+ T cell differentiation. J Immunol 188:4323–4333. doi:10.4049/jimmunol.1101439
122. Gattinoni L, Ji Y, Restifo NP (2010) Wnt/beta-catenin signaling in T-cell immunity and cancer immunotherapy. Clin Cancer Res 16:4695–4701. doi:10.1158/1078-0432.CCR-10-0356
123. Xue HH, Zhao DM (2012) Regulation of mature T cell responses by the Wnt signaling pathway. Ann N Y Acad Sci 1247:16–33. doi:10.1111/j.1749-6632.2011.06302.x
124. Cohen P, Goedert M (2004) GSK3 inhibitors: development and therapeutic potential. Nat Rev Drug Discov 3:479–487. doi:10.1038/nrd1415
125. Zhao Y, Butler EB, Tan M (2013) Targeting cellular metabolism to improve cancer therapeutics. Cell Death Dis 4, e532. doi:10.1038/cddis.2013.60
126. Hinrichs CS, Borman ZA, Gattinoni L et al (2011) Human effector CD8+ T cells derived from naive rather than memory subsets possess superior traits for adoptive immunotherapy. Blood 117:808–814. doi:10.1182/blood-2010-05-286286
127. Zippelius A, Batard P, Rubio-Godoy V et al (2004) Effector function of human tumor-specific CD8 T cells in melanoma lesions: a state of local functional tolerance. Cancer Res 64:2865–2873
128. Ahmadzadeh M, Johnson LA, Heemskerk B et al (2009) Tumor antigen-specific CD8 T cells infiltrating the tumor express high levels of PD-1 and are functionally impaired. Blood 114:1537–1544. doi:10.1182/blood-2008-12-195792
129. Gros A, Robbins PF, Yao X et al (2014) PD-1 identifies the patient-specific CD8+ tumor-reactive repertoire infiltrating human tumors. J Clin Invest 124:2246–2259. doi:10.1172/JCI73639
130. Baitsch L, Baumgaertner P, Devevre E et al (2011) Exhaustion of tumor-specific CD8(+) T cells in metastases from melanoma patients. J Clin Invest 121:2350–2360. doi:10.1172/JCI46102
131. Crompton JG, Clever D, Vizcardo R et al (2014) Reprogramming antitumor immunity. Trends Immunol 35:178–185. doi:10.1016/j.it.2014.02.003
132. Staerk J, Dawlaty MM, Gao Q et al (2010) Reprogramming of human peripheral blood cells to induced pluripotent stem cells. Cell Stem Cell 7:20–24. doi:10.1016/j.stem.2010.06.002
133. Zhang XB (2013) Cellular reprogramming of human peripheral blood cells. Genomics Proteomics Bioinformatics 11:264–274. doi:10.1016/j.gpb.2013.09.001
134. Seki T, Yuasa S, Oda M et al (2010) Generation of induced pluripotent stem cells from human terminally differentiated circulating T cells. Cell Stem Cell 7:11–14. doi:10.1016/j.stem.2010.06.003
135. Schmitt TM, de Pooter RF, Gronski MA et al (2004) Induction of T cell development and establishment of T cell competence from embryonic stem cells differentiated in vitro. Nat Immunol 5:410–417. doi:10.1038/ni1055
136. Schmitt TM, Zuniga-Pflucker JC (2002) Induction of T cell development from hematopoietic progenitor cells by delta-like-1 in vitro. Immunity 17:749–756
137. Clark RA, Yamanaka K, Bai M et al (2005) Human skin cells support thymus-independent T cell development. J Clin Invest 115:3239–3249. doi:10.1172/JCI24731
138. Zhao Y, Parkhurst MR, Zheng Z et al (2007) Extrathymic generation of tumor-specific T cells from genetically engineered human hematopoietic stem cells via Notch signaling. Cancer Res 67:2425–2429. doi:10.1158/0008-5472.CAN-06-3977
139. Lei F, Haque R, Weiler L et al (2009) T lineage differentiation from induced pluripotent stem cells. Cell Immunol 260:1–5. doi:10.1016/j.cellimm.2009.09.005
140. Vizcardo R, Masuda K, Yamada D et al (2013) Regeneration of human tumor antigen-specific T cells from iPSCs derived from mature CD8(+) T cells. Cell Stem Cell 12:31–36. doi:10.1016/j.stem.2012.12.006
141. Nishimura T, Kaneko S, Kawana-Tachikawa A et al (2013) Generation of rejuvenated antigen-specific T cells by reprogramming to pluripotency and redifferentiation. Cell Stem Cell 12:114–126. doi:10.1016/j.stem.2012.11.002

142. Themeli M, Kloss CC, Ciriello G et al (2013) Generation of tumor-targeted human T lymphocytes from induced pluripotent stem cells for cancer therapy. Nat Biotechnol 31:928–933. doi:10.1038/nbt.2678
143. Obokata H, Wakayama T, Sasai Y et al (2014) Stimulus-triggered fate conversion of somatic cells into pluripotency. Nature 505:641–647. doi:10.1038/nature12968
144. Cyranoski D (2014) Stem-cell scientist found guilty of misconduct. Nature. doi:10.1038/nature.2014.14974
145. Vierbuchen T, Ostermeier A, Pang ZP et al (2010) Direct conversion of fibroblasts to functional neurons by defined factors. Nature 463:1035–1041. doi:10.1038/nature08797
146. Marro S, Pang ZP, Yang N et al (2011) Direct lineage conversion of terminally differentiated hepatocytes to functional neurons. Cell Stem Cell 9:374–382. doi:10.1016/j.stem.2011.09.002
147. Sekiya S, Suzuki A (2011) Direct conversion of mouse fibroblasts to hepatocyte-like cells by defined factors. Nature 475:390–393. doi:10.1038/nature10263
148. Ieda M, Fu JD, Delgado-Olguin P et al (2010) Direct reprogramming of fibroblasts into functional cardiomyocytes by defined factors. Cell 142:375–386. doi:10.1016/j.cell.2010.07.002
149. Szabo E, Rampalli S, Risueno RM et al (2010) Direct conversion of human fibroblasts to multilineage blood progenitors. Nature 468:521–526. doi:10.1038/nature09591
150. Li W, Nakanishi M, Zumsteg A et al (2014) In vivo reprogramming of pancreatic acinar cells to three islet endocrine subtypes. Elife 3, e01846. doi:10.7554/eLife.01846
151. Chen YJ, Finkbeiner SR, Weinblatt D et al (2014) De novo formation of insulin-producing "neo-beta cell islets" from intestinal crypts. Cell Rep 6:1046–1058. doi:10.1016/j.celrep.2014.02.013
152. Donia M, Junker N, Ellebaek E et al (2011) Characterization and comparison of "Standard" and "Young" tumor infiltrating lymphocytes for adoptive cell therapy at a Danish Translational Research Institution. Scand J Immunol 75:157–167. doi:10.1111/j.1365-3083.2011.02640.x
153. Itzhaki O, Hovav E, Ziporen Y et al (2011) Establishment and large-scale expansion of minimally cultured "young" tumor infiltrating lymphocytes for adoptive transfer therapy. J Immunother 34:212–220. doi:10.1097/CJI.0b013e318209c94c
154. Tran KQ, Zhou J, Durflinger KH et al (2008) Minimally cultured tumor-infiltrating lymphocytes display optimal characteristics for adoptive cell therapy. J Immunother 31:742–751. doi:10.1097/CJI.0b013e31818403d5
155. Dudley ME, Gross CA, Langhan MM et al (2010) CD8+ enriched "young" tumor infiltrating lymphocytes can mediate regression of metastatic melanoma. Clin Cancer Res 16:6122–6131. doi:10.1158/1078-0432.CCR-10-1297
156. Dudley ME, Gross CA, Somerville RP et al (2013) Randomized selection design trial evaluating CD8+–enriched versus unselected tumor-infiltrating lymphocytes for adoptive cell therapy for patients with melanoma. J Clin Oncol 31:2152–2159. doi:10.1200/JCO.2012.46.6441
157. Ullenhag GJ, Sadeghi AM, Carlsson B et al (2012) Adoptive T-cell therapy for malignant melanoma patients with TILs obtained by ultrasound-guided needle biopsy. Cancer Immunol Immunother 61:725–732. doi:10.1007/s00262-011-1182-4
158. Butler MO, Friedlander P, Milstein MI et al (2011) Establishment of antitumor memory in humans using in vitro-educated CD8+ T cells. Sci Transl Med 3:80ra34. doi:10.1126/scitranslmed.3002207
159. Chapuis AG, Ragnarsson GB, Nguyen HN et al (2013) Transferred WT1-reactive CD8+ T cells can mediate antileukemic activity and persist in post-transplant patients. Sci Transl Med 5:174ra127. doi:10.1126/scitranslmed.3004916
160. Berger C, Jensen MC, Lansdorp PM et al (2008) Adoptive transfer of effector CD8+ T cells derived from central memory cells establishes persistent T cell memory in primates. J Clin Invest 118:294–305. doi:10.1172/JCI32103
161. Wang X, Berger C, Wong CW et al (2011) Engraftment of human central memory-derived effector CD8+ T cells in immunodeficient mice. Blood 117:1888–1898. doi:10.1182/blood-2010-10-310599

162. Wang A, Chandran S, Shah SA et al (2012) The stoichiometric production of IL-2 and IFN-gamma mRNA defines memory T cells that can self-renew after adoptive transfer in humans. Sci Transl Med 4:149ra120. doi:10.1126/scitranslmed.3004306
163. Wang X, Naranjo A, Brown CE et al (2012) Phenotypic and functional attributes of lentivirus-modified CD19-specific human CD8+ central memory T cells manufactured at clinical scale. J Immunother 35:689–701. doi:10.1097/CJI.0b013e318270dec7
164. Hinrichs CS, Borman ZA, Cassard L et al (2009) Adoptively transferred effector cells derived from naive rather than central memory CD8+ T cells mediate superior antitumor immunity. Proc Natl Acad Sci U S A 106:17469–17474. doi:10.1073/pnas.0907448106
165. Gattinoni L, Restifo NP (2013) Moving T memory stem cells to the clinic. Blood 121:567–568. doi:10.1182/blood-2012-11-468660
166. Nat Immunol. 2008 May;9(5):513-21. doi: 10.1038/ni.1603. Epub 2008 Apr 6. Erratum in: Nat Immunol. 2008 Jun;9(6):692. PMID: 18391955

Chapter 9
T Cell Blockade Immunotherapy Against Cancer and Abscopal Effect in Combination Therapy

Giuseppe V. Masucci, Luigi De Petris, Andreas Lundqvist, Rolf Kiessling, and Rolf Lewensohn

Abstract Stereotactic ablative body radiotherapy (SABR) has been described to induce abscopal effects. The "abscopal effect" is the occurrence of objective tumour regressions induced following irradiation at sites outside the irradiated field. However, this effect is limited and occurs in about 5 % of patients. This phenomenon has been reported in various tumour forms but mainly as singular events. Recent evidence that radiation induces immunogenic tumour cell death and alters the tumour microenvironment to enhance recruitment of antitumor T cells supports the hypothesis that radiation can enhance both the priming and the effector-phase of the antitumor immune response. SABR treatment of inoperable renal cancer results in local tumour sterilization with release of tumour cell fragments containing molecules that may be immunogenic. For instance, apoptosis and necroptosis, forms of cell death, have recently been demonstrated to be immunogenic if induced by drugs like anthracyclines or by ionizing radiation. These tumour antigens are taken up locally and systemically by antigen presenting cells, particularly the dendritic cells (DC), which have the potential to stimulate de novo production of specific immune responses. Cells and molecules, however, regulate anti-tumoral immuneresponse, with the ability to inhibit particularly response to "self" tumour antigens. The intervention at the checkpoint level through the administration of anti-Cytotoxic T-Lymphocyte Antigen 4 (CTLA-4) antibodies or anti-programmed cell death protein 1 (PD-1) can arrest these immune regulatory mechanisms, resulting in activation of the anti-tumour responses to the antigens released by SABR, and thus acting in synergy with SABR. Additional activation of anti-tumour immunity can also be obtained by administration of autologous DC "pulsed" ex vivo, with tumour-derived material; adoptively transfer tumour specific T cells, derived either from Tumour Infiltrating T cells (TIL) or from autologous T cells retrovirally transduced with tumour specific T cell receptors (TCRs).

G.V. Masucci (✉) • L. De Petris • A. Lundqvist • R. Kiessling • R. Lewensohn
Department of Oncology and Pathology, Karolinska Institutet, Stockholm 17176, Sweden
e-mail: giuseppe.masucci@ki.se

P.A. Ascierto et al. (eds.), *Developments in T Cell Based Cancer Immunotherapies*, Cancer Drug Discovery and Development,
DOI 10.1007/978-3-319-21167-1_9

Keywords Stereotactic ablative radiotherapy • Abscopal effect • Immunotherapy • Cytotoxic T-lymphocyte antigen 4 • Programmed cell death protein 1

Acronyms

AE	Adverse events
AHH	Ancestral HLA Haplotypes
CBR	Clinical benefit rate
CD	Cluster of differentiation
CEA	Carcinoembryonic antigen
COX-2	Cyclooxygenase type 2
CT	Computerized tomography
CTLA-4	Cytotoxic T-lymphocyte antigen 4
DAMPs	Damage-associated molecular patterns
DC	Dendritic cells
EGFR	Endothelial growth factor
Flt3	Fms-like tyrosine kinase 3
GM-CSF	Granulocytes, macrophages colony stimulating factor
GVAX	Vaccine with irradiated tumour cells engineered to secrete GM-CSF
HLA	Human leukocytes antigen
HMGB1	High-mobility group protein B1
ICAM-1	Adhesion molecules
IFN	Interferon
iNOS	Nitric oxide synthase
MDSC	Myeloid-derived suppressor cells
MHC	Major histocompatibility complex
MIC A/B	MHC class I polypeptide-related sequence A or B
MoAB	Monoclonal antibody
NK	Natural Killer
NKG2D	Natural killer group 2, member D
NSCLC	Non-small cell lung cancer
PAMPs	Pathogen-associated molecular patterns
PD-1	Programmed cell death protein 1
PD-L1/2	Ligands for PD1
PET	Positron emission tomography
RCC	Renal cell carcinoma
SABR	Stereotactic ablative body radiotherapy
TCRs	T cell receptors
TGF	Tumour growth factor
TIL	Tumour infiltrating T cells
Treg	Regulatory T cells

Introduction

Combination strategies in the treatment of human diseases in general and of cancer in particular are necessary. The possibility to harmonize surgery, radiotherapy, and chemotherapy has been the standard care for treating several tumours [1, 2]. The area of application of radiotherapy and in particular SABR has increased during the last decade [3–7]. Previous experience of SABR in the treatment of metastases from renal cell carcinoma (RCC) showed a high local control (90 %) at different tumour locations. This has been documented by several Institutions [5–9]. Nonetheless, the occurrence of distant micro-metastases not visualized on Computerized Tomography (CT) and Positron emission tomography (PET) remains a treatment problem. Even if the local control is high after SABR in various tumour diagnoses a large fraction of the patients will have recurrences with new metastases due to the occurrence of distant micro-metastases not visualized on CT or PET/CT [3, 10].

In renal cancer there have been reports on abscopal effects on distant metastases where non-irradiated tumours have regressed temporarily or seemingly permanently after treatment with SABR of either the primary tumour or other metastatic lesions [11]. The term "abscopal effect" covers occurrence of objective tumour regressions induced following irradiation of sites outside the irradiated field.

Recent evidence that radiation induces immunogenic tumour cell death and alters the tumour microenvironment to enhance recruitment of anti-tumour T cells supports the hypothesis that radiation can enhance both the priming and the effector-phase of the anti-tumour immune response [12, 13]. The leukocytes phenotype relates to the underlying mechanisms of these abscopal effects. SABR treatment of inoperable renal cancer results in local tumour sterilization with release of tumour cell fragments containing molecules that may be immunogenic. For instance, apoptosis, as a form of cell death, was originally considered as non-immunogenic and non-inflammatory. However, recently it has been demonstrated to be immunogenic when induced by drugs like anthracyclines or by ionizing radiation [14, 15].

Apoptosis is not the main type of cell death caused by radiotherapy on epithelial tumour cells., On the other hand hyperthermia combined with radiation-induced DNA damage has shown to stimulate another mechanism of cell death, namely necroptosis [16, 17]. The residuals from radiotherapy induced cell destruction are strong candidates to spread potential antigens recognized by the immune system. Tumour antigens are taken up locally and systemically by antigen presenting cells, particularly the dendritic cells, which have the potential to stimulate de novo production of specific immune responses (either cellular or humoral) or enhance, or recall already existing immune competent cells. Induction of immune response to these tumour related antigens is regulated, however, by a complex mechanism based on cells and molecules that modulate the strength of the responses blocking a possible anti-tumoral effect. Only through the blocking of the negative feed-back on cytotoxic T cells there is a possibility to restore the effector arm of the anti-tumoral response. For instance, through the administration of checkpoint inhibitors such as anti-CTLA-4 antibodies these immune regulatory mechanisms can be halted or diminished in activity, resulting in activation of the anti-tumour responses to the antigens released by SABR and thus acting in synergy with SABR [18].

SABR and Abscopal Effect

SABR Treatment

The SABR technique is based on a few key features (Fig. 9.1).

1. *High precision and reproducibility*. The patient is accurately positioned on the treatment table with the help of fixation tools, such as the stereotactic box or vacuum pillows. These aim at limiting the movements of the target between the subsequent radiotherapy fractions. In addition, the movement of the tumor within each treatment session, due for example to breathing movements for targets in the thorax and upper abdomen, can be monitored with diverse gating techniques, These include for example irradiating only when the tumor is under the beam or keep track of the tumor with the help fiducials. Finally, with the help of imaging devices, such as the cone-beam CT, coupled orthogonally to the radiation beam, the tumor is visualized before each treatment and patient position accordingly modified in order not to miss the target. The result is that the treatment can be planned with very small margins around the tumor, limiting unwanted doses to the surrounding normal tissues.
2. *High dose per fraction*. Taking advantage of the high precision and small margins around the tumor, SABR allows for the delivery of very high radiation doses per fraction, which are repeated few times. A typical SABR fractionation schedule is 3 times 15 Gy, as opposed to the conventionally fractionated radiotherapy, which instead is typically administered with about 2 Gy per daily fraction, throughout several weeks of treatment. SABR ablative doses can be safely delivered because the target usually is a small lesion, with e.g. a maximal diameter of approximately 5 cm. In addition, the tumor must not lay too close to sensitive structures, as for example the bowel or the central airways.

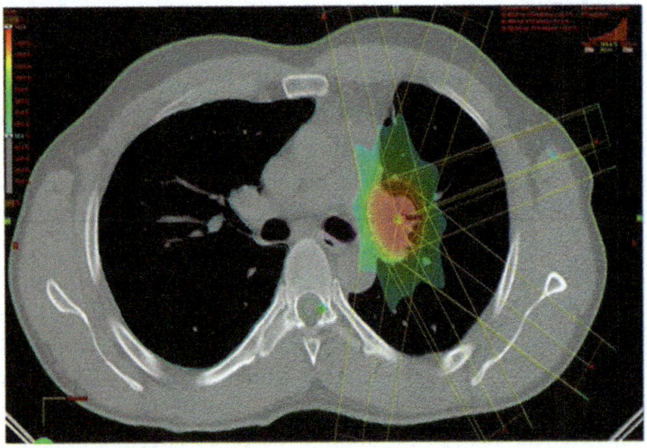

Fig. 9.1 SABR dose planning of a lung metastasis. Performed at the Radiotherapy Unit, Department of Oncology, Karolinska University Hospital, Stockholm, Sweden

3. *Inhomogeneous dose distribution*. One of the major characteristics of the conventional 3Dimension conformal radiotherapy planning is that the entire target is covered by the same dose. In contrast, with the SABR technique the radiation dose in not homogenously distributed on the target, but by a gradient increased to the isocenter to reach a total dose of 150–200 Gy, where the cells are anticipated to be more radiation resistant, as related to e.g. hypoxia.

Altogether, all these features makes SABR to a treatment that generates a large variety of cellular and tissue responses, and is easy to combine with immunotherapy strategies as induction phase, because of the very short total treatment time. Yet, it must be noted that these considerations are pure empirical, since a radiobiological modeling, that specifically predict the dose- and fractionation-effects in SABR as opposed to conventional radiotherapy, has not been validated yet. Although radiotherapy has been generally considered as an immunosuppressive intervention, since T cells are very radiosensitive, recent evidence suggests that a number or reactions are triggered in the tissues exposed to ionizing radiation, resulting in potentially vivid immune responses.

Firstly, radiotherapy induces any kind of cell death, including apoptosis, necrosis, autophagy and mitotic catastrophe. Dying cells secrete ATP and high-mobility group protein B1 (HMGB1). The latter interacts with the Toll-like receptor factor 4 (TLR4) on dendritic cells activating them. In addition, the up regulation of calreticulin on the tumor cell surface facilitates phagocytosis by the DC [19, 20]. Furthermore, radiation induced damage in tumor and in bystander tissues triggers a stress signaling that results in the over-expression of MHC class I, adhesion molecules (ICAM-1) and death receptors, and in increased secretion of pro-inflammatory cytokines and chemokines (IL-1beta, TNF-alpha, CXCL16) as well as heat-shock proteins. Similarly, even immunosuppressive mediators such as TGF-beta are released. The latter, could constitute a possible target for inhibitory strategies to further enhance immune response after radiation [21, 22]. Finally, as compared to conventional radiotherapy, SABR induces more endothelial and micro vascular damage, mediated by the ceramide pathway [23], resulting in increased cell death, high T-cell infiltration and effective T-cell priming and expansion.

The Abscopal Effect

The "abscopal effect" is the occurrence of objective tumour regressions induced following irradiation in sites outside the irradiated field. This phenomenon has been reported in various tumour forms but mainly as singular events [24]. There have been reports on abscopal effects on distant metastases where non-irradiated tumours have regressed temporarily or seemingly permanently after treatment with SABR of either the primary tumour or other metastatic lesions [11]. Typically, this occurs several months after radiotherapy, suggesting mechanisms probably triggered by radiotherapy, but sustained and developed by other factors, primarily the immune system. The abscopal effect was firstly described as systemic tumor regression following the local intra-tumoral injection of Staphylococcus, causing a subsequent acute inflammatory response [25]. During infection, the release of pathogenic signals induces so-called pathogen-associated molecular patterns (PAMPs), which triggers a potent immune reaction. Similarly, the signals resulting from tumor cell death caused by anti-tumoral agents generate so-called damage-associated molecular patterns (DAMPs), which in tumours lead to a variable level of tumor antigen presentation to immune cells [26]. The intensity and efficacy of the immune response generated by DAMPs depends on multiple factors, including the timing, amount and possible combination of the anti-tumoral agents implemented. For radiotherapy, such mechanisms have not been completely elucidated yet (see above).

Abscopal effect following radiotherapy for diverse tumor types has been reported in a sporadic number of clinical cases, summarized in detail by Siva et al. [27]. In brief, in a few articles it has been described a spontaneous regression of metastasis induced by conventional radiotherapy, against either the primary site or other metastasis. Of note, in the majority of cases the total dose of radiotherapy was far below curative or ablative thresholds (mean total dose 36 Gy), indicating that potent systemic responses after a local intervention may be triggered even by sub-lethal stimuli. On the other hand, a report from our Institution describes long term abscopal effects in 4 out of 28 consecutive cases with renal cell carcinoma treated with SABR [28]. Notwithstanding the limitations associated to these communications, that describe clinical events without exploring mechanisms, still they represent relevant hypotheses-generating observations on how to take advantage of ionizing radiation as adjuvant to immune treatments (Fig. 9.2).

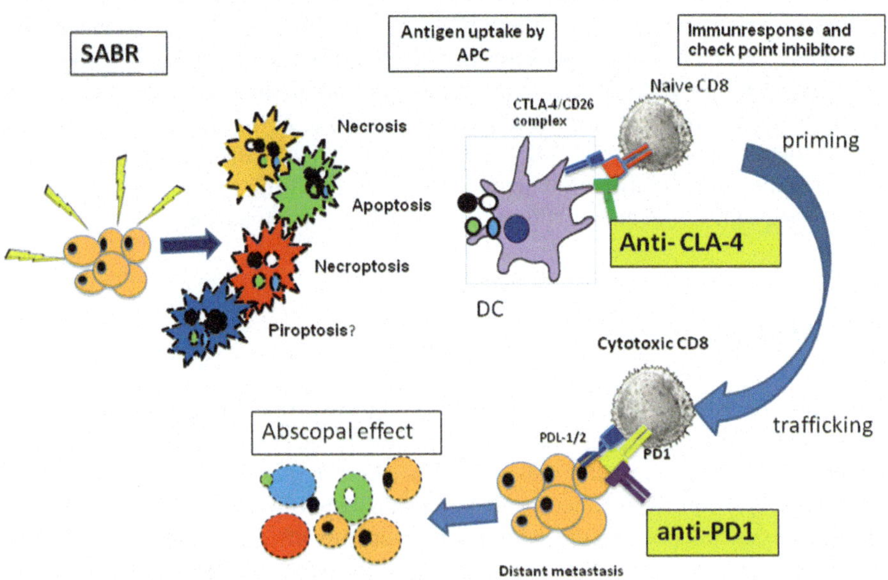

Fig. 9.2 Summary chart presenting the possible steps involved in the cascade of reactions when combining the effect of auto-vaccination induced by SABR with consequent abscopal effect and immuneresponse enhancing strategies by check-point inhibitors

Checkpoint Inhibitors and Therapy

Among the most promising approaches to activate therapeutic anti-tumour immunity is the blockade of immune checkpoints that bloc the anti-tumour cytolytic effect by the immune system. They refer to several inhibitory pathways involved in the immune system. They are crucial for maintaining self-tolerance and modulating the duration and amplitude of physiological immune responses in peripheral tissues in order to minimize unnecessary tissue damage. Tumours choose certain immune-checkpoint pathways as a major mechanism of immune resistance, particularly against cytotoxic specific T cells to tumour antigens. Many of the immune checkpoints are initiated by ligand–receptor interactions, permitting their inhibition by antibodies or modulated by recombinant forms of ligands or receptors [29].

CTLA-4

CTLA-4, also known as CD152 (Cluster of differentiation 152), is a protein receptor down regulating the immune system. It is found on the surface of T cells involved in the efferent arm of the immunesystem. The T cell attack can be turned on by stimulating the CD28 receptor on the T cell. On the other hand, their attack can be turned off by stimulating the CTLA-4 receptor acting as an "off" switch. In humans, the CTLA-4 protein is encoded by the *CTLA-4* gene [30, 31].

Ipilimumab is the first monoclonal antibody used in clinic that blocks CTLA-4 ligand, negative regulator of T-cells, and thereby augments T cell activation and proliferation. it should be also mentioned the introduction of tremelimumab in phase III clinical trial for melanoma treatment [32]. Ipilimumab was approved and regarded as the standard of care treatment for patients with unresectable and metastatic melanoma that are either previously untreated (approved with a first line indication in EU and USA) or treated with other therapies (approved with a second line indication in EU and USA) [29].

PD-1

The programmed cell death 1 (PD-1) pathway represents a major immune control switch that may be engaged by tumour cells to overcome active T-cell immune surveillance [33]. The ligands for PD-1 (PD-L1 and PD-L2) are constitutively expressed or can be induced in various tumours as well as they can be reinvigorated by IL-2 in exhausted T cells [34–38]. High expression of PD-L1 on tumour cells (and to a lesser extent of PD-L2) has been found to correlate with poor prognosis and survival in various cancer types, including RCC [39], pancreatic carcinoma [40], hepatocellular carcinoma [41], ovarian carcinoma [42] and non-small cell lung cancer (NSCLC) [43]. Furthermore, PD-1 has been suggested to regulate tumour-specific T-cell expansion in patients with malignant melanoma [44].

Clinical Findings

Clinical trials using Ipilimumab have shown an overall survival benefit in malignant melanoma patients. Grade 3/4 drug-related adverse events were reported in 22.8 % of subjects treated with ipilimumab alone, the majority being immune-related events (14.5 % of the entire patient cohort) [43, 45]. Recently the group of Ascierto et al. described the effect induced by anti CTLA-4 therapy sequenced with radiotherapy (SABR). In the preliminary analysis the abscopal response has been detected in 11/21 (52 %) patients. They observed 9 abscopal partial response (42, 8 %), 2 abscopal stable disease (9, 6 %), and 10 cases with progressive disease (47, 6 %). The median overall survival (OS) for all 21 patients was 13 months (range 6–26). The median OS for patients with and without abscopal responses was 22.4 months (range 2, 5–50, 3) and 8, 3 months (range 7, 6–9.0) respectively as assessed in 11 (84.6 %) out of 13 patients with local response showing an abscopal effect [46]. Preclinical in vitro and in vivo experiments have shown that PD-1 and/or PD-L1 blockade using monoclonal antibodies (mAb) enhances tumour-cell specific T-cell activation, cytokine production, anti-tumour effector mechanisms, and clearance of tumour cells by the immune system [40, 47–51]. Recently, 135 patients with advanced melanoma were treated, with anti-PD-1 (pembrolizumab) including those who had had disease progression while they had been receiving ipilimumab [52]. Treatment resulted in a high rate of sustained tumor regression, with mainly grade 1 or 2 toxic effects. The response rate did not differ significantly between patients who had received prior ipilimumab treatment and those who had not. Responses were durable in the majority of patients (Median follow-up was 11 months among patients who had a response) and 81 % of the responders (42 of 52) were still receiving treatment at the time of analysis. The overall median progression-free survival among the 135 patients was longer than 7 months and the updated 1-year overall survival rate was 82 % [53]).

Recent data with nivolumab (BMS-936558), an IgG4 antibody against PD-1, have validated PD-1 as an attractive target for clinical therapeutic intervention [54]. In a recent report of the clinical trial data with nivolumab a total of 296 patients with advanced melanoma, NSCLC, castration -resistant prostate cancer, renal-cell carcinoma or colorectal cancer were treated at a dose of 0.1, 0.3, 1, 3 or 10 mg/kg every 2 weeks. Among 236 evaluable patients, cumulative response rates were 18 % among patients with NSCLC, 28 % among patients with melanoma, and 27 % among patients with renal-cell cancer. Responses were reported to be durable: 20 of 31 responses lasted 1 year or more in patients with 1 year or more of follow-up. Grade 3 or 4 adverse events were observed in 49 % of patients, while 14 % of patients had treatment related Grade 3 or 4 adverse events.

Concurrent therapy with nivolumab and ipilimumab had a manageable safety profile and provided clinical activity that appears to be distinct from that in published data on monotherapy, with rapid and deep tumor regression in a substantial proportion of patients [55].

Cell Therapy and Cancer Vaccines

Active immunotherapy in the form of vaccination with tumour-loaded DC has been evaluated in patients with cancer in the past two decades. Monocytes or CD34 positive hematopoietic stem cells are isolated from patients and differentiated to DC in cytokine cocktails typically containing interleukin-4 and GM-CSF. Prior to vaccination, DC are pulsed with tumour antigens. The infused DC migrates to the lymph node and presents the antigens to T cells. The first clinical studies showed promising results. In a pilot study where four patients with were follicular B-cell lymphoma were vaccinated with tumour-specific idiotype protein-loaded DC resulted in measurable anti-tumour T cell responses. Importantly, tumour regression was observed in three of four vaccinated patients [56]. Similarly, a clinical study of patients with advanced melanoma vaccinated with DC loaded with tumour lysate or melanoma-specific peptides showed induction of DTH reactions to peptide-pulsed DC. In this study, objective responses were evident in 5 out of 16 patients including two complete responders [57]. Until 2010, more than 900 cancer patients have been enrolled in clinical trials of DC-based vaccination [58]. As a 'proof of principle' a statistically significant effect of DC-mediated cellular immune response and of DC dose on clinical benefit rate (CBR) could be demonstrated. However, advances in basic immunology and pre-clinical studies have identified several crucial hurdles that need to be overcome in order to generate improved clinical responses in patients undergoing DC-based vaccines. Tumour-induced immune suppression plays a major role in limiting the efficacy of immunotherapy including DC-based vaccination. Strategies to overcome immune suppression in patients vaccinated with DC include targeting of myeloid-derived suppressor cells (MDSC) and regulatory T cells (Treg). We recently showed that MDSC impair the quality of DC vaccines. Changes in DC characteristics were most notable when MDSC frequencies of >50 % CD14+HLA-DR-cells were present in the starting culture. MDSC frequencies did not affect yield or viability of the produced DC, but induced a dose-dependent decrease in DC maturation, ability to take up antigen, migrate and induce T-cell IFN-gamma production [59]. In a randomized clinical trial of patients with metastatic small cell lung cancer, targeting MDSC with all-trans-retinoic acid improved the induction of immune responses by a DC vaccine [60]. Depletion of Treg can enhance antigen-specific immune responses to cancer vaccines. Targeting Treg with the immunotoxin, denileukin diftitox followed by vaccination with DC loaded with carcinoembryonic antigen results in selective depletion of Treg leading to improved T-cell response to CEA [61]. Other efforts to improve the clinical outcome of DC-based vaccination include combination with standard therapy like chemotherapy, surgery and radiation therapy. Retrospective analysis of a phase II trial in patients with stage IV melanoma receiving DC-based vaccination showed that therapeutic DC vaccination could favor clinical response in patients after more than one line of therapy. Patients receiving DC-vaccine alone had a median overall survival of 15 months whereas patients receiving DC-vaccine and other treatments after stopping vaccination had a median overall survival of 34 months [62]. The rationale behind combining radiation therapy with DC-based vaccination is that: radiation induces apoptosis of tumour cells and infused DCs phagocytose apoptotic tumour cells and processes the antigenic material to present these antigens to T cells. In addition, radiation causes inflammation at the tumour site leading to enhanced activation and ability of DC to present these antigens. Several studies have shown safety and feasibility of incorporating radiation therapy to DC vaccination strategies in patients with cancer [63–65]. A phase I trial investigated the immunologic response induced by direct injection of autologous immature DC into the tumour under radiotherapy in advanced hepatoma patients. Induction of alpha-fetoprotein specific immune responses was observed in 8 of 14 patients. In this trial, there were two partial responses and four minor responses. Interestingly, six patients showed an increased NK cell cytotoxic activity after vaccination [63]. In another study, patients with newly diagnosed glioblastoma who underwent surgery and standard conformal external beam radiotherapy, received thereafter three consecutive intra dermal vaccinations with autologous mature DCs pulsed with an EGFRvIII-specific peptide conjugated to keyhole limpet hemocyanin. Induction of EGFRvIII-specific immune responses was observed in most patients and overall median survival from time of histological diagnosis was 22.8 months [63]. Similarly, ten patients with glioblastoma received vaccination with autologous tumour lysate-loaded DC after radiation therapy and temozolomide. Immune responses were induced in all patients and all patients were alive at 6 months after diagnosis and the median overall survival was 28 months [64].

Rational and Strategies of Combining Therapies to Increase the Abscopal Effect

Irradiation can convert a tumour into a vaccine in situ, hence generating T-cell responses that, upon priming in the context of CTLA-4 blockade, reject tumours also at non-irradiated metastatic sites [13, 18, 66]. Demaria et al. showed that T cell-mediated elimination of cancer cells that survived irradiation by mounting a stress response, which are marked by the expression of NKG2D ligands, contributes to therapeutic success in response to anti-CTLA-4 antibodies [67] The critical role played by NKG2D ligands in determining the response of poorly immunogenic tumours to anti-CTLA-4 antibodies has important clinical implications. The up-regulation of NKG2D ligands on the surface of tumour cells by some chemotherapy drugs or by radiotherapy may indeed provide a biomarker that predicts the success of combination regimens involving anti-CTLA-4 antibodies. Recently, two groups have demonstrated that PD-L1 was up regulated in the tumor microenvironment after SABR. Administration of anti–PD-L1 [68] or anti PD1 (Demaria S, February 2014, personal communication) enhanced the efficacy of radiotherapy through a cytotoxic T cell–dependent mechanism [69]. As mentioned earlier, ionizing radiation can reduce tumour burden outside the field of radiation, known as the abscopal effect. Several pre-clinical studies have demonstrated that dendritic cells are central in augmenting the abscopal effect. In mice bearing mammary carcinoma, treatment with Flt3-ligand increased the number of DC. Irradiation of the primary tumour followed by treatment with Flt3-ligand resulted in reduced tumour growth of secondary non-irradiated tumours. No growth delay of non-irradiated tumours was observed when T cell deficient mice were treated with radiation plus Flt3-L concluding that the abscopal effect is dependent on DC activation of tumour-specific T cells [12]. In a murine model of squamous cell carcinoma intra-tumoral administration of DC and radiation therapy, tumour growth was markedly suppressed compared with mice treated with either modality alone. Growth of non-treated tumours was also suppressed, indicating that the combination therapy of DC and radiation therapy stimulated an abscopal effect [70]. Since the approval of the immune checkpoint inhibitor ipilimumab, several studies have evaluated its combination with cancer vaccines. Although, as of today, no reports on combining DC vaccination and ipilimumab have been published, several clinical trials are being planned. The combination, vaccination with irradiated tumour cells engineered to secrete GM-CSF (GVAX) and ipilimumab has been evaluated in patients with melanoma, prostate cancer and pancreatic cancer. Periodic infusions of ipilimumab after GVAX vaccination generated clinically meaningful anti-tumour immunity without grade 3 or 4 toxicity in a majority of metastatic melanoma patients [71]. A phase I/II dose escalation/expansion trial of combined GVAX and ipilimumab treatment showed that treatment-induced increases in absolute CD4 and CD8 T cell count was associated with prolonged overall survival [71]. In patients with pancreatic ductal adenocarcinoma surviving longer than 4.3 months after treatment with ipilimumab and GVAX vaccine, there was an increase in the frequency of tumour specific T cells was detected, suggesting that checkpoint blockade in combination with GVAX has the potential for clinical benefit [72]. In contrast, one report on combining ipilimumab with a peptide-based vaccine in patients with advanced melanoma showed that vaccine combination was not associated with additional benefit [73]. A study in prostate cancer explored the safety and tolerability of ipilimumab in combination with a poxviral-based vaccine targeting prostate-specific antigen. No dose-limiting toxic effects were observed and randomised trials are needed to further assess clinical outcomes of the combination of ipilimumab and viral-based vaccines in prostate [74].

Future Strategies

Additional activation of anti-tumour immunity can also be obtained by administration of autologous DC, produced ex vivo from autologous leukapheresis derived monocytes, with the capacity to take up the circulating tumour antigens released by SABR for efficient priming of T cells. Alternatively, autologous DCs can be "pulsed" ex vivo with tumour derived material and provided as a tumour vaccine with the capacity to re-activate the patient's anti-tumour response. Yet another principle of anti-tumour treatment, which can be applied to these patients, would be to adoptively transfer tumour specific T cells, derived either from Tumour Infiltrating T cells

(TIL) or from autologous T cells retroviral transduced with tumour specific TCRs. SABR creates an inflammatory environment that may augment the activity of adoptively infused TILs. Consequently, we consider that the rationale beyond these strategies is: (a) to improve the clinical outcome of metastatic solid tumours (renal cell cancer, malignant melanoma, lung cancer etc) by combining the SABR with indirect induced immune re-activation; (b) to elicit an "abscopal effect" by sequential treatment with immunotherapeutic principles such as (anti CTLA-4 and/or anti PD1 MoABs autologous dendritic cells and/or TIL cells adoptive cell transfer, cytokines GM-CSF, Interleukins etc.).

Recruitment of a patient cohort could include in a first stage patients with metastatic tumours (Fig. 9.2). These patients would have at least one metastatic lesion available for SABR (15Gy × 3) and one marker lesion for evaluation of effect without SABR. In light of the recent knowledge a phase-I study could be planned for these patients for a combination treatment of anti-CTLA4+ SABR + anti PD1 (or anti PDL-1) sequentially [46, 68], SABR would be delivered to at least one metastatic or primary lesion with a standard dose of 15 Gy × 3 prescribed to the 67 % isodose. This dose has been documented to result in high local control in various tumour types and it is in accordance with experimental evidence for induction of the abscopal effect.

Search for the best and most effective individualized treatment strategy is absolutely dependent on a complete analysis of the predictive and prognostics markers that can scrutinize the capacity to response at the individual level. Prediction should be oriented to base line and during treatment. This relates to determination of immune-markers that best reveal the specificity of the treatment. The individualized treatment against cancer has to be defined more specifically, in order to understand which, are the markers that indicates individual benefit. These may be HLA, gene profile in general and the markers that identify the tumour at the cellular, microenvironment and genetic level (e.g. Infiltrating inflammatory cells cytokines, oncogenes, miRNA).

The Relevance of Predictive and Prognostic Biomarkers in Relation to Combined Therapies

Treatment with checkpoint inhibitors such as anti CTLA-4, anti PD-1 or anti PD-1 L1, needs biomarkers to identify those patients that will respond to treatment and also those who will show severe side effects. Since both the treatment and the side effects are immune mediated, there should be a high likelihood to find such biomarkers by monitoring the patient's immune system. The analysis of the immune response should be focused at each single patient at treatment start and during its consequent manipulation as related to treatment with checkpoints inhibitors. This is crucial for the determination of efficacy and also for predicting clinical outcome. Immune biomarkers can be divided into effector type (they give the ability of the immune system to react to the presence of cancer cells) in priming markers (they measure the characteristics of the tumour itself and the microenvironment) and meaning the genetic fingerprinting of the individual (i.e. HLA genotype). A possible panel of markers might be also presented as positive or negative predictors in relation to clinical outcome. The immune related biomarkers are not necessarily bound to the tumour histology but rather to the individual immune response state. It is known that the tumour microenvironment can induce recruitment and expansion of suppressive immune cells and immune suppressive factors to the tumour site. In patients treated with checkpoint inhibitors, it is relevant to investigate the potential value of immune-markers. Studies on ipilimumab have suggested a possible predictive role for a number of biomarkers, including the early increase of the absolute lymphocyte count (ALC) or the early decrease of lactate dehydrogenase (LDH) levels, the maintained expression of the inducible co-stimulator molecule (ICOS) and the detection of humoral immune response against NY-ESO-1 antigen [75].

Levels of Tregs are increased in different cancer types [76]. This accumulation of Tregs may result from aberrant proliferation and trafficking and, as we recently have described, greater resilience to oxidative stress compared with conventional T cells [77]. This enhanced antioxidative capacity of Tregs possibly serves as feedback inhibition during inflammation and prevents uncontrolled immune reactions by favoring survival of suppressor rather than effector cells. Also Myeloid Derived Suppressor Cells (MDSC) accumulate in cancer patients. Thus, we have recently described the presence of increased numbers of MDSC in the peripheral blood of patients with stage III-IV melanoma [78]. These melanoma MDSC:s mediate strong suppression of CD8 and CD4 T-cell proliferation and inhibit T-cell IFN-gamma production and can also impair the quality of Dendritic cell tumour vaccines [59]. However, their presence appears to be strongly tumour-dependent, as they rapidly disappear once the tumour is surgically removed or successfully diminished by other treatments. Both nutrient starvation as mediated by the enzyme Arginase-1 and oxidative stress contributed to T cell suppression, but additional unknown mechanisms must also be at work. It is therefore conceivable that in patients with advanced malignant melanoma, MDSC and their products can have a major influence on their ability to respond to antibodies to checkpoint molecules. Several other immune factors contribute to the immune suppressive environment and can constitute important prognostic markers for therapy. Inducible nitric oxide synthase (iNOS) and cyclooxygenase type 2 (COX-2) have been described to correlate with progression of tumour tissue [79].

NK cells are of potential importance in inhibiting the spreading of malignant cells [80, 81]. For instance several of the ligands for the NK cell receptors and for cytotoxic T cells (CTLs) can be detected on malignant melanoma cells [82]. There is pre-clinical evidence that the NKG2D Ligands (NKG2DL), which belong either to the MIC A/B or the ULPB 1–6 molecule families, are involved in immune surveillance, and NKG2DL can sensitize tumour cells to NK and CTL killing [83]. Others have described that NKG2DL in the sera of cancer patients are strong independent predictors of poor prognosis [82]. Therefore, NK cells, their receptors and their ligands are of potential interest as predictive factors for therapies.

As a prognostic factor, HLA has been studied in association with lung [84] and head and neck tumours [85]. HLA has been studied as a factor that may predict the response to different immune therapies, such as vaccine or cytokine therapy, in e.g. melanoma [10, 86–93] and in chronic myelogenous leukaemia [94]. There is an association of the HLA-A2 allele with severe prognosis in serous adenocarcinoma of the ovary in stage III-IV and malignant melanoma [95]. There is also a correlation of two so called Ancestral HLA Haplotypes (AHH) 8.1 (A1-B6-C7-DRB-1*03) and 62.1 (A2-B15(62)-C3-DRB1*04) to a rapid tumour progression in patients with ovarian-, prostate cancer and malignant melanoma [96, 97]. The presence or absence of Major histocompatibility molecule in tumour cells is relevant for the tumour cell to be recognized by the immune system and lack of these is a negative prognostic factor 1 [98]. HLA-G and -E are a new series of tumour surface markers that can predict the aggressiveness of the tumour tissue and has a inhibitory activity toward the efferent arm of the immune cells as reviewed by Rebmann et al. [99]. Challenges exist that need to be resolved to facilitate development of innovative approaches such as the utilization of the radiotherapy induced abscopal effect and checkpoint inhibitor therapy. There is a need for development and validation of tools to identify patients who can benefit from these particular forms of combined immunotherapy. For example, only a fraction of patients are eligible for adoptive tumour-infiltrating lymphocyte (TIL) cell transfer [59] only a fraction of patients can achieve durable regressions in response to cell or antigen vaccination [79], or antibody therapies. The problem is that we do not know how to select for such cases because we lack deep knowledge on the responsible mechanisms responsible. Despite substantial efforts from many groups, we do not know which parameters of immune responses, and which assays used to assess these parameters, are optimal for efficacy analysis. Indeed, the tumour-specific cellular immune response promoted by immunization often has not correlation with clinical cancer regression despite the induced cytotoxic T cells detected in "**in vitro**" assays [82, 83, 100, 101].

Conclusions

Radiation therapy for cancer is usually considered simply a local treatment modality; however, there is clinical evidence of longer-range effects that render this conceptual simplicity misleading. It is clear that abscopal effects are beneficial in terms of tumor control. The increasing use of high dose per-fraction radiotherapeutic approaches (SABR) offers the possibility that novel combinations with current systemic strategies could enhance systemic anti-tumor effects. Challenges exist that need to be resolved to facilitate development of innovative approaches such as the utilization of the radiotherapy induced abscopal effect and checkpoint inhibitor therapy. There is a need for development and validation of tools to identify patients who can benefit from these particular forms of combined immunotherapy.

References

1. Ascierto PA, Marincola FM (2011) Combination therapy: the next opportunity and challenge of medicine. J Transl Med 9:115
2. Masucci GV, Mansson-Brahme E, Ragnarsson-Olding B et al (2006) Alternating chemo-immunotherapy with temozolomide and low-dose interleukin-2 in patients with metastatic melanoma. Melanoma Res 16(4):357–363
3. Baumann P, Nyman J, Hoyer M et al (2009) Outcome in a prospective phase II trial of medically inoperable stage I non-small-cell lung cancer patients treated with stereotactic body radiotherapy. J Clin Oncol 27(20):3290–3296
4. Baumann P, Nyman J, Lax I et al (2006) Factors important for efficacy of stereotactic body radiotherapy of medically inoperable stage I lung cancer. A retrospective analysis of patients treated in the Nordic countries. Acta Oncol 45(7):787–795
5. Svedman C, Karlsson K, Rutkowska E et al (2008) Stereotactic body radiotherapy of primary and metastatic renal lesions for patients with only one functioning kidney. Acta Oncol 47(8):1578–1583
6. Svedman C, Sandstrom P, Pisa P et al (2006) A prospective phase II trial of using extracranial stereotactic radiotherapy in primary and metastatic renal cell carcinoma. Acta Oncol 45(7):870–875
7. Wersall PJ, Blomgren H, Lax I et al (2005) Extracranial stereotactic radiotherapy for primary and metastatic renal cell carcinoma. Radiother Oncol 76:S37–S37
8. Siomin V, Lin JL, Marko NF et al (2011) Stereotactic radiosurgical treatment of brain metastases to the choroid plexus. Int J Radiat Oncol Biol Phys 80(4):1134–1142
9. Zelefsky MJ, Greco C, Motzer R et al (2012) Tumor control outcomes after hypofractionated and single-dose stereotactic image-guided intensity-modulated radiotherapy for extracranial metastases from renal cell carcinoma. Int J Radiat Oncol Biol Phys 82(5):1744–1748
10. Onishi T, Ohishi Y, Iizuka N et al (1996) Phenotype frequency of human leukocyte antigens in Japanese patients with renal cell carcinoma who responded to interferon-alpha treatment. Int J Urol 3(6):435–440
11. Severin AC, Heijbel M, Alvevik P et al (2006) Comparison of implanted gold markers and cone beam CT for high precision on-line image-guided radiotherapy of prostate. Radiother Oncol 81:S214–S214
12. Demaria S, Ng B, Devitt ML et al (2004) Ionizing radiation inhibition of distant untreated tumors (abscopal effect) is immune mediated. Int J Radiat Oncol Biol Phys 58(3):862–870
13. Postow MA, Callahan MK, Barker CA et al (2012) Immunologic correlates of the abscopal effect in a patient with melanoma. N Engl J Med 366(10):925–931
14. Apetoh L, Obeid M, Tesniere A et al (2007) Immunogenic chemotherapy: discovery of a critical protein through proteomic analyses of tumor cells. Cancer Genomics Proteomics 4(2):65–70
15. Tesniere A, Panaretakis T, Kepp O et al (2008) Molecular characteristics of immunogenic cancer cell death. Cell Death Differ 15(1):3–12
16. Mantel F, Frey B, Haslinger S et al (2010) Combination of ionising irradiation and hyperthermia activates programmed apoptotic and necrotic cell death pathways in human colorectal carcinoma cells. Strahlenther Onkol 186(11):587–599
17. Schildkopf P, Frey B, Mantel F et al (2010) Application of hyperthermia in addition to ionizing irradiation fosters necrotic cell death and HMGB1 release of colorectal tumor cells. Biochem Biophys Res Commun 391(1):1014–1020
18. Dewan MZ, Galloway AE, Kawashima N et al (2009) Fractionated but not single-dose radiotherapy induces an immune-mediated abscopal effect when combined with anti-CTLA-4 antibody. Clin Cancer Res 15(17):5379–5388
19. Demaria S, Formenti SC (2012) Radiation as an immunological adjuvant: current evidence on dose and fractionation. Front Oncol 2:153
20. Demaria S, Formenti SC (2012) Role of T lymphocytes in tumor response to radiotherapy. Front Oncol 2:95

21. Finkelstein SE, Timmerman R, McBride WH et al (2011) The confluence of stereotactic ablative radiotherapy and tumor immunology. Clin Dev Immunol 2011:439752
22. Zeng J, Harris TJ, Lim M et al (2013) Immune modulation and stereotactic radiation: improving local and abscopal responses. Biomed Res Int 2013:658126
23. Truman JP, Garcia-Barros M, Kaag M et al (2010) Endothelial membrane remodeling is obligate for anti-angiogenic radiosensitization during tumor radiosurgery. PLoS One 5(9)
24. Snyder AR (2004) Review of radiation-induced bystander effects. Hum Exp Toxicol 23(2):87–89
25. Hoption Cann SA, van Netten JP, van Netten C (2003) Dr William Coley and tumour regression: a place in history or in the future. Postgrad Med J 79(938):672–680
26. Ludgate CM (2012) Optimizing cancer treatments to induce an acute immune response: radiation abscopal effects, PAMPs, and DAMPs. Clin Cancer Res 18(17):4522–4525
27. Siva S, Macmanus MP, Martin RF et al (2015) Abscopal effects of radiation therapy: a clinical review for the radiobiologist. Cancer Lett 356:82–90
28. Wersall PJ, Blomgren H, Pisa P et al (2006) Regression of non-irradiated metastases after extracranial stereotactic radiotherapy in metastatic renal cell carcinoma. Acta Oncol 45(4):493–497
29. Pardoll DM (2012) The blockade of immune checkpoints in cancer immunotherapy. Nat Rev Cancer 12(4):252–264
30. Dariavach P, Mattéi M-G, Golstein P et al (1988) Human Ig superfamily CTLA-4 gene: chromosomal localization and identity of protein sequence between murine and human CTLA-4 cytoplasmic domains. Eur J Immunol 18(12):1901–1905
31. Grosso JF, Jure-Kunkel MN (2013) CTLA-4 blockade in tumor models: an overview of preclinical and translational research. Cancer Immun 13:5
32. Ribas A, Kefford R, Marshall MA et al (2013) Phase III randomized clinical trial comparing tremelimumab with standard-of-care chemotherapy in patients with advanced melanoma. J Clin Oncol 31(5):616–622
33. Disis ML (2010) Immune regulation of cancer. J Clin Oncol 28(29):4531–4538
34. Brown JA, Dorfman DM, Ma FR et al (2003) Blockade of programmed death-1 ligands on dendritic cells enhances T cell activation and cytokine production. J Immunol 170(3): 1257–1266
35. Dong H, Strome SE, Salomao DR et al (2002) Tumor-associated B7-H1 promotes T-cell apoptosis: a potential mechanism of immune evasion. Nat Med 8(8):793–800
36. Francisco LM, Sage PT, Sharpe AH (2010) The PD-1 pathway in tolerance and autoimmunity. Immunol Rev 236:219–242
37. Sharpe AH, Freeman GJ (2002) The B7-CD28 superfamily. Nat Rev Immunol 2(2):116–126
38. West EE, Jin HT, Rasheed AU et al (2013) PD-L1 blockade synergizes with IL-2 therapy in reinvigorating exhausted T cells. J Clin Invest 123(6):2604–2615
39. Thompson RH, Dong H, Lohse CM et al (2007) PD-1 is expressed by tumor-infiltrating immune cells and is associated with poor outcome for patients with renal cell carcinoma. Clin Cancer Res 13(6):1757–1761
40. Nomi T, Sho M, Akahori T et al (2007) Clinical significance and therapeutic potential of the programmed death-1 ligand/programmed death-1 pathway in human pancreatic cancer. Clin Cancer Res 13(7):2151–2157
41. Gao Q, Wang XY, Qiu SJ et al (2009) Overexpression of PD-L1 significantly associates with tumor aggressiveness and postoperative recurrence in human hepatocellular carcinoma. Clin Cancer Res 15(3):971–979
42. Hamanishi J, Mandai M, Iwasaki M et al (2007) Programmed cell death 1 ligand 1 and tumor-infiltrating CD8+ T lymphocytes are prognostic factors of human ovarian cancer. Proc Natl Acad Sci U S A 104(9):3360–3365
43. Topalian SL, Hodi FS, Brahmer JR et al (2012) Safety, activity, and immune correlates of anti-PD-1 antibody in cancer. N Engl J Med 366(26):2443–2454
44. Hino R, Kabashima K, Kato Y et al (2010) Tumor cell expression of programmed cell death-1 ligand 1 is a prognostic factor for malignant melanoma. Cancer 116(7):1757–1766

45. Hodi FS, O'Day SJ, McDermott DF et al (2010) Improved survival with ipilimumab in patients with metastatic melanoma. N Engl J Med 363(8):711–723
46. Grimaldi A, Simeone E, Giannarelli D et al (2013) The abscopal effect: efficacy of radiotherapy in patients on progression after treatment with ipilimumab 3 mg/kg. J Immunother Cancer 1(Suppl 1):O23
47. Blank C, Mackensen A (2007) Contribution of the PD-L1/PD-1 pathway to T-cell exhaustion: an update on implications for chronic infections and tumor evasion. Cancer Immunol Immunother 56(5):739–745
48. Cai G, Karni A, Oliveira EM et al (2004) PD-1 ligands, negative regulators for activation of naive, memory, and recently activated human CD4+ T cells. Cell Immunol 230(2):89–98
49. Fourcade J, Kudela P, Sun Z et al (2009) PD-1 is a regulator of NY-ESO-1-specific CD8+ T cell expansion in melanoma patients. J Immunol 182(9):5240–5249
50. Iwai Y, Ishida M, Tanaka Y et al (2002) Involvement of PD-L1 on tumor cells in the escape from host immune system and tumor immunotherapy by PD-L1 blockade. Proc Natl Acad Sci U S A 99(19):12293–12297
51. Tsushima F, Tanaka K, Otsuki N et al (2006) Predominant expression of B7-H1 and its immunoregulatory roles in oral squamous cell carcinoma. Oral Oncol 42(3):268–274
52. Hamid O, Robert C, Daud A et al (2013) Safety and tumor responses with lambrolizumab (anti-PD-1) in melanoma. N Engl J Med 369(2):134–144
53. Robert C, Ribas A, Wolchok JD et al (2014) Anti-programmed-death-receptor-1 treatment with pembrolizumab in ipilimumab-refractory advanced melanoma: a randomised dose-comparison cohort of a phase 1 trial. Lancet 384(9948):1109–1117
54. Homsi J, Kashani-Sabet M, Messina JL et al (2005) Cutaneous melanoma: prognostic factors. Cancer Control 12(4):223–229
55. Wolchok JD, Kluger H, Callahan MK et al (2013) Nivolumab plus Ipilimumab in advanced melanoma. N Engl J Med 369(2):122–133
56. Hsu FJ, Benike C, Fagnoni F et al (1996) Vaccination of patients with B-cell lymphoma using autologous antigen-pulsed dendritic cells. Nat Med 2(1):52–58
57. Nestle FO, Alijagic S, Gilliet M et al (1998) Vaccination of melanoma patients with peptide- or tumor lysate-pulsed dendritic cells. Nat Med 4(3):328–332
58. Draube A, Klein-Gonzalez N, Mattheus S et al (2011) Dendritic cell based tumor vaccination in prostate and renal cell cancer: a systematic review and meta-analysis. PLoS One 6(4), e18801
59. Poschke I, Mao Y, Adamson L et al (2012) Myeloid-derived suppressor cells impair the quality of dendritic cell vaccines. Cancer Immunol Immunother 61(6):827–838
60. Iclozan C, Antonia S, Chiappori A et al (2013) Therapeutic regulation of myeloid-derived suppressor cells and immune response to cancer vaccine in patients with extensive stage small cell lung cancer. Cancer Immunol Immunother 62(5):909–918
61. Morse MA, Hobeika AC, Osada T et al (2008) Depletion of human regulatory T cells specifically enhances antigen-specific immune responses to cancer vaccines. Blood 112(3):610–618
62. Ridolfi L, Petrini M, Fiammenghi L et al (2010) Unexpected high response rate to traditional therapy after dendritic cell-based vaccine in advanced melanoma: update of clinical outcome and subgroup analysis. Clin Dev Immunol 2010:504979
63. Chi KH, Liu SJ, Li CP et al (2005) Combination of conformal radiotherapy and intratumoral injection of adoptive dendritic cell immunotherapy in refractory hepatoma. J Immunother 28(2):129–135
64. Fadul CE, Fisher JL, Hampton TH et al (2011) Immune response in patients with newly diagnosed glioblastoma multiforme treated with intranodal autologous tumor lysate-dendritic cell vaccination after radiation chemotherapy. J Immunother 34(4):382–389
65. Sampson JH, Archer GE, Mitchell DA et al (2009) An epidermal growth factor receptor variant III-targeted vaccine is safe and immunogenic in patients with glioblastoma multiforme. Mol Cancer Ther 8(10):2773–2779
66. Demaria S, Kawashima N, Yang AM et al (2005) Immune-mediated inhibition of metastases after treatment with local radiation and CTLA-4 blockade in a mouse model of breast cancer. Clin Cancer Res 11(2 Pt 1):728–734

67. Demaria S, Pilones KA, Formenti SC et al (2013) Exploiting the stress response to radiation to sensitize poorly immunogenic tumors to anti-CTLA-4 treatment. Oncoimmunology 2(3), e23127
68. Deng L, Liang H, Burnette B et al (2014) Irradiation and anti-PD-L1 treatment synergistically promote antitumor immunity in mice. J Clin Invest 124(2):687–695
69. Zeng J, See AP, Phallen J et al (2013) Anti-PD-1 blockade and stereotactic radiation produce long-term survival in mice with intracranial gliomas. Int J Radiat Oncol Biol Phys 86(2):343–349
70. Akutsu Y, Matsubara H, Urashima T et al (2007) Combination of direct intratumoral administration of dendritic cells and irradiation induces strong systemic antitumor effect mediated by GRP94/gp96 against squamous cell carcinoma in mice. Int J Oncol 31(3):509–515
71. Hodi FS, Butler M, Oble DA et al (2008) Immunologic and clinical effects of antibody blockade of cytotoxic T lymphocyte-associated antigen 4 in previously vaccinated cancer patients. Proc Natl Acad Sci U S A 105(8):3005–3010
72. Le DT, Lutz E, Uram JN et al (2013) Evaluation of ipilimumab in combination with allogeneic pancreatic tumor cells transfected with a GM-CSF gene in previously treated pancreatic cancer. J Immunother 36(7):382–389
73. Sarnaik AA, Yu B, Yu D et al (2011) Extended dose ipilimumab with a peptide vaccine: immune correlates associated with clinical benefit in patients with resected high-risk stage IIIc/IV melanoma. Clin Cancer Res 17(4):896–906
74. Madan RA, Mohebtash M, Arlen PM et al (2012) Ipilimumab and a poxviral vaccine targeting prostate-specific antigen in metastatic castration-resistant prostate cancer: a phase 1 dose-escalation trial. Lancet Oncol 13(5):501–508
75. Ascierto PA, Kalos M, Schaer DA et al (2013) Biomarkers for immunostimulatory monoclonal antibodies in combination strategies for melanoma and other tumor types. Clin Cancer Res 19(5):1009–1020
76. Mougiakakos D, Choudhury A, Lladser A et al (2010) Regulatory T cells in cancer. Adv Cancer Res 107:57–117
77. Mougiakakos D, Johansson CC, Jitschin R et al (2011) Increased thioredoxin-1 production in human naturally occurring regulatory T cells confers enhanced tolerance to oxidative stress. Blood 117(3):857–861
78. Poschke I, Mao Y, Nyström M, Hansson J, Masucci G, Kiessling R (2013) Title Ipilimumab treatment results in an early decrease in the frequency of circulating granulocytic myeloid derived suppressor cells as well as their arginase1 production. J Cancer Immunol Res 1:158–62
79. Johansson CC, Egyhazi S, Masucci G et al (2009) Prognostic significance of tumor iNOS and COX-2 in stage III malignant cutaneous melanoma. Cancer Immunol Immunother 58(7):1085–1094
80. Purdy AK, Campbell KS (2009) Natural killer cells and cancer: regulation by the killer cell Ig-like receptors (KIR). Cancer Biol Ther 8(23):2211–2220
81. Srivastava S, Lundqvist A, Childs RW (2008) Natural killer cell immunotherapy for cancer: a new hope. Cytotherapy 10(8):775–783
82. Paschen A, Sucker A, Hill B et al (2009) Differential clinical significance of individual NKG2D ligands in melanoma: soluble ULBP2 as an indicator of poor prognosis superior to S100B. Clin Cancer Res 15(16):5208–5215
83. Smyth MJ, Swann J, Kelly JM et al (2004) NKG2D recognition and perforin effector function mediate effective cytokine immunotherapy of cancer. J Exp Med 200(10):1325–1335
84. So T, Takenoyama M, Sugaya M et al (2001) Unfavorable prognosis of patients with non-small cell lung carcinoma associated with HLA-A2. Lung Cancer 32(1):39–46
85. Tisch M, Kyrberg H, Weidauer H et al (2002) Human leukocyte antigens and prognosis in patients with head and neck cancer: results of a prospective follow-up study. Laryngoscope 112(4):651–657
86. Hoon DS, Okamoto T, Wang HJ et al (1998) Is the survival of melanoma patients receiving polyvalent melanoma cell vaccine linked to the human leukocyte antigen phenotype of patients? J Clin Oncol 16(4):1430–1437

87. Le YS, Kim TE, Kim BK et al (2002) Alterations of HLA class I and class II antigen expressions in borderline, invasive and metastatic ovarian cancers. Exp Mol Med 34(1):18–26
88. Lee JE, Abdalla J, Porter GA et al (2002) Presence of the human leukocyte antigen class II gene DRB1*1101 predicts interferon gamma levels and disease recurrence in melanoma patients. Ann Surg Oncol 9(6):587–593
89. Marincola FM, Venzon D, White D et al (1992) HLA association with response and toxicity in melanoma patients treated with interleukin 2-based immunotherapy. Cancer Res 52(23):6561–6566
90. Mitchell MS, Harel W, Groshen S (1992) Association of HLA phenotype with response to active specific immunotherapy of melanoma. J Clin Oncol 10(7):1158–1164
91. Sosman JA, Unger JM, Liu PY et al (2002) Adjuvant immunotherapy of resected, intermediate-thickness, node-negative melanoma with an allogeneic tumor vaccine: impact of HLA class I antigen expression on outcome. J Clin Oncol 20(8):2067–2075
92. Bain C, Merrouche Y, Puisieux I et al (1997) Correlation between clinical response to interleukin 2 and HLA phenotypes in patients with metastatic renal cell carcinoma. Br J Cancer 75(2):283–286
93. Franzke A, Buer J, Probst-Kepper M et al (2001) HLA phenotype and cytokine-induced tumor control in advanced renal cell cancer. Cancer Biother Radiopharm 16(5):401–409
94. Cortes J, Fayad L, Kantarjian H et al (1998) Association of HLA phenotype and response to interferon-alpha in patients with chronic myelogenous leukemia. Leukemia 12(4):455–462
95. Gamzatova Z, Villabona L, Dahlgren L et al (2006) Human leucocyte antigen (HLA) A2 as a negative clinical prognostic factor in patients with advanced ovarian cancer. Gynecol Oncol 103(1):145–150
96. Gamzatova Z, Villabona L, van der Zanden H et al (2007) Analysis of HLA class I-II haplotype frequency and segregation in a cohort of patients with advanced stage ovarian cancer. Tissue Antigens 70(3):205–213
97. Helgadottir H, Andersson E, Villabona L et al (2009) The common Scandinavian human leucocyte antigen ancestral haplotype 62.1 as prognostic factor in patients with advanced malignant melanoma. Cancer Immunol Immunother 58(10):1599–1608
98. Andersson E, Villabona L, Bergfeldt K et al (2012) Correlation of HLA-A02* genotype and HLA class I antigen down-regulation with the prognosis of epithelial ovarian cancer. Cancer Immunol Immunother 61(8):1243–1253
99. Rebmann V, Wagner S, Grosse-Wilde H (2007) HLA-G expression in malignant melanoma. Semin Cancer Biol 17(6):422–429
100. Luo W, Liao WJ, Huang YT et al (2011) Normalization of T cell receptor repertoire diversity in patients with advanced colorectal cancer who responded to chemotherapy. Cancer Sci 102(4):706–712
101. Luo W, Liao WJ, Huang YT et al (2011) Cancer of the gastrointestinal tract results in a restricted T-cell repertoire dependent upon tumor differentiation. Cell Immunol 270(1):47–52

Chapter 10
T Cell Modulation: Anti-PD-1 Antibodies for the Treatment of Cancer

Patrick A. Ott and F. Stephen Hodi

Abstract The expression of inhibitory receptors on tumor specific T cells leads to compromised effector function such as decreased proliferation, cytokine secretion, and tumor cell lysis. These receptors can be targeted therapeutically using monoclonal antibodies, an approach that was termed "checkpoint blockade". The improved survival of advanced melanoma patients treated with the anti-CTLA-4 antibody ipilimumab validates this new treatment concept. Inhibition of another inhibitory pathway, PD-1/PD-L1, using monoclonal antibodies has recently shown much promise in the treatment of melanoma, renal cell cancer, non-small cell lung cancer, among other tumor types. Durable anti-tumor activity with a favorable safety profile has lead to fast paced clinical development of many compounds targeting both PD-1 and PD-L1.

Keywords T cell • PD-1 • PD-L1 • CTLA-4 • Immune checkpoint blockade • Cancer

Introduction

Immune checkpoint blockade was proven as a successful strategy for cancer therapy based on the improved overall survival of advanced melanoma patients who were treated with the anti-Cytotoxic T-Lymphocyte Antigen 4 (CTLA-4) antibody ipilimumab in phase 3 trials [1, 2]. Programmed Death-1 (PD-1) is a member of the immunoglobulin superfamily; it was initially discovered as an upregulated gene in a T cell hybridoma undergoing apoptosis [3]. PD-1 interacts with two ligands: PD-L1 (B7-H1) and PD-L2 (B7-DC) (Fig. 10.1). PD-1 is expressed on T cells, B cells, natural killer (NK) T cells, activated monocytes, and myeloid CD11c^{+} dendritic cells (DCs). PD-L1 is expressed in many immune cells including T and B cells, monocytes, dendritic cells (DCs) as well as non-hematopoietic tissues and many different tumor types [4–6]. The expression of PD-L1 is constitutive in many

P.A. Ott, M.D., Ph.D. (✉) • F.S. Hodi
Dana Farber Cancer Institute, Harvard Medical School, Melanoma Disease Center and Center for Immuno-Oncology, 450 Brookline Avenue, Boston, MA 02215-5450, USA
e-mail: Patrick_Ott@DFCI.harvard.edu

P.A. Ascierto et al. (eds.), *Developments in T Cell Based Cancer Immunotherapies*, Cancer Drug Discovery and Development,
DOI 10.1007/978-3-319-21167-1_10

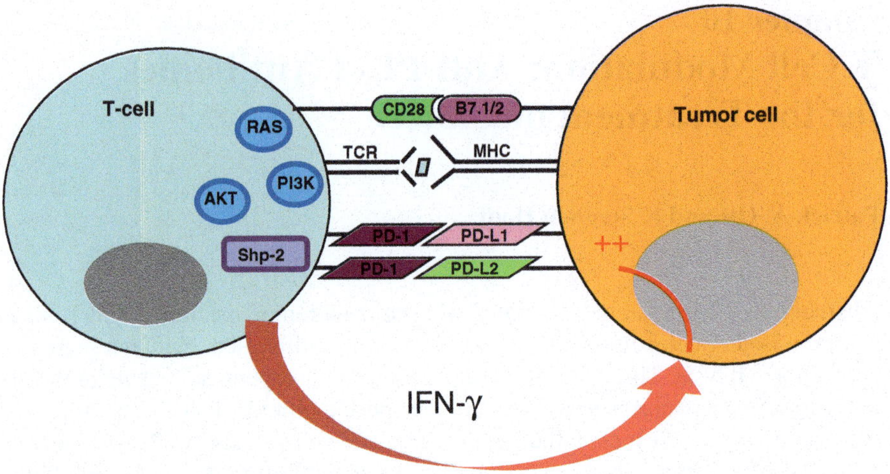

Fig. 10.1 PD-1 is upregulated on antigen- experienced memory T cells in peripheral tissues, thereby protecting them from collateral damage during an inflammatory response. Tumor cells can evade T cell mediated destruction by upregulating PD-L1 and PD-L2. IFN-γ production by T cells leads to upregulation of both PD-1 and PD-L1

cell types, however it can also be induced by type I and type II interferons [7, 8]. In contrast to the fairly wide expression of PD-L1 including non-hematopoietic tissues, PD-L2 is primarily expressed upon induction in hematopoietic cells [9, 10]. The MyD88-, TRAF6-, and MEK and AKT dependent pathways have been implicated in mediating upregulation of PD-L1; loss of phosphatase and tensin homolog (PTEN) leads to increased PD-L1 expression in tumors [11]. The engagement of PD-1 on T cells mediates inhibitory signals, affecting cytokine production, cell proliferation, apoptosis and cytolysis [12–14].

CTLA-4 and PD-1: Different Functionality

CTLA-4 modulates T cell function primarily during early activation in the lymph node. In contrast, regulation of T cells through the PD-1/PD-L1 pathways takes place both in the activation and effector phases of the T cell response in peripheral tissues e.g. during chronic inflammation, thereby protecting tissues as described above. This distinct functionality is reflected in the phenotype of CTLA-4 and PD-1 knockout (KO) models: CTLA-4 KO mice succumb to massive systemic lymph node hyperproliferation several weeks after birth, whereas PD-1 KO is manifested in autoimmune syndromes such as autoimmune diabetes, dilated cardiomyopathy, or glomerulonephritis in sensitive mouse strains [15–17].

The presence of PD-L1 on peripheral tissues represents a protective mechanism against T cell mediated tissue damage such as during chronic tissue inflammation/ infection. PD-L1 expression on CD8+ cells correlates with an "exhaustive" phenotype in viral infection models resulting in impaired or lost effector function such as cytokine secretion (e.g. IFN-γ, IL-2, TNF-α) and cytolytic function [18–20]. Consistent with this immunosuppressive effect in the periphery, PD-1/PD-L1 interaction was found to be critical in the control of self-reactive T cells during both initial activation and re-encounter in the periphery in animal models of type 1 diabetes and multiple sclerosis (Experimental Allergic Encephalomyelitis) [21–24]. In PD-1 and PD-L1 double knock out mouse models, different types of spontaneous autoimmunity have been observed, including glomerulonephritis and dilated cardiomyopathy [15, 25]. In the non obese diabetes (NOD) model, PD-1/PD-L1 blockade reversed anergy in islet-antigen-specific T cells, whereas inhibition of CTLA-4 blockade did not, suggesting a unique function for PD-1:PD-L1 interactions in maintaining peripheral tolerance [26].

Complexity of PD-1:PD-L1 pathway interactions is added by the finding that PD-L1 exerts an additional inhibitory signal to T cells through B7.1, independent of PD-1, which also modulates T cell activation and proliferation [27]. As a result, PD-1 blockade, which only suppresses interactions between PD-1 and its two ligands, B7-H1 and B7-DC, does not block the inhibitory signal through B7.1. Similarly, PD-L1 blockade affects the inhibitory signals through PD-1 and B7.1, but does not impact the B7-DC-PD-1 pathway. These different sets of receptor-ligand interactions need to be taken into account when assessing antibodies with specificities to PD-1 versus B7-H1 in the clinical context [28].

Tumors can co-opt the PD-1/PD-L1 pathway as a defense mechanism against attack by tumor infiltrating lymphocytes (TIL). PD-L1 expression has been reported in a broad spectrum of tumor types including melanoma, glioblastoma, non small cell lung cancer (NSCLC), urothelial cancer, ovarian cancer, breast cancer, cervical cancer, colon cancer, pancreatic cancer, and gastric cancer [29–34]. PD-L1 and PD-L2 expression has also been found on hematologic malignancies such as multiple myeloma, acute myeloid leukemia chronic lymphocytic leukemia, adult T-cell leukemia/lymphoma, Hodgkin lymphoma, primary mediastinal B cell lymphoma, and angioimmunoblastic T-cell lymphoma [35–39]. Tumor PD-L1 expression has been associated with poor clinical outcome in many retrospective series, which seems intuitive and consistent with the immunosuppressive role of the PD-1/PD-L1 pathway in the tumor microenvironment. Intriguingly however, tumor PD-L1 expression can also correlate with favorable outcome in melanoma patients [40]. In a recent study, PD-L1 expression, IFN-γ production, and TIL were found to be co-localized in melanoma specimens. In this context, the expression of PD-L1 was interpreted as an adaptive homeostasis reaction to the infiltration with TILs since IFN-γ production by T cells can directly mediate upregulation of PD-L1 expression locally as a resistance mechanism of the tumor, thus explaining the improved clinical outcomes [40].

PD-1/PD-L1 Inhibition in Cancer Patients

Clinical activity of PD-1 blockade in cancer patients was first reported in 2008 [41]. In subsequent large phase 1 studies, encouraging objective tumor response rates in patients with different tumor types including melanoma, renal cell carcinoma

Table 10.1 Tumor activity of monoclonal antibodies targeting PD-1/PD-L1 (efficacy population, all doses)

	Melanoma	RCC	NSCLC
Anti-PD-1			
Nivolumab (BMS)			
Number of patients	94	33	76
Objective response rate (%)	26 (28)	9 (27)	14 (18)
Stable disease (24+ weeks)	6 (6)	9 (27)	5 (7)
MK3475 (Merck)			
Number of patients	117	NA	NA
Objective response rate (%)	44 (38)	NA	NA
Stable disease (24+ weeks)	NA	NA	NA
Anti-PD-L1			
MDX-1105 (BMS)			
Number of patients	52	17	49
Objective response rate (%)	9 (17)	2 (12)	5 (10)
Stable disease (24+ weeks)	14 (27)	7 (41)	6 (12)
MPDL3280A (Roche/Genentech)			
Number of patients	38	NA	53
Objective response rate (%)	11 (29)	NA	12 (23)
Stable disease (24+ weeks)	NA	NA	2 (5)

(RCC), and NSCLC were reported with both PD-1 and PD-L1 blocking monoclonal antibodies [42, 43] (Table 10.1). The tremendous promise of this treatment strategy from these early studies prompted the initiation of intense and fast paced clinical development programs of monoclonal antibodies directed at the PD-1/PD-L1 pathway in cancer patients (Table 10.2).

PD-1 Inhibition

Nivolumab

Nivolumab (also known as MDX-1106; ONO-4538) is a fully human monoclonal IgG4 antibody directed at the PD-1 receptor. In the dose range of 0.1–10 mg, the pharmacokinetics are linear with dose-proportional increases in C_{max} and $AUC_{0–14\ days}$. Because of its IgG4 Fc-domain, the drug does not exhibit antibody dependent cell death (ADCC). Patients with melanoma, RCC, NSCLC, prostate cancer, and colorectal cancer (n=296) who had tumor progression on prior anti-cancer therapies were treated with nivolumab every 2 weeks at doses of 0.1, 0.3, 1, 3, and 10 mg/kg. Objective tumor responses were observed in 26 of 94 (28 %) patients with melanoma, 14 of 76 (18 %) patients with NSCLC, and 9 of 33 (27 %) of patients with RCC [43]. Many of the responses were durable: Thirteen of 18 (72 %) melanoma patients and 5 of 8 (63 %) RCC patients who had received nivolumab for more than 1 year had responses that lasted for 1 year or longer; 8 of 14 (57 %) NSCLC patients who were

Table 10.2 PD-1/PD-L1 directed monoclonal antibodies in clinical development

	Phase I (monotherapy)	Phase I (combo)	Phase II (monotherapy)	Phase II (combo)	Phase III (monotherapy)
Anti-PD-1					
Nivolumab (BMS)	Solid tumors (2 studies)[a]	RCC (+Pazopanib/Sunitinib)	RCC, NSCLC		NSCLC, RCC, Melanoma
	Hematological Malignancies	NSCLC (+chemo/Erlotinib/ Bevacizumab)			
	HCC	Melanoma (+ Ipilimumab)			
		Solid tumors (+ Ipilimumab)			
		Melanoma (+ vaccine)			
		Solid tumors (+ IL-21)			
		Solid tumors (+ anti-LAG-3)			
		Solid tumors (+ anti-KIR)			
CT-011 (CureTech)	Hematological malignancies[a]		Melanoma, DLBCL	Multiple tumor types[b]	
MK-3475 (Merck)	Solid tumors, Hematologic malignancies, Melanoma, NSCLC	NSCLC	Melanoma (vs chemotherapy), Microsatellite unstable tumors		Melanoma
Anti-PD-L1					
BMS-936559 (BMS)	Solid tumors[a]				
MPDL3280A	Solid tumors	Melanoma (+Vemurafenib)	RCC, NSCLC	RCC	
		Solid tumor (+Bevacizumab +/– chemo)			
		Solid tumor (+Cobimetinib)			
MEDI4736 (MedImmune)	Melanoma, RCC, NSCLC, CRC	Solid tumors (+Tremelimumab)			
MSB0010718C (EMD Serono)	Solid tumors				

RCC renal cell cancer, *NSCLC* non-small cell lung cancer, *DLBCL* diffuse large B cell lymphoma, *HCC* hepatocellular carcinoma, *Cobimetinib* MEK inhibitor, *Vemurafenib* BRAF inhibitor, *BMS* Bristol Myers Squibbs

[a]Completed and reported

[b]Separate studies in Pancreatic Cancer, CRC, Prostate Ca, Myeloma, RCC, Follicular lymphoma, AML

treated for 24+ weeks had tumor responses that lasted 24 weeks or longer. Updated results from this study were presented for the melanoma, NSCLC, and RCC cohorts at ASCO 2013 [44]. In melanoma patients (all dose cohorts), the median duration of response was 104 weeks (18.4–117.0+ weeks). Thirty-three of 107 (31 %) melanoma patients in the updated dataset had an objective tumor response, in addition to 7 patients (7 %) with stable disease (SD) lasting ≥24 weeks and 4 patients with an unconventional (immune-related) pattern. The onset of RECIST responses was rapid: 15 of 33 (45 %) of patients had a CR or PR at the time of the first tumor assessment at 8 weeks. The median OS of melanoma patients was 16.8 months whereas 1 year OS and 2 year OS were 62 % and 43 %, respectively. Twenty of 122 (16 %) of NSCLC patients had objective tumor response in the updated data set; median OS was 9.6 months. One-year OS and 2-year OS were 43 % and 32 %, respectively. In RCC, median OS was >22 months with 70 % and 52 % 1-year and 2-year OS, respectively. Nivolumab was remarkably well tolerated in this trial. Eighteen of 296 (6 %) patients experienced a treatment related grade 3 or 4 adverse event and 122 of 296 (41 %) patients had adverse events associated with an immune-related etiology. Adverse events observed in more than 5 % of patients were rash (12 %), diarrhea (11 %), and pruritus (9 %). Other, infrequently observed inflammatory adverse events included pneumonitis, vitiligo, hepatitis, and hypo/hyperthyroidism. Most of these toxicities were reversible with treatment interruption, treatment discontinuation, or the administration of glucocorticoids. Three deaths due to pneumonitis related to nivolumab occurred on the study. Algorithms are now employed to guide treatment of inflammatory events associated with PD-1/PD-L1 blockade.

MK3475

MK3475 is a fully human monoclonal anti-PD-1 antibody of the IgG4-κ isotype. In a multi-institutional, phase 1 trial, 135 patients with metastatic melanoma, many of whom had visceral metastases, received MK-3575 at doses of 10 mg every 2 weeks, 2 mg every 3 weeks, or 10 mg every 3 weeks [45]. The treatment was well tolerated; grade 3 or 4 related adverse events attributed to treatment were seen in 17 of the 135 (13 %) patients. Toxicities included fatigue, fever, chills, myalgias, and headaches, which were almost exclusively grade 1 or 2. Treatment-related pneumonitis was seen in 6/135 (4 %) of the patients. Forty-four of 117 (38 %) of the patients had an objective response by RECIST 1.1 criteria. Eight additional patients had an unconfirmed response, resulting in a RR including confirmed and unconfirmed responses of 44 %. In the highest dose cohort, 27 of 52 (52 %) patients experienced a confirmed tumor response. Interestingly, response rates did not differ between patients who had been previously treated with ipilimumab (anti-CTLA-4) versus those who had not. Most tumor responses were observed when the tumors were first assessed after 12 weeks of treatment. Moreover, the median duration of responses and median OS had not been reached after 11 months of follow-up, indicating that the responses may be durable in many patients. In an update presented at the annual meeting of the Society of Melanoma Research (SMR), the one year overall survival was 82 %

and median OS had still not been reached. Phase 2 and 3 trials with MK3475 are currently underway in melanoma and NSCLC; other tumor types are under investigation in phase 1 and 2 studies.

PD-L1 Inhibition

BMS936559

In a multi-institutional phase 1 trial with dose escalation cohorts and tumor type specific expansion cohorts, patients with NSCLC, melanoma, RCC, ovarian cancer, gastric cancer, pancreatic cancer, and colorectal cancer were treated with the anti-PD-L1 monoclonal antibody BMS936559 (also known as MDX-1105) [42]. Most of these patients were refractory to one or more lines of prior systemic treatment. The treatment was well tolerated: 39 % of patients had grades 1 and 2 inflammatory toxicities such as rash, diarrhea, hypothyroidism, and hepatitis, whereas 9 % of patients had treatment-related grade 3 or 4 AEs. In this study, objective tumor responses were seen in 9 of 52 (17 %) melanoma patients, 5 of 49 (10 %) NSCLC patients, 2 of 17 (12 %) RCC patients, and one of 17 ovarian cancer patients. In addition to the objective tumor responses, between 12 and 41 % of patients in these cohorts had SD ≥ 24 weeks.

MPDL3280A

MPDL3280A is a fully human monoclonal IgG4 antibody directed at PD-L1. Safety and efficacy data in patients with melanoma and NSCLC have been reported from a phase 1A dose escalation study including patients with multiple different tumor types [46]. Six of 38 (14 %) advanced melanoma patients had treatment-related grade 3 and 4 adverse events and two patients (5 %) had immune-related grade 3 and 4 events. Eleven of 38 (29 %) melanoma patients had a CR or PR while 2 of 38 (5 %) patients had SD lasting ≥24 weeks. In a cohort of 53 patients with NSCLC the RR was 23 % (12/53) [47]. Notably, 11 of the 12 responders in the NSCLC cohort were smokers, for a RR of 26 % (11/43) in this subset of patients. Objective tumor responses were also observed in patients with colorectal cancer, gastric cancer, and squamous cell head and neck cancer treated with MPDL3280A [48].

MEDI4736

MEDI4736 is an anti-PD-L1 human monoclonal antibody that lacks the Fc domain and therefore has no Fc-mediated effector function. Preliminary data from a phase 1, multi-center, open label study in patients with advanced solid tumors suggest that the antibody is safe; a signal for clinical activity was detected as evident by tumor regression and prolonged stable disease in several tumor types [49].

Other Immune Modulating Antibodies and Opportunities for Combinatorial Approaches

The efficacy of PD-1/PD-L1 and anti-CTLA-4 blockade given without vaccination indicates that pre-existing, endogenous tumor specific T cells, once uncoupled from the inhibitory signaling, can eradicate tumor cells. It may be surprising that targeting a single inhibitory receptor such as PD-1 or CTLA-4 can induce tumor responses given the multitude of immune regulatory mechanisms, which also include T regulatory cells, myeloid derived suppressor cells [50], and immunosuppressive mediators such as IDO (indoleamine 2,3-dioxygenase), arginase and prostaglandin E2 (PGE2), IL-6, IL-10, and vascular endothelial growth factor (VEGF).

Multiple inhibitory molecules expressed on T cells such as LAG-3 [51], Tim-3 [52], BTLA [53] as well as a number of co-stimulatory members of the CD28/CTLA-4 family (ICOS) and the TNF superfamily (CD40, CD27, CD137, Glucocorticoid-induced TNFR-related protein (GITR), and OX40) can be targeted with monoclonal antibodies. Some of these antibodies (directed at CD27, CD137, GITR, CD40, LAG-3) are already under clinical investigation. In addition, inhibitory receptors in other immune cell populations, such as Killer-cell immunoglobulin-like receptor (KIR) expressed by natural killer (NK) cells, are suitable targets for immune intervention and being investigated in clinical trials.

Co-expression of PD-1 and LAG-3

There is preclinical evidence of co-expression of multiple inhibitory receptors and synergistic effects achieved with combined checkpoint blockade. In patients with ovarian cancer, TILs co-expressing LAG-3 and PD-1 were found to be more functionally impaired than TILs expressing either of the receptors individually. This CD8exhaustion could be rescued in vitro by double blockade of PD-1 and LAG-3, but not by inhibition of either of the receptors alone [54].

Co-expression of PD-1 and Tim-3

Tumor infiltrating lymphocytes in mouse models of acute myeloid leukemia (AML) and colon cancer were found to co-express PD-1 and Tim-3, leading to impaired proliferation and decreased IL-2, TNF-α, and IFN-γ secretion in both models. Of note, co-inhibition of PD-1 and Tim-3 revealed a synergistic anti-tumor effect [55, 56]. In metastatic melanoma patients, NY-ESO-1–specific CD8+ cells present in PBMC were found to predominantly co-express Tim-3 and PD-1, resulting in decreased IFN-γ, TNF-α, and IL-2 production; Tim-3 and PD-1 blockade was synergistic in rescuing the PD1+ Tim3+NY-ESO-1–specific CD8+ cells from their dysfunctional state [57].

Co-expression of PD-1 and CTLA-4

In the B16 melanoma mouse, CTLA-4 blockade induces PD-1 expression in TILs whereas PD-1 blockade induces upregulation of CTLA-4 on TIL [58] indicating the complexity and redundancy of the compensatory T cell regulation network and providing one possible explanation for the unpredictable response rates. In this model, both CTLA-4 and PD-1 inhibition individually are compromised by leaving the other inhibitory pathway unopposed, even leading to its upregulation. As discussed earlier in this chapter, it is recognized that CTLA-4 and PD-1 play distinct and complementary roles in the regulation of T cell responses. PD-1 mediates exhaustion of antigen experienced T cells in peripheral/tumor tissue, whereas CTLA-4 dampens early T cell activation in lymphoid tissues. Combined CTLA-4 and PD-1 blockade has been explored in patients with advanced melanoma in a phase 1 study.

Nivolumab Plus Ipilimumab

In a phase 1 study, 53 patients with metastatic melanoma received nivolumab (anti-PD-1) and ipilimumab in combination [59]. The two antibodies were given concurrently every 3 weeks for 4 doses, followed by nivolumab monotherapy every 3 weeks for 4 doses. Subsequently, nivolumab and ipilimumab were administered concurrently every 12 weeks for up to 8 doses. A separate cohort of 33 patients previously treated with ipilimumab received nivolumab every 2 weeks for up to 48 doses. Grade 3 or 4 treatment-related adverse events were observed in 28 of 53 (53 %) patients and included elevated transaminases (11–13 %), diarrhea (6 %), rash (4 %), elevated lipase (13 %), and emesis (2 %). Common grade 1 and 2 events (>20 %, all grades) included rash, pruritus, fatigue, diarrhea, fever, nausea, and elevated transaminases. Twenty of 52 (42 %) patients receiving the concurrent regimen had a confirmed objective tumor response according to modified WHO criteria. Four additional patients had a response by immune-related response criteria and two patients had an unconfirmed response. Sixteen of 52 (31 %) patients experienced rapid and deep responses as defined as ≥80 % tumor reduction at 12 weeks, including 5 patients with a CR. In the 17 patients who received concurrent ipilimumab and nivolumab at 3 mg/kg and 1 mg/kg, respectively (the maximal dose levels associated with acceptable toxicity), 9 patients (53 %) had an objective response, 7 of which (41 %) achieved ≥80 % tumor reduction. Notably, objective responses were seen in patients with extensive and bulky disease. Tumor responses may be durable as 19 of 21 responses were ongoing at the time of data-cutoff, with response duration ranging between 6.1 and 72 weeks.

These data suggest that combined CTLA-4 and PD-1 blockade mediates more rapid and deeper tumor responses in a higher proportion of melanoma patients compared to monotherapy with either CTLA-4 or PD-1 inhibition. Larger, comparative trials to confirm these observations are ongoing.

TumorPD-L1 Expression

PD-L1 expression by the tumor is actively being pursued as a predictive marker for tumor activity of PD-1 and PD-L1 directed antibodies. In a retrospective analysis of a subset of 101 melanoma and NSCLC patients treated with nivolumab, none of the patients with tumors lacking PD-L1 expression had a tumor response [43, 60]. This observation has lead to the selection of patients based on tumor PD-L1 expression in some of the ongoing anti-PD-1 and anti-PD-L1 clinical development programs. It is important to emphasize that the data from the nivolumab study are preliminary and that only tumor responses (and not disease stabilization) was captured as a clinical benefit endpoint. Of note, objective responses were seen in a small subset of advanced melanoma patients with PD-L1 negative tumors who were treated with the anti-PD-L1 antibody MPDL3280A [61].

Furthermore, PD-L1 expression may undergo changes driven by alterations in the tumor microenvironment, e.g. infiltrations with immune cells, potentially limiting its use as a clear-cut predictive biomarker as opposed to, for example, $BRAF^{V600}$ mutational status. Furthermore, reliable PD-L1 specific antibodies have been difficult to develop and an agreement on acceptable criteria for the definition of PD-L1 positivity such as appropriate staining patterns (membranous versus cytosolic) and cutoffs for percentages of staining cells have so far been elusive. Prospective randomized trials using validated immunohistochemical assessment of PD-L1 expression are needed for further exploration.

Conclusions and Outlook

The B7-H1/PD-1 pathway is an important mechanism cancers utilize to evade destruction by the host immune response; the anti-tumor activity achieved with PD-1/PD-L1 inhibition across a spectrum of different cancers indicates that the pathway is a critical target for immunotherapy. Particularly encouraging is the observation that PD-1/PD-L1 blockade can be effective in tumor types that were previously not considered amenable to immune intervention such as NSCLC, the favorable toxicity profile of PD-1 and B7L1 inhibition, and the potential for durability of tumor responses. These attributes of PD-1/PD-L1 blockade have led to ongoing immunotherapy development programs that are unparalleled in scope and tempo. Preclinical studies suggest that many other inhibitory and stimulatory receptors such as LAG-3, Tim-3, CD40, CD137, and others are potential targets for immune modulation and the remarkable efficacy of combined PD-1 and CTLA-4 blockade in melanoma patients indicates that synergy can be achieved clinically with combination of immune modulatory antibodies. Many other combinatorial approaches including combinations of checkpoint blockade with novel vaccines, angiogenesis inhibition, molecular targeted therapies, and direct targeting of other immunosuppressive pathways are ongoing.

References

1. Hodi FS, O'Day SJ, McDermott DF et al (2010) Improved survival with ipilimumab in patients with metastatic melanoma. N Engl J Med 363:711–723
2. Robert C, Thomas L, Bondarenko I et al (2011) Ipilimumab plus dacarbazine for previously untreated metastatic melanoma. N Engl J Med 364:2517–2526
3. Ishida Y, Agata Y, Shibahara K, Honjo T (1992) Induced expression of PD-1, a novel member of the immunoglobulin gene superfamily, upon programmed cell death. EMBO J 11:3887–3895
4. Freeman GJ, Long AJ, Iwai Y et al (2000) Engagement of the PD-1 immunoinhibitory receptor by a novel B7 family member leads to negative regulation of lymphocyte activation. J Exp Med 192:1027–1034
5. Dong H, Zhu G, Tamada K, Chen L (1999) B7-H1, a third member of the B7 family, co-stimulates T-cell proliferation and interleukin-10 secretion. Nat Med 5:1365–1369
6. Keir ME, Butte MJ, Freeman GJ, Sharpe AH (2008) PD-1 and its ligands in tolerance and immunity. Annu Rev Immunol 26:677–704
7. Eppihimer MJ, Gunn J, Freeman GJ et al (2002) Expression and regulation of the PD-L1 immunoinhibitory molecule on microvascular endothelial cells. Microcirculation 9:133–145
8. Schreiner B, Mitsdoerffer M, Kieseier BC et al (2004) Interferon-beta enhances monocyte and dendritic cell expression of B7-H1 (PD-L1), a strong inhibitor of autologous T-cell activation: relevance for the immune modulatory effect in multiple sclerosis. J Neuroimmunol 155:172–182
9. Latchman Y, Wood CR, Chernova T et al (2001) PD-L2 is a second ligand for PD-1 and inhibits T cell activation. Nat Immunol 2:261–268
10. Tseng SY, Otsuji M, Gorski K et al (2001) B7-DC, a new dendritic cell molecule with potent costimulatory properties for T cells. J Exp Med 193:839–846
11. Parsa AT, Waldron JS, Panner A et al (2007) Loss of tumor suppressor PTEN function increases B7-H1 expression and immunoresistance in glioma. Nat Med 13:84–88
12. Carter L, Fouser LA, Jussif J et al (2002) PD-1:PD-L inhibitory pathway affects both CD4(+) and CD8(+) T cells and is overcome by IL-2. Eur J Immunol 32:634–643
13. Bennett F, Luxenberg D, Ling V et al (2003) Program death-1 engagement upon TCR activation has distinct effects on costimulation and cytokine-driven proliferation: attenuation of ICOS, IL-4, and IL-21, but not CD28, IL-7, and IL-15 responses. J Immunol 170:711–718
14. Fife BT, Bluestone JA (2008) Control of peripheral T-cell tolerance and autoimmunity via the CTLA-4 and PD-1 pathways. Immunol Rev 224:166–182
15. Nishimura H, Nose M, Hiai H, Minato N, Honjo T (1999) Development of lupus-like autoimmune diseases by disruption of the PD-1 gene encoding an ITIM motif-carrying immunoreceptor. Immunity 11:141–151
16. Okazaki T, Tanaka Y, Nishio R et al (2003) Autoantibodies against cardiac troponin I are responsible for dilated cardiomyopathy in PD-1-deficient mice. Nat Med 9:1477–1483
17. Wang J, Yoshida T, Nakaki F, Hiai H, Okazaki T, Honjo T (2005) Establishment of NOD-Pdcd1–/– mice as an efficient animal model of type I diabetes. Proc Natl Acad Sci U S A 102:11823–11828
18. Flies DB, Sandler BJ, Sznol M, Chen L (2011) Blockade of the B7-H1/PD-1 pathway for cancer immunotherapy. Yale J Biol Med 84:409–421
19. Wherry EJ (2011) T cell exhaustion. Nat Immunol 12:492–499
20. Barber DL, Wherry EJ, Masopust D et al (2006) Restoring function in exhausted CD8 T cells during chronic viral infection. Nature 439:682–687
21. Liang SC, Latchman YE, Buhlmann JE et al (2003) Regulation of PD-1, PD-L1, and PD-L2 expression during normal and autoimmune responses. Eur J Immunol 33:2706–2716
22. Keir ME, Liang SC, Guleria I et al (2006) Tissue expression of PD-L1 mediates peripheral T cell tolerance. J Exp Med 203:883–895
23. Ansari MJ, Salama AD, Chitnis T et al (2003) The programmed death-1 (PD-1) pathway regulates autoimmune diabetes in nonobese diabetic (NOD) mice. J Exp Med 198:63–69

24. Zhu B, Guleria I, Khosroshahi A et al (2006) Differential role of programmed death-ligand 1 [corrected] and programmed death-ligand 2 [corrected] in regulating the susceptibility and chronic progression of experimental autoimmune encephalomyelitis. J Immunol 176:3480–3489
25. Nishimura H, Okazaki T, Tanaka Y et al (2001) Autoimmune dilated cardiomyopathy in PD-1 receptor-deficient mice. Science 291:319–322
26. Fife BT, Guleria I, Gubbels Bupp M et al (2006) Insulin-induced remission in new-onset NOD mice is maintained by the PD-1-PD-L1 pathway. J Exp Med 203:2737–2747
27. Butte MJ, Keir ME, Phamduy TB, Sharpe AH, Freeman GJ (2007) Programmed death-1 ligand 1 interacts specifically with the B7-1 costimulatory molecule to inhibit T cell responses. Immunity 27:111–122
28. Topalian SL, Drake CG, Pardoll DM (2012) Targeting the PD-1/B7-H1(PD-L1) pathway to activate anti-tumor immunity. Curr Opin Immunol 24:207–212
29. Thompson RH, Gillett MD, Cheville JC et al (2004) Costimulatory B7-H1 in renal cell carcinoma patients: indicator of tumor aggressiveness and potential therapeutic target. Proc Natl Acad Sci U S A 101:17174–17179
30. Hamanishi J, Mandai M, Iwasaki M et al (2007) Programmed cell death 1 ligand 1 and tumor-infiltrating CD8+ T lymphocytes are prognostic factors of human ovarian cancer. Proc Natl Acad Sci U S A 104:3360–3365
31. Inman BA, Sebo TJ, Frigola X et al (2007) PD-L1 (B7-H1) expression by urothelial carcinoma of the bladder and BCG-induced granulomata: associations with localized stage progression. Cancer 109:1499–1505
32. Nakanishi J, Wada Y, Matsumoto K, Azuma M, Kikuchi K, Ueda S (2007) Overexpression of B7-H1 (PD-L1) significantly associates with tumor grade and postoperative prognosis in human urothelial cancers. Cancer Immunol Immunother 56:1173–1182
33. Nomi T, Sho M, Akahori T et al (2007) Clinical significance and therapeutic potential of the programmed death-1 ligand/programmed death-1 pathway in human pancreatic cancer. Clin Cancer Res 13:2151–2157
34. Wu C, Zhu Y, Jiang J, Zhao J, Zhang XG, Xu N (2006) Immunohistochemical localization of programmed death-1 ligand-1 (PD-L1) in gastric carcinoma and its clinical significance. Acta Histochem 108:19–24
35. Liu J, Hamrouni A, Wolowiec D et al (2007) Plasma cells from multiple myeloma patients express B7-H1 (PD-L1) and increase expression after stimulation with IFN-{gamma} and TLR ligands via a MyD88-, TRAF6-, and MEK-dependent pathway. Blood 110:296–304
36. Dorfman DM, Brown JA, Shahsafaei A, Freeman GJ (2006) Programmed death-1 (PD-1) is a marker of germinal center-associated T cells and angioimmunoblastic T-cell lymphoma. Am J Surg Pathol 30:802–810
37. Rosenwald A, Wright G, Leroy K et al (2003) Molecular diagnosis of primary mediastinal B cell lymphoma identifies a clinically favorable subgroup of diffuse large B cell lymphoma related to Hodgkin lymphoma. J Exp Med 198:851–862
38. Tamura H, Dan K, Tamada K et al (2005) Expression of functional B7-H2 and B7.2 costimulatory molecules and their prognostic implications in de novo acute myeloid leukemia. Clin Cancer Res 11:5708–5717
39. Xerri L, Chetaille B, Serriari N et al (2008) Programmed death 1 is a marker of angioimmunoblastic T-cell lymphoma and B-cell small lymphocytic lymphoma/chronic lymphocytic leukemia. Hum Pathol 39:1050–1058
40. Taube JM, Anders RA, Young GD et al (2012) Colocalization of inflammatory response with B7-h1 expression in human melanocytic lesions supports an adaptive resistance mechanism of immune escape. Sci Transl Med 4:127ra37.
41. Brahmer JR, Topalian S, Wollner I, Powderly JD, Picus J, Drake C, Covino J, Korman A, Pardoll D, Lowy I (2008) Safety and activity of MDX-1106 (ONO-4538), an anti-PD-1 monoclonal antibody, in patients with selected refractory or relapsed malignancies. J Clin Oncol 26(15S):3000
42. Brahmer JR, Tykodi SS, Chow LQ et al (2012) Safety and activity of anti-PD-L1 antibody in patients with advanced cancer. N Engl J Med 366:2455–2465

43. Topalian SL, Hodi FS, Brahmer JR et al (2012) Safety, activity, and immune correlates of anti-PD-1 antibody in cancer. N Engl J Med 366:2443–2454
44. Sznol M, Kluger H, Hodi F, David F, McDermott RDC, Carvajal RD, Lawrence DP, Topalian SL, Atkins MB, Powderly JD, Sharfman WH, Puzanov I, Smith DC, Wigginton JM, Kollia G, Gupta AK, Sosman JA (2013) Survival and long-term follow-up of safety and response in patients (pts) with advanced melanoma (MEL) in a phase I trial of nivolumab (anti-PD-1; BMS-936558; ONO-4538). J Clin Oncol 31 (Suppl; abstr CRA9006)
45. Hamid O, Robert C, Daud A et al (2013) Safety and tumor responses with lambrolizumab (anti-PD-1) in melanoma. N Engl J Med 369:134–144.
46. Hamid O, Sosman JA, Lawrence DP, Sullivan RJ, Ibrahim N, Kluger HM, Boasberg PD, Flaherty K, Hwu P, Ballinger M, Mokatrin A, Kowanetz M, Chen DS, Stephen Hodi F (2013) Clinical activity, safety, and biomarkers of MPDL3280A, an engineered PD-L1 antibody in patients with locally advanced or metastatic melanoma (mM). J Clin Oncol 31 (Suppl; abstr CRA9006^); 31, 2013 (suppl; abstr 9010)
47. Soria JEA (2013) Clinical activity, safety and biomarkers of PD-L1 blockade in non-small cell lung cancer (NSCLC): additional analyses from a clinical study of the engineered antibody MPDL3280A (anti-PDL1). In: The 2013 European Cancer Congress is the 17th congress of the European CanCer Organisation (ECCO), the 38th congress of the European Society for Medical Oncology (ESMO) and the 32nd congress of European Society for Therapeutic Radiology and Oncology (ESTRO) 2013, Amsterdam
48. Tabernero JPJ, Hamid O, Gordon MS, Fisher GA, Braiteh FS, Garbo LE, Fine GD, Kowanetz M, McCall B, Shen X, Chen DS, Kohrt HE (2013) Clinical activity, safety, and biomarkers of MPDL3280A, an engineered PD-L1 antibody in patients with locally advanced or metastatic CRC, gastric cancer (GC), SCCHN, or other tumors. In: ASCO Annual Meeting 2013, Chicago
49. Khleif SN, Lutzky J, Segal NH, Antonia S, Blake-Haskins A, Stewart R et al (2013) MEDI4736, an anti-PD-L1 antibody with modified Fc domain: preclinical evaluation and early clinical results from a phase I study in patients with advanced solid tumors. In: Proceedings from the European Cancer Congress 2013, 2013 September 27–October 1, Amsterdam. Abstract No. 802
50. Gabrilovich DI, Nagaraj S (2009) Myeloid-derived suppressor cells as regulators of the immune system. Nat Rev Immunol 9:162–174
51. Goldberg MV, Drake CG (2011) LAG-3 in cancer immunotherapy. Curr Top Microbiol Immunol 344:269–278
52. Anderson AC (2012) Tim-3, a negative regulator of anti-tumor immunity. Curr Opin Immunol 24:213–216
53. Watanabe N, Gavrieli M, Sedy JR et al (2003) BTLA is a lymphocyte inhibitory receptor with similarities to CTLA-4 and PD-1. Nat Immunol 4:670–679
54. Matsuzaki J, Gnjatic S, Mhawech-Fauceglia P et al (2010) Tumor-infiltrating NY-ESO-1-specific CD8+ T cells are negatively regulated by LAG-3 and PD-1 in human ovarian cancer. Proc Natl Acad Sci U S A 107:7875–7880
55. Sakuishi K, Apetoh L, Sullivan JM, Blazar BR, Kuchroo VK, Anderson AC (2010) Targeting Tim-3 and PD-1 pathways to reverse T cell exhaustion and restore anti-tumor immunity. J Exp Med 207:2187–2194
56. Zhou Q, Munger ME, Veenstra RG et al (2011) Coexpression of Tim-3 and PD-1 identifies a CD8+ T-cell exhaustion phenotype in mice with disseminated acute myelogenous leukemia. Blood 117:4501–4510
57. Fourcade J, Sun Z, Benallaoua M et al (2010) Upregulation of Tim-3 and PD-1 expression is associated with tumor antigen-specific CD8+ T cell dysfunction in melanoma patients. J Exp Med 207:2175–2186
58. Curran MA, Montalvo W, Yagita H, Allison JP (2010) PD-1 and CTLA-4 combination blockade expands infiltrating T cells and reduces regulatory T and myeloid cells within B16 melanoma tumors. Proc Natl Acad Sci U S A 107:4275–4280
59. Wolchok JD, Kluger H, Callahan MK et al (2013) Nivolumab plus ipilimumab in advanced melanoma. N Engl J Med 369:122–133

60. Grosso J, Horak CE, Inzunza D, Cardona DM, Simon JS, Gupta AK, Sankar V, Park J-S, Kollia G, Taube JM, Anders R, Jure-Kunkel M, Novotny J, Jr., Taylor CR, Zhang X, Phillips T, Simmons P, Cogswell J (2013) Association of tumor PD-L1 expression and immune biomarkers with clinical activity in patients (pts) with advanced solid tumors treated with nivolumab (anti-PD-1; BMS-936558; ONO-4538). In: 2013 ASCO Annual Meeting
61. Hamid O, Sosman JA, Lawrence DP, Sullivan RJ, Ibrahim N, Kluger HM, Boasberg PD, Flaherty K, Hwu P, Ballinger M, Mokatrin A, Kowanetz M, Chen DS, Stephen Hodi F (2013) Clinical activity, safety, and biomarkers of MPDL3280A, an engineered PD-L1 antibody in patients with locally advanced or metastatic melanoma (mM). J Clin Oncol 31 (Suppl; abstr CRA9006^) 2013;31, 2013 (suppl; abstr 9010)

Chapter 11
Enhancing T Cell Performance Against Cancer in Combination Treatment Strategies

Maria Libera Ascierto, Salvador Martin Algarra, Ignacio Melero, and Paolo A. Ascierto

Abstract Increasing evidence suggests that some patients with cancer can mount an antitumor immune response capable of controlling cancer. Although both innate (natural killer cells, dendritic cells) and adaptive (T cells) immunity play important roles in cancer immune surveillance, most emphasis has been placed on the exploitation of adaptive immune responses for cancer immunotherapy. In particular, several preclinical and clinical reports have shown that both cytotoxic CD8+ T cells (CTLs) and CD4+ Th1 cells are able to control and even completely reject tumors through the secretion of cytokines such as interferon gamma (IFN-γ) and tumor necrosis factor alfa (TNF-α). These observations have spurred a revolution in cancer therapeutics, as several treatment strategies are clinically tested and developed in an iterative fashion. In melanoma, multiple immunotherapeutic approaches aimed at manipulating and optimizing the activation of CTLs and CD4+ Th1 cells and their recruitment to malignant tissue are being assessed. Here, we provide an overview of the main immune-oncology treatment strategies that, either alone or in combination, are undergoing clinical development at different stages. Namely, we will refer to those immunotherapeutic strategies that include cancer vaccines, adoptive transfer of ex vivo activated T cells and immunomodulatory monoclonal antibodies.

Keywords Immunotherapy • Vaccination • Adoptive T cell therapy • Immunomodulatory antibodies • Combination therapy

---

M.L. Ascierto
Department of Oncology, Johns Hopkins University, Baltimore, MD, USA

S.M. Algarra
Department of Oncology, University of Navarra, Pamplona, Spain

I. Melero (✉)
Department of Oncology, Centro de Investigación Médica Aplicada (CIMA), Clinica Universidad de Navarra, Avenida de Pio XII, 55, Pamplona 31008, Spain
e-mail: imelero@unav.es

P.A. Ascierto (✉)
Unit of Melanoma, Cancer Immunotherapy and Innovative Therapy, Istituto Nazionale Tumori Fondazione "G. Pascale", Via Mariano Semmola, Naples 80131, Italy
e-mail: paolo.ascierto@gmail.com

P.A. Ascierto et al. (eds.), *Developments in T Cell Based Cancer Immunotherapies*, Cancer Drug Discovery and Development,
DOI 10.1007/978-3-319-21167-1_11

Introduction

Anti-cancer therapies have traditionally targeted transformed cancer cells. However, the contribution of non-transformed host immune cells to control the progression of cancer has also been extensively explored over the last six decades. Clinical observations during the past 10 years provide compelling evidence that some patients with cancer can mount an antitumor immune response capable of controlling their disease. These finding emphasize that the roles played by host immune cells in the tumor stroma are critical determinants of cancer biology and key factors for the success or failure of cancer therapy. Such discoveries have changed the field of tumor immunology so that, as remarked by Hanahan and Weinberg [1] in a recent review, 'avoiding immune destruction' is now considered an emerging hallmark of cancer.

Although both innate (natural killer cells, macrophages and dendritic cells) and adaptive (T cells) immunity arms are postulated to play coordinated roles in cancer immune surveillance [2, 3], most emphasis in immunotherapy has been placed on the adaptive immune response. In support of the critical role of T lymphocytes in human against cancer, tumor infiltrating T cell numeration estimated by tissue microarray have revealed a positive prognostic association between high density of CD8+ effector memory cells and overall survival (OS) of patients with colon and other cancer types [4, 5]. Moreover, analyses conducted on pretreatment melanoma biopsies using gene expression analysis revealed the exist of activation of interferon signal transduction pathway signified by STAT1, IFNg, CCR5 CXCR3 and their ligand are positively correlated with favorable clinic outcome in response to therapy [6]. Furthermore, colon, melanoma and breast cancer specimens have shown a positive correlation between upregulation of genes involved in the CD4+Th1 adaptive immune response and reduced risk of relapse [3, 7].

To date, there is limited evidence with regard to the antigen specificity and understand of the tumor tropic mechanism of such spontaneous tumor-infiltrating T (TIL) cells. However, these observations have given rise to a conceptual model in which an adaptive T cell response composed of both cytotoxic CD8+ T cells (CTLs) and CD4+ Th1 cells controls cancer progression. In this immunosurveillance interpretation, the ability to produce cytokines such as interferon (IFN)-γ and tumor necrosis factor (TNF)-α, together with the expansion and activation of cytotoxic CD8+ T cells, is considered critical [8]. Immunosurveillance also predicts the evolutionary adaptation of cancer through genetic or epigenetic changes that determine the appearance of immune-evasive variants, a phenomenon shown in mouse models of sarcoma and termed immunoediting [9].

Because of these observations, attempts to exploit CTLs, TILs and CD4+ Th1 cells have predominated in immunotherapeutic approaches for cancer. Here, we provide an overview of the major strategies in immunotherapy, involving single agents as well as potentially synergistic treatment combinations. These include cancer vaccines, adoptive transfer of ex vivo activated T cells and immunomodulatory monoclonal antibodies. Challenges, advantages and disadvantages associated with each of these approaches will be discussed.

Cancer Vaccines

Tumors are antigenic as a result of somatic genetic alterations, viral oncogenesis and deregulated expression of genes. However, their immunogenicity is generally weak due to constant immunosurveillance selection and therefore tumor antigens need proper formulation in vaccines in order to elicit a strong enough and sufficient anti-cancer response. Among the different immunotherapeutic strategies that have been introduced, cancer vaccination is the one most often investigated in a variety of pre-clinical and clinical settings [10]. The first immunotherapy to show efficacy in large controlled phase III studies was sipuleucel-T, which consists of a preparation of autologous antigen-presenting cells activated and pulsed with a fusion protein consisting of prostatic acid phosphate (PAP) and granulocyte-macrophage colony-stimulating factor (GM-CSF). Treatment with sipuleucel-T in patients with metastatic prostate cancer improved median survival by approximately 4 months and it was approved by the US Food and Drug Administration (FDA) in 2009 as the first antigen-specific cancer immunotherapy [11]. This complex cell product, which involves three sequential leukoapheresis leukocyte suspensions being cultured for 48 h with the PAP-GM-CSF fusion protein, contains both antigen-presenting cells and activated T cells and thus the exact mechanism of action is undefined. However, its use has been shown to prolong OS without improving progression-free survival (PFS).

Also in prostate cancer, a randomized phase IIb study in patients with metastatic disease revealed that vaccination with prostate-specific antigen (PSA) encoding poxviral vector (Prostvac) improved OS by more than 8 months compared with the control vector group [12]. Prostvac is a prostate cancer vaccine regimen consisting of a recombinant vaccinia vector as a primary vaccination, followed by multiple booster vaccinations employing a recombinant fowlpox vector. Both vectors contain the transgenes for PSA and multiple T-cell costimulatory molecules. Its clinical benefit is currently being further evaluated in an ongoing phase III trial.

In melanoma, vaccination with a peptide derived from melanoma-associated antigen gp100, applied together with interleukin (IL)-2, showed in a controlled phase III trial significantly improved PFS in a cohort of patients with metastatic disease [13]. In this case, the antigen is a tissue differentiation protein expressed in melanosomes. Moreover, in a fully individualized and cancer specific approach (BiovaxID), the B cell rearranged receptors expressed on the malignant clones of B cell lymphoma cells were used in a phase III trial of a heterohybrioma-derived idiotype (Id) vaccine in patients with follicular lymphoma (FL) resulting in prolonged duration of clinical remission [14]. These idiotypic vaccines are formulated with GM-CSF and a mollusc protein to enhance immunogenicity. However, insufficient accrual in the phase III trial, together with the advent of CD20-specific monoclonal antibodies regimens, have prevented it being approved to date, although it is currently being reviewed by the regulatory authority in Europe for the treatment of follicular lymphoma in patients who have achieved a first complete remission. Another phase III trial with anti-idiotype vaccination has rendered negative results, although the results remain unpublished.

Currently, a series of promising clinical phase II and III trials applying various technical approaches and targeting different tumor indications are being conducted and are expected to release results over the upcoming years. Importantly, some of these vaccines are targeting minimal residual disease following surgical resection.

The concept of therapeutic vaccination against cancer is based on the idea that there is a repertoire of functionally competent effector and memory T cells with specificity for tumor cell antigens at sufficient frequencies to control tumor progression [10]. However, at least three major hurdles need to be considered. Firstly, tumor antigens used in vaccination are self-proteins and T cells with high affinity receptors recognizing these antigens might have been deleted through thymic and peripheral negative selection. Secondly, tumors deploy panoply of molecules that are locally and systemically immunosuppressive, meaning mechanisms of immune tolerance against cancer-associated antigens need to be overcome. Lastly, tumor cells are notoriously heterogenic and their antigen expression of a given tumor can range from completely negative to highly expressed and changing upon microenvironment and systemic conditions. Hence, it should be considered that cancer vaccines not only induce a newly primed T cell repertoire de novo but may also reactivate and potentially reeducate pre-existing tumor reactive memory T cells. As such, the presence or absence of preexisting effector/memory T cell responses represent an important prognostic biomarker for response to vaccination.

In order to have a successful vaccination a high quality CD4 and CD8 effector and memory T cell responses against tumor-associated antigens need to be achieved and circumvent the problems of immune tolerance and the selection of immune escape variants of cancer cells become extremely important. Several vaccination strategies have currently been adopted including whole proteins, peptides, mRNA encoding for antigens, and viral vectors encoding tumor antigens. Vaccines consist of an antigen recognized in tumor cells or critical components of the tumor stroma together with adjuvant biomolecules that enhance the immunogenicity of the vaccination antigens. In this regard, viral and microbe mimicry play a crucial role. Accordingly, vaccines are usually particulated, encompass bacterial agonists of toll-like receptors (TLRs), or contain viral or viral-like nucleic acids. Immune system cells such as dendritic cells bear surface and endosomal receptors that recognize these microbial-associated patterns.

Short peptide vaccines consist of one or more HLA-class I restricted epitopes derived from tumor antigens [15, 16]. These peptides can bind to HLA antigen-presenting molecules at the binding groove on the cell surface with no need to be processed. Short peptides have the advantage of selecting only the immune dominant epitopes and facilitating the precise assessment of subsequent immune response, e.g. with fluorescent HLA-peptide multimers. Disadvantages derive from the high risk of selection of immune escape variants if a single antigen is used and from the lack of CD4 T cell help if the vaccine is based only on class I HLA restricted epitopes. Recent strategies have aimed to overcome these limitations by combining multiple HLA-I and II restricted short peptides from many tumor antigens [16].

Another approach is the use of synthetic long overlapping peptides of 20–30 amino acids. Long peptides are avidly taken up by antigen-presenting cells, processed, and presented by HLA-class I as well as HLA-class II molecules. Therefore, they provide CD4 T cell help that supports the induction of long-lived memory T cells [17]. Long peptide antigens have the advantage of being selectively presented by professional antigen-presenting cells, resulting in better quality T lymphocytes [18]. Using this approach with the human papillomavirus (HPV)-16 viral oncoproteins E6 and E7, therapeutic effects have been reported in patients with premalignant vulvar lesions [19]. Similar to the use of long overlapping peptides is the application of recombinant full-length proteins. Such approaches are currently being investigated in controlled clinical trials by various pharmaceutical companies [20].

Alternatively, antigen expression in an immunogenic fashion can be induced by means of gene therapy employing DNA or mRNA encoding for single or multiple tumor antigens. After injection, not all the naked nucleic acids are completely degraded but some are locally expressed by a variety of somatic cells, including antigen-presenting cells. Local inflammation and immunity is fostered through recognition of the naked nucleic acids by innate endosomal and cytoplasmic receptors in leukocytes. Stimulation of TLR9, stimulator of IFN genes (STING), TLR3, TLR7, RIG-I, and MDA-5 by DNA is vital for the mechanism of action. TLR stimulation increases antigen processing and presentation of dendritic cells, their migration into regional lymph nodes and the expression of costimulatory receptors and cytokines favoring T cell activation. Because of some proportions of applied nucleic acids can be rapidly degraded by intra- and extracellular nucleases, many protective modifications have been chemically introduced [10]. Major advantages of using DNA and RNA vaccines are related to ease of production and high stability in vitro. The first cancer immunotherapy that has been approved using this approach is a DNA-based tyrosinase vaccine for the treatment of malignant melanoma in dogs [21]. In humans, an ongoing phase III trial is testing an mRNA vaccine encoding for PSA in patients with castration-resistant prostate cancer, with the aim being to improve OS [22].

No matter which vaccine formulation is used, the strength and type of the vaccine-induced immune response is truly determined by the amount of antigen presented by the activation/maturation of dendritic cells induced through the stimulation of danger-associated molecular pattern (DAMP) receptors [10].

A cumbersome but actively tested approach is to culture patients dendritic cells ex vivo, loading their antigen-presentation pathways with sources of tumor antigens to induce activation before being reinfused into the patient [23]. Such a complex and individualized approach is difficult to test in large-scale randomized trials. However, the use of dendritic cells differentiated in culture with monocytes and loaded with whole cell tumor lysates is being tested in phase III trials in glioblastoma to prevent postsurgical relapse.

Cancer vaccines have been a source of disappointment in the past due to repeated failures. It is important to recognize that it is likely that the critical antigens are not shared differentiation antigens but rather are individual mutations of each tumor [24].

Technical advances in biotechnology may permit cost-efficient and rapid identification of such mutations by means of whole exome sequencing and bioinformatic predictions for HLA binding of ensuing peptides. Alternatively, HLA-eluted peptides can be sequenced by proteomic approaches. Individualized peptide or mRNA vaccines are predicted to be a major line of progress in the near future.

Adoptive T Cell Therapy

Adoptive cellular immunotherapy involves administering autologous or allogeneic tumor-reactive T cells to patients in order to achieve tumor regression. Today, adoptive T therapy represents one of the most promising therapies in the field of cancer treatment, shown promising results in patients with transplant-related malignancies, leukemia and melanoma [25]. Adoptive T cell therapy involves the isolation of lymphocytes with high affinity for tumor antigens which can be selected ex vivo, stimulated, expanded and infused back into the patient. The feasibility of generating T cells ex vivo is limited by the initial frequency of tumor antigen-specific T cells, which can be very low. Initially, tumor-specific T cells were selected from peripheral blood mononucleated cell (PBMCs) and subjected to cloning. However, cultures of monoclonal and polyclonal T cells do not show great efficiency in T cell expansion nor in induce tumor rejection. The recent advance in tumor-specific T cells have been generated from enriched sources such as tumor-infiltrating lymphocyte (TILs) populations and tumor-draining lymph nodes. In melanoma, it has been shown that numerous tumor antigen-specific T cells can be isolated from excisional the tumor mass, dissociating cells into single cell suspensions and adding the T-cell growth factor interleukin-2 (IL-2) [26]. Several clinical trials using this approach have provided highly promising results, especially in melanoma. For instance, a recent study involving 93 patients with stage IV melanoma showed objective clinical response rates, ranged from 49 to 72 % [27]. Of particular note, 22 % of patients had a complete tumor regression and some of these patients have been alive and disease-free for more than 8 years [26]. To achieve these results patients must be heavily conditioned by total body irradiation and lymphodepleting chemotherapy. This aggressive procedure is supposed to eliminate regulatory T cells and eliminate competition for homeostatic cytokines sustaining T cell proliferation and survival. Recent reports document that several other groups have showed similar results with objective response rates of up to 48 % [28].

Despite the successes of T adoptive therapy with TILs, a clinical response is still not guaranteed for all patients. Indeed, not all patients respond to this type of therapy in the same way and the reasons behind this differential response to TIL T cell transfer remain unknown. It has been shown that tumors escape TIL transfer by several mechanisms and there is considerable evidence that TILs are blocked in vivo by many immunosuppressive molecules, such as programmed cell death protein-1 (PD1) and cytotoxic T cells antigen-4 (CTLA-4). Thus, increased expression of molecules with immunosuppressive properties such as NOS1 [29] might block TIL activity in tumors and decrease response to adoptive T cell transfer therapy [5]. As such, a clear portrait of the phenomena associated with the lack of response observed in some patients is necessary in order to prevent both costly regimens and side effects associated with the therapy.

According to immune repertoire selection established by immunosuppressive tumors in vivo, the affinity of the repertoire of endogenous T cell receptor (TCRs) for tumor-specific antigens can be very low. Thus, several approaches have been developed to genetically engineer T cells with high affinity TCRs and confer strong effector functions upon recognition of tumor-associated antigens. High avidity human TCRs are isolated either from in vitro cultures of naïve T cells with allogeneic peptide pulsed antigen presenting cells [30] or following vaccination of humanized mice expressing human HLA alleles together with the human TCR alpha and beta genetic loci [31]. Candidate TCRs are sequenced, cloned and inserted into retrovirus or lentiviruses which can be then used to transduce autologous T cells from other patients with matching HLA restriction elements [32]. The specificity of the TCR-transduced T cells is not altered compared to the parental T cell clone [33].

Another problem associated with the transfer of genetically modified TCRs is the restriction to single HLA alleles. This limitation is overcome by the use of chimeric antigen receptors (CARs) in which a single chain antibody (artificially linked light and heavy chains) is coupled to the transmembrane and cytoplasmic signaling domains of the TCR complex and costimulatory molecules, thus inducing a lymphocyte activation upon antigen encounter. Genetic modification of T cells can also involve inserting genes to improve the efficacy of the T cells or induce co-stimulation and inhibit apoptosis of modified T cells. Risks associated with modifying T cells include the emergence of serious toxicity, such as cytokine release syndrome [26] or organ damage due to overexpression of the recognized antigen or cross-reactivity with other self-antigens. However, safety can be improved by identifying those antigens that can be targeted to destroy cancer cells without toxic effects in normal tissues.

Monoclonal Antibodies for Cancer Therapy

The use of monoclonal antibodies (mAbs) in cancer therapy is one of the great clinical successes of the past decade. The fact that these antibodies can be used in the effective treatment of cancer is based on many years of comprehensive research into their complex structure, physical, chemical and biological properties, specificity to targeted antigens, and therapeutic activity in vivo alone and in combination. Following the discovery of mAbs by Cesar Milstein and George Kohler immortalizing antibody-producing rodent B cells, their therapeutic use has relied on strategies to humanize mouse sequences such that their immune potential is much reduced when administered to humans.

In cancer therapy, an antibody can impact tumor regression through various mechanisms. Firstly, direct action of the antibody on tumor cells through binding to a tumor cell surface receptor can induce receptor activation or an antagonist activity leading to tumor cell apoptosis. Secondly, the antibody may not act directly on the tumor but instead induce an immune response which mediates tumor cell death by phagocytosis, complement activation (complement-dependent cytotoxicity [CDC]) or antibody-dependent cellular cytotoxicity (ADCC). Thirdly, the antibody may have a specific effect on the tumor stroma, such as toxin delivery to the stromal cell or the blockade of angiogenesis by an antagonizing vasculature receptor or growth factor [34].

In case of a direct effect or effect on tumor stroma, it is important to consider that the efficacy of therapeutic mAbs is based on the target antigen which, ideally, should be very abundant and mostly or only expressed on cancer cells or selectively expressed in tumor stroma. If possible, the antigen should be functionally related to the biology of cancer development and progression. In solid tumors, some of the most successful antibodies directly targeting tumor cells are those that block the ErbB family (which include epidermal growth factor receptor [EGFR]) or vascular endothelial growth factor (VEGF). For example, much evidence has shown that the EGFR-specific antibody cetuximab has improved response and survival in patients with colon cancer, breast cancer and glioma. In colon cancer, cetuximab has been indicated for first line of treatment of metastatic disease in combination with chemotherapy [35].

Similarly, trastuzumab is specifically targeted against ERBB2 and has been shown to be effective in patients with breast, ovarian and prostatic cancer with high expression of ERBB2 [36]. Trastuzumab is a good example of how the genetic background of the patient plays an important role in therapy response and thus urges the need for biomarkers that identify patients most likely to benefit from treatment.

Several mAbs have also been approved for the treatment of hematological malignancies. The most well known and widely used of these is rituximab, which is directed against the CD20 receptor and has shown considerable success in patients with B cell non-Hodgkin's lymphoma (NHL) and chronic lymphocytic leukemia [37]. A major advance in this field of anti-CD20 mAbs is the advent of glyco-engineered antibodies with higher avidity for the CD16 receptor that ignites ADCC on NK cells. In addition, antibody-conjugated drugs or toxins have been approved by FDA. These include brentuximab vedotin for patients with CD30-positive Hodgkin's lymphoma [38] which provided the first proof-in-principle that antibodies can selectively deliver active drug to cancer cells.

Antibodies have also been devised to enhance cellular immune responses by activating or antagonizing immunological receptors important for cancer immunosurveillance. These are thus referred to as immunomodulatory or immunostimulatory mAbs. The concept behind the usage of these immunomodulatory mAbs is based on the knowledge that the immune system, and in particular T lymphocyte activity, is regulated by a balance of costimulatory and co-inhibitory signals known as immune checkpoints. Under physiological conditions, these immune checkpoints are crucial for avoiding autoimmunity and protecting tissue during an immune response to infection. However, during cancer development, the balance can shift towards a reduced immune response, thereby promoting unchecked progression of the tumor. The two molecules involved in the immune checkpoint regulation that have been most actively studied in the context of cancer immunotherapy are the CTLA-4 and PD-1 receptors. These are believed to inhibit immune responses at different levels and by different mechanisms [39]. The CTLA-4 receptor antagonizes binding between the T cell receptor co-stimulatory molecule CD28 and the ligand CD86, thereby mediating a down modulation of T cell activation [40]. A mAb able to block CTLA-4 was designed by Allison et al. [41], who used preclinical models to show that a significant antitumor response without overt immune toxicity was achieved when mice were treated with anti CTLA-4 antibody. These preclinical findings encouraged the production and testing of two fully humanized CTLA-4 antibodies, ipilimumab and tremelimumab, which began clinical trials in 2000. Both antibodies produced a clinical response in patients with melanoma.

However, ipilimumab was more successful, being associated with a 17 % survival benefit in patients with advanced melanoma [42, 43] and opening a window of hope for long-term survival in these patients [44]. Compelling evidence led to the FDA approval of ipilimumab for patients with advanced melanoma in 2010. However, tremelimumab also produced an OS benefit of 12.6 % in a phase III melanoma trial [45] and its use in the treatment of melanoma and other cancer types, possibly in combination with other therapies, should not be completely disregarded [46].

The immunostimulatory mechanism of action of ipilimumab has meant immune-related adverse effects can occur. However, no correlation between efficacy and toxicity has been observed in ipilimumab-treated melanoma patients [47]. Efficacy of ipilimumab has also been reported in patients with advanced uveal and mucosal melanoma, who generally have a poor prognosis and otherwise have limited treatment options [48, 49].

In physiologic conditions, PD-1 interacts with PD-1 ligand 1 or 2 (PD-L1 and PD-L2) to limit and regulate T cell activity in peripheral tissues during the inflammatory or autoimmune processes. This co-inhibitory system has probably evolved to minimize collateral damage to healthy tissue and non-infected cells during clearance of viral and bacterial intracellular infections. Recently, a fully human PD-1 blocking antibody, nivolumab, has been developed and a durable clinical response reported in a large phase I trial of patients with different tumor types, including melanoma, renal cell carcinoma, colorectal cancer and non-small-cell lung cancer [50]. Some studies have also suggested that the response to nivolumab might correlate with the expression of PD-L1 [51], but further confirmation is required from prospective studies with larger number of patients. The use of nivolumab has been also tested in combination with other treatments, such as ipilimumab, with an objective response rate of 53 % and a tumor reduction of 80 % or more in a study of 53 patients with advanced melanoma [52]. Importantly, responses to the nivolumab and ipilimumab combination are both rapid and dramatic.

Other antibodies that target the PD-1/PD-L1 pathway are also under clinical development. These include the anti-PD-1 antibody, MK-3475 (previously known as lambrolizumab), which showed a high rate of sustained tumor regression with mainly grade 1 or 2 toxic effects in 135 patients with advanced melanoma, including some with previous disease progression on ipilimumab. Moreover, the response rate did not significantly differ between patients who had received prior ipilimumab treatment (38 % [95 % CI: 23–55]) and those who had not (37 % [95 % CI: 26–49]) [53]. The future of immunostimulatory antibody therapies in cancer immunotherapy is rapidly advancing [54] and is dependent on better understanding of the underlying biology as well as innovative clinical trial design.

Enhance Trafficking of T Cells

Infiltration of immune system cells into healthy tissues is limited. However, inflammation due to microbial invasion or injury results in dramatic histological changes. At the vascular cell level, this is controlled by chemokines that attract leukocytes and adhesion molecules that arrest leukocytes from circulation and extravasate them into tissue. Recirculation of memory lymphocytes via afferent lymphatic vessels is also relevant to ensure systemic immunity.

Effector and potentially tumorocidal T lymphocytes are recruited into the tumor by comparable mechanisms. However, the aberrant vasculature of tumors in not prone to the entrance of T lymphocytes. Vascular normalization antagonizing VEGF is known to enhance entrance of T lymphocytes [53]. The tumor and the tumor vasculature also attract factors that downregulate the expression of adhesion molecules on endothelial cells in the tumor microvasculature. Tumors often show an infiltrate that is mainly composed of immunosuppressive leukocyte subsets such as myeloid-derived suppressor cells (MDSCs), M2 macrophages and regulatory T cells. These subsets are attracted by defined and undefined chemo-attractants. Counteracting chemokines that attract immunosuppressive populations is another strategy currently under scrutiny in experimental models.

Enhancing T cell migration to tumors is a neglected field in cancer immunotherapy although it is envisioned to be a major factor in limiting tumor growth. Local approaches including radiotherapy and intratumoral injection of proinflammatory agents including TLR agonists may be effective. MAbs can also be used to directly upregulate chemokines and adhesion molecules in the tumor endothelium, such as those directed to CD40 and CD137.

Recent research suggests that T cells retain a memory of the tissue and draining lymphoid tissue of the organ in which they were primed. This is due to imprinted epigenetic changes induced by the specific dendritic cells in each tissue and their draining lymph nodes. Hence, memory T lymphocytes primed in the skin, gut or respiratory system tend to migrate home to the tissue where originally primed. This molecular mechanism relies on selective adhesion molecules and chemokine receptors and has profound implications for the route of vaccination [55] and the in vitro activation of T cells for adoptive T cell therapy.

Conclusions

Recently, an improved understanding of cancer pathogenesis has given rise to new treatment options for cancer patients, including targeted agents and cancer immunotherapy which endeavors to stimulate a host immune response that effectuates long-lived tumor destruction. Among the multiple immune therapeutic approaches that have been discovered and tested, some have been clearly showed to increase the capacity of the immune system to attack and destroy cancer cells thus increasing the survival of patients with cancer. We predict that these novel immune therapeutic approaches will help revolutionize, either alone or in combinations, the management of cancer.

Already some trials conducted in mice have showed that the combination of vaccines and PD1 blockade increased overall survival and decreased tumor growth [56]. Additionally, combining blockade of multiple inhibitory pathways decreases T cell anergy and improves T cell responsiveness [57].

In humans, the combination of ipilimumab with plus bevacizumab (anti VEGF) in a phase I study of 22 patients, has shown interesting results with one-year survival rate of 72 % [58]. A similar combination approach of bevacizumab plus MPDL3280A (an anti-PD-L1) is currently being evaluated (NCT01633970). The combination of ipilimumab and GM-CSFhas also been recently investigated. In a phase II trial, 245 patients were randomized to receive ipilimumab and GM-CSF in combination or ipilimumab alone [59]. The survival rate after one year of treatment in the combination arm was 68.9 % compared with 52.9 % in the monotherapy arm, while the median OS in the combination arm was 17.5 months compared with 12.7 months in the group of patients that received ipilimumab alone.

Taken altogether the above observations showed that the usage, either alone or in combinations of immunomodulatory approaches holds an absolutely unprecedented hope for a robust impact on the survival of cancer patients and may represent a decisive turning for cancer therapy.

References

1. Hanahan D, Weinberg RA (2011) Hallmarks of cancer: the next generation. Cell 144(5):646–674
2. Ascierto ML, Idowu MO, Zhao Y, Khalak H, Payne KK, Wang XY, Dumur CI, Bedognetti D, Tomei S, Ascierto PA et al (2013) Molecular signatures mostly associated with NK cells are predictive of relapse free survival in breast cancer patients. J Transl Med 11(1):145
3. Ascierto ML, Kmieciak M, Idowu MO, Manjili R, Zhao Y, Grimes M, Dumur C, Wang E, Ramakrishnan V, Wang XY et al (2012) A signature of immune function genes associated with recurrence-free survival in breast cancer patients. Breast Cancer Res Treat 131(3):871–880
4. Galon J, Costes A, Sanchez-Cabo F, Kirilovsky A, Mlecnik B, Lagorce-Pages C, Tosolini M, Camus M, Berger A, Wind P et al (2006) Type, density, and location of immune cells within human colorectal tumors predict clinical outcome. Science 313(5795):1960–1964
5. Ascierto ML, De Giorgi V, Liu Q, Bedognetti D, Spivey TL, Murtas D, Uccellini L, Ayotte BD, Stroncek DF, Chouchane L et al (2011) An immunologic portrait of cancer. J Transl Med 9:146
6. El-Hamamy E, Wright A, B-Lynch C (2009) The B-Lynch suture technique for postpartum haemorrhage: a decade of experience and outcome. J Obstet Gynaecol 29(4):278–283
7. Disis ML (2010) Immune regulation of cancer. J Clin Oncol 28(29):4531–4538
8. Wang E, Uccellini L, Marincola FM (2012) A genetic inference on cancer immune responsiveness. Oncoimmunology 1(4):520–525
9. Schreiber RD, Old LJ, Smyth MJ (2011) Cancer immunoediting: integrating immunity's roles in cancer suppression and promotion. Science 331(6024):1565–1570
10. Beckhove P (2013) Cancer immunotherapy: overview in brief. In: The tumor immunoenvironment. Springer
11. Higano CS, Small EJ, Schellhammer P, Yasothan U, Gubernick S, Kirkpatrick P, Kantoff PW (2010) Sipuleucel-T. Nat Rev Drug Discov 9(7):513–514
12. Kantoff PW, Higano CS, Shore ND, Berger ER, Small EJ, Penson DF, Redfern CH, Ferrari AC, Dreicer R, Sims RB et al (2010) Sipuleucel-T immunotherapy for castration-resistant prostate cancer. N Engl J Med 363(5):411–422
13. Schwartzentruber DJ, Lawson DH, Richards JM, Conry RM, Miller DM, Treisman J, Gailani F, Riley L, Conlon K, Pockaj B et al (2011) gp100 peptide vaccine and interleukin-2 in patients with advanced melanoma. N Engl J Med 364(22):2119–2127
14. Schuster SJ, Neelapu SS, Gause BL, Janik JE, Muggia FM, Gockerman JP, Winter JN, Flowers CR, Nikcevich DA, Sotomayor EM et al (2011) Vaccination with patient-specific

tumor-derived antigen in first remission improves disease-free survival in follicular lymphoma. J Clin Oncol 29(20):2787–2794
15. Rammensee HG (2006) Peptides made to order. Immunity 25(5):693–695
16. Dutoit V, Herold-Mende C, Hilf N, Schoor O, Beckhove P, Bucher J, Dorsch K, Flohr S, Fritsche J, Lewandrowski P et al (2012) Exploiting the glioblastoma peptidome to discover novel tumour-associated antigens for immunotherapy. Brain 135(Pt 4):1042–1054
17. Bonertz A, Weitz J, Pietsch DH, Rahbari NN, Schlude C, Ge Y, Juenger S, Vlodavsky I, Khazaie K, Jaeger D et al (2009) Antigen-specific Tregs control T cell responses against a limited repertoire of tumor antigens in patients with colorectal carcinoma. J Clin Invest 119(11):3311–3321
18. Melief CJ (2008) Cancer immunotherapy by dendritic cells. Immunity 29(3):372–383
19. Kenter GG, Welters MJ, Valentijn AR, Lowik MJ, Berends-van der Meer DM, Vloon AP, Essahsah F, Fathers LM, Offringa R, Drijfhout JW et al (2009) Vaccination against HPV-16 oncoproteins for vulvar intraepithelial neoplasia. N Engl J Med 361(19):1838–1847
20. Powell E, Chow LQ (2008) BLP-25 liposomal vaccine: a promising potential therapy in non-small-cell lung cancer. Expert Rev Respir Med 2(1):37–45
21. Grosenbaugh DA, Leard AT, Bergman PJ, Klein MK, Meleo K, Susaneck S, Hess PR, Jankowski MK, Jones PD, Leibman NF et al (2011) Safety and efficacy of a xenogeneic DNA vaccine encoding for human tyrosinase as adjunctive treatment for oral malignant melanoma in dogs following surgical excision of the primary tumor. Am J Vet Res 72(12): 1631–1638
22. Dayyani F, Gallick GE, Logothetis CJ, Corn PG (2011) Novel therapies for metastatic castrate-resistant prostate cancer. J Natl Cancer Inst 103(22):1665–1675
23. Palucka K, Ueno H, Roberts L, Fay J, Banchereau J (2010) Dendritic cells: are they clinically relevant? Cancer J 16(4):318–324
24. Sahin U, Tureci O, Chen YT, Seitz G, Villena-Heinsen C, Old LJ, Pfreundschuh M (1998) Expression of multiple cancer/testis (CT) antigens in breast cancer and melanoma: basis for polyvalent CT vaccine strategies. Int J Cancer 78(3):387–389
25. Brenner MK, Heslop HE (2010) Adoptive T cell therapy of cancer. Curr Opin Immunol 22(2):251–257
26. Restifo NP, Dudley ME, Rosenberg SA (2012) Adoptive immunotherapy for cancer: harnessing the T cell response. Nat Rev Immunol 12(4):269–281
27. Rosenberg SA, Yang JC, Sherry RM, Kammula US, Hughes MS, Phan GQ, Citrin DE, Restifo NP, Robbins PF, Wunderlich JR et al (2011) Durable complete responses in heavily pretreated patients with metastatic melanoma using T-cell transfer immunotherapy. Clin Cancer Res 17(13):4550–4557
28. Besser MJ, Shapira-Frommer R, Itzhaki O, Treves AJ, Zippel DB, Levy D, Kubi A, Shoshani N, Zikich D, Ohayon Y et al (2013) Adoptive transfer of tumor-infiltrating lymphocytes in patients with metastatic melanoma: intent-to-treat analysis and efficacy after failure to prior immunotherapies. Clin Cancer Res 19(17):4792–4800
29. Liu QZ, Tomei S, Ascierto ML, De Giorgi V, Bedognetti D, Dai CL, Uccellini L, Spivey T, Pos Z, Thomas J et al (2014) Melanoma NOS1 expression promotes dysfunctional IFN signaling. J Clin Invest 124(5):2147–2159
30. Wilde S, Geiger C, Milosevic S, Mosetter B, Eichenlaub S, Schendel DJ (2012) Generation of allo-restricted peptide-specific T cells using RNA-pulsed dendritic cells: a three phase experimental procedure. Oncoimmunology 1(2):129–140
31. Li LP, Lampert JC, Chen X, Leitao C, Popovic J, Muller W, Blankenstein T (2010) Transgenic mice with a diverse human T cell antigen receptor repertoire. Nat Med 16(9):1029–1034
32. Voss RH, Thomas S, Pfirschke C, Hauptrock B, Klobuch S, Kuball J, Grabowski M, Engel R, Guillaume P, Romero P et al (2010) Coexpression of the T-cell receptor constant alpha domain triggers tumor reactivity of single-chain TCR-transduced human T cells. Blood 115(25): 5154–5163
33. Rosenberg SA, Restifo NP, Yang JC, Morgan RA, Dudley ME (2008) Adoptive cell transfer: a clinical path to effective cancer immunotherapy. Nat Rev Cancer 8(4):299–308

34. Scott AM, Wolchok JD, Old LJ (2012) Antibody therapy of cancer. Nat Rev Cancer 12(4):278–287
35. Van Cutsem E, Rivera F, Berry S, Kretzschmar A, Michael M, DiBartolomeo M, Mazier MA, Canon JL, Georgoulias V, Peeters M et al (2009) Safety and efficacy of first-line bevacizumab with FOLFOX, XELOX, FOLFIRI and fluoropyrimidines in metastatic colorectal cancer: the BEAT study. Ann Oncol 20(11):1842–1847
36. Hudis CA (2007) Trastuzumab—mechanism of action and use in clinical practice. N Engl J Med 357(1):39–51
37. Cheson BD, Leonard JP (2008) Monoclonal antibody therapy for B-cell non-Hodgkin's lymphoma. N Engl J Med 359(6):613–626
38. Younes A, Bartlett NL, Leonard JP, Kennedy DA, Lynch CM, Sievers EL, Forero-Torres A (2010) Brentuximab vedotin (SGN-35) for relapsed CD30-positive lymphomas. N Engl J Med 363(19):1812–1821
39. Pardoll DM (2012) The blockade of immune checkpoints in cancer immunotherapy. Nat Rev Cancer 12(4):252–264
40. Linsley PS, Clark EA, Ledbetter JA (1990) T-cell antigen CD28 mediates adhesion with B cells by interacting with activation antigen B7/BB-1. Proc Natl Acad Sci U S A 87(13): 5031–5035
41. Leach DR, Krummel MF, Allison JP (1996) Enhancement of antitumor immunity by CTLA-4 blockade. Science 271(5256):1734–1736
42. Melero I, Grimaldi AM, Perez-Gracia JL, Ascierto PA (2013) Clinical development of immunostimulatory monoclonal antibodies and opportunities for combination. Clin Cancer Res 19(5):997–1008
43. Prieto PA, Yang JC, Sherry RM, Hughes MS, Kammula US, White DE, Levy CL, Rosenberg SA, Phan GQ (2012) CTLA-4 blockade with ipilimumab: long-term follow-up of 177 patients with metastatic melanoma. Clin Cancer Res 18(7):2039–2047
44. Hodi FS, O'Day SJ, McDermott DF, Weber RW, Sosman JA, Haanen JB, Gonzalez R, Robert C, Schadendorf D, Hassel JC et al (2010) Improved survival with ipilimumab in patients with metastatic melanoma. N Engl J Med 363(8):711–723
45. Ribas A, Kefford R, Marshall MA, Punt CJ, Haanen JB, Marmol M, Garbe C, Gogas H, Schachter J, Linette G et al (2013) Phase III randomized clinical trial comparing tremelimumab with standard-of-care chemotherapy in patients with advanced melanoma. J Clin Oncol 31(5):616–622
46. Ascierto PA, Marincola FM, Ribas A (2011) Anti-CTLA4 monoclonal antibodies: the past and the future in clinical application. J Transl Med 9:196
47. Ascierto PA, Kalos M, Schaer DA, Callahan MK, Wolchok JD (2013) Biomarkers for immunostimulatory monoclonal antibodies in combination strategies for melanoma and other tumor types. Clin Cancer Res 19(5):1009–1020
48. Maio M, Danielli R, Chiarion-Sileni V, Pigozzo J, Parmiani G, Ridolfi R, De Rosa F, Del Vecchio M, Di Guardo L, Queirolo P et al (2013) Efficacy and safety of ipilimumab in patients with pre-treated, uveal melanoma. Ann Oncol 24(11):2911–2915
49. Del Vecchio M, Di Guardo L, Ascierto PA, Grimaldi AM, Sileni VC, Pigozzo J, Ferraresi V, Nuzzo C, Rinaldi G, Testori A et al (2014) Efficacy and safety of ipilimumab 3 mg/kg in patients with pretreated, metastatic, mucosal melanoma. Eur J Cancer 50(1):121–127
50. Topalian SL, Hodi FS, Brahmer JR, Gettinger SN, Smith DC, McDermott DF, Powderly JD, Carvajal RD, Sosman JA, Atkins MB et al (2012) Safety, activity, and immune correlates of anti-PD-1 antibody in cancer. N Engl J Med 366(26):2443–2454
51. Taube JM, Anders RA, Young GD, Xu H, Sharma R, McMiller TL, Chen S, Klein AP, Pardoll DM, Topalian SL et al (2012) Colocalization of inflammatory response with B7-h1 expression in human melanocytic lesions supports an adaptive resistance mechanism of immune escape. Sci Transl Med 4(127):127ra137
52. Wolchok JD, Kluger H, Callahan MK, Postow MA, Rizvi NA, Lesokhin AM, Segal NH, Ariyan CE, Gordon RA, Reed K et al (2013) Nivolumab plus ipilimumab in advanced melanoma. N Engl J Med 369(2):122–133

53. Chiang CL, Kandalaft LE, Coukos G (2011) Adjuvants for enhancing the immunogenicity of whole tumor cell vaccines. Int Rev Immunol 30(2–3):150–182
54. Grimaldi AM, Cassidy PB, Leachmann S, Ascierto PA (2014) Novel approaches in melanoma prevention and therapy. Cancer Treat Res 159:443–455
55. Sandoval F, Terme M, Nizard M, Badoual C, Bureau MF, Freyburger L, Clement O, Marcheteau E, Gey A, Fraisse G et al (2013) Mucosal imprinting of vaccine-induced CD8(+) T cells is crucial to inhibit the growth of mucosal tumors. Sci Transl Med 5(172):172ra120
56. Li B, Van Roey M, Wang CY, Chen THT, Korman A, Jooss K (2009) Anti-programmed death-1 synergizes with granulocyte macrophage colony-stimulating factor-secreting tumor cell immunotherapy providing therapeutic benefit to mice with established tumors. Clin Cancer Res 15(5):1623–1634
57. Curran MA, Montalvo W, Yagita H, Allison JP (2010) PD-1 and CTLA-4 combination blockade expands infiltrating T cells and reduces regulatory T and myeloid cells within B16 melanoma tumors. Proc Natl Acad Sci U S A 107(9):4275–4280
58. Hodi FS, Friedlander PA, Atkins MB, McDermott DF, Lawrence DP, Ibrahim N, Wu X, Zhou J, Giobbie-Hurder A, Murphy G et al (2011) A phase I trial of ipilimumab plus bevacizumab in patients with unresectable stage III or stage IV melanoma. J Clin Oncol 29(15)
59. Hodi FS, Lee SJ, McDermott DF, Rao UNM, Butterfield LH, Tarhini AA, Leming PD, Puzanov I, Kirkwood JM, Eastern Cooperative Oncology Group (2013) Multicenter, randomized phase II trial of GM-CSF (GM) plus ipilimumab (Ipi) versus Ipi alone in metastatic melanoma: E1608. J Clin Oncol 31(18)

Chapter 12
Chimeric Antigen Receptor (CAR) T Cells

Daniel W. Lee and Alan S. Wayne

Abstract Recent advances in cellular engineering techniques coupled with modern chimeric antigen receptors (CARs) now permit the efficient targeting and killing of malignant cells using patients' own T cells. Freedom from MHC restriction by relying most commonly on single chain variable fragments of monoclonal antibodies for antigen recognition rather than T cell receptors expands the list of potential targets. Unlike small molecule inhibitors, the targets of CAR T cells are not required to play a critical function in the tumor cell. CAR targets must be extracellular and ideally should have limited expression on normal, vital tissues. By incorporating primary and co-stimulatory domains, CAR T cells possess a substantial proliferative capacity allowing for small cell doses, which reduces the manufacturing burden. This therapeutic approach allows a potent yet customized in vivo response. The potential of CAR T cells to contribute to the overall treatment of cancer is exemplified by the impressive clinical responses with predominantly reversible toxicities seen in early phase clinical trials targeting the B-cell antigen, CD19, in B-lineage hematologic malignancies.

Early CARs

Immunotherapists have struggled for years to re-program the immune system to attack cancer. Cancer vaccines have proven difficult to translate into positive patient outcomes [1, 2]. Advances in retroviral systems used to genetically engineer immune cells first showed promise with T cell receptors (TCRs) designed to target cancer peptides. Responses have been seen particularly in synovial sarcoma and melanoma

D.W. Lee (✉)
Center for Cancer Research, National Cancer Institute, Building 10 – Hatfield CRC, Room 1-3750, Bethesda, MD 20892-1104, USA
e-mail: leed3@mail.nih.gov

A.S. Wayne
Division of Hematology, Oncology and Blood & Marrow Transplantation, Children's Center for Cancer and Blood Diseases, Children's Hospital Los Angeles, Los Angeles, CA, USA

P.A. Ascierto et al. (eds.), *Developments in T Cell Based Cancer Immunotherapies*, Cancer Drug Discovery and Development,
DOI 10.1007/978-3-319-21167-1_12

[3–5], but TCR-based therapies are limited by major histocompatibility complex (MHC) restriction, which reduces the number of patients that can benefit.

Recognizing this limitation, Zelig Eshhar while on sabbatical at the National Cancer Institute conceptualized and cloned the first T-body, later to be known as a chimeric antigen receptor (CAR) [6, 7]. The T-body incorporated the CD3 zeta signaling domain, like the TCR, but replaced the α and β peptide-MHC recognition subunits with a single chain variable fragment (scFv) from an antibody. This single maneuver, after more than a decade of study and improvements, has translated into the first overwhelmingly successful targeted immunotherapy for relapsed and refractory malignancies.

Mechanism of Action

The most basic components of a CAR include an extracellular antigen-recognition sequence, most often an scFv, transmembrane domain and the CD3-zeta signaling domain from the T cell receptor (Fig. 12.1a, b). This antigen-recognition strategy overcomes MHC restriction inherent in TCR-based therapies thereby allowing for the production of one receptor that can be applied to all patients.

Upon antigen recognition, the CAR signals through its CD3-zeta providing Signal 1 required for T cell activation. While these first-generation CARs showed activity in tumor-bearing mice [8] early human trials of such CARs expressed in autologous T cells showed minimal activity. Perhaps the best responses were seen by Pule, et al who treated 11 children with neuroblastoma with a GD2-specific first-generation CAR in EBV-specific cytotoxic T lymphocytes [9]. In that study, 4 of 8 evaluable patients had tumor necrosis or regressions including one complete remission. Interestingly, in this trial PCR-based evidence of CAR T cell persistence extended for several years in survivors [10]. Specific factors responsible for the long-term persistence of CAR T cells remain elusive.

Co-stimulation

The Pule experience provided key insights into the limitations of first-generation CARs. After infusion, circulating CAR T cells were detected and disease burden appeared to stabilize for a period of time in 5 of 11 patients [9]. This suggests that in some patients CAR T cells were successfully able to target and interact with the malignant cells but fell short of becoming completely activated. Many tumors, especially acute lymphocytic leukemia (ALL), evade immune surveillance in part by downregulating or interfering with potential co-stimulatory signals [11, 12].

Second-generation CARs (Fig. 12.1c) sought to circumvent this limitation by incorporating a co-stimulatory domain in the receptor. There are multiple candidate signaling domains, but the most commonly used are CD28 and 4-1BB (CD137).

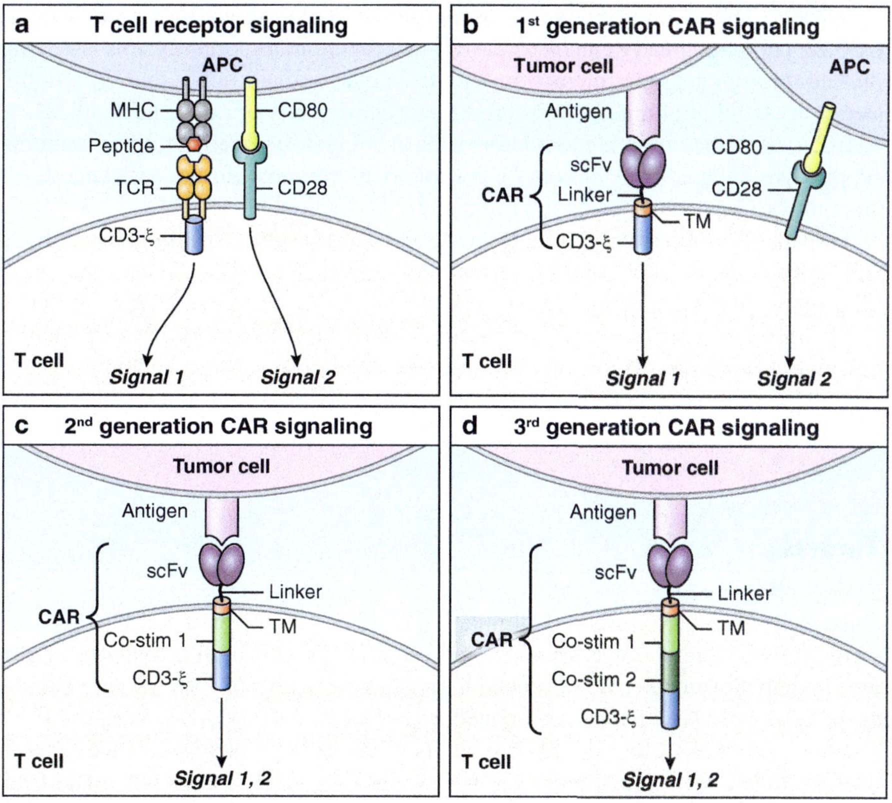

Fig. 12.1 Basic components of CAR T cell signaling. Two activating signals are required for full T cell activation. (**a**) The endogenous TCR upon encountering its cognate peptide-MHC complex on an antigen presenting cell (APC) provides Signal 1 via the CD3-zeta signaling domain. Full activation is not achieved until Signal 2 is provided by a subsequent interaction with a co-stimulatory signal (e.g., CD80-CD28 interaction). (**b**) 1st generation CAR T cells rely on MHC-independent interactions (e.g., single chain variable fragment, scFv) to provide Signal 1, but still depend on an additional co-stimulation step for complete activation. The scFv is joined with a linker sequence, transmembrane (TM) domain and CD3-zeta. (**c**) Integrating CD3-zeta and a co-stimulatory domain in the same receptor provides 2nd generation CARs with the capacity to fully activate T cells in a single step. (**d**) 3rd generation CARs incorporate two co-stimulatory domains with CD3-zeta

CD28 interacts with B7 family members on antigen presenting cells and in the presence of TCR-peptide:MHC complex induces IL-2, IL-6 and other inflammatory cytokine production. Like CD28, 4-1BB activation induces T-cell proliferation, cytokine production and inhibits apoptotic signals produced by T cell activation, but appears to play a more important role in CD8+ cytolytic T cells than CD4+ T cells [13–15]. Whether CD28 or 4-1BB signaling in the context of a CAR leads to superior anti-tumor response, proliferation and/or CAR T cell persistence remains to be determined and is an active area of investigation.

What is clear at this point is that the addition of a co-stimulatory molecule to the CD3 zeta domain greatly enhances CAR T cell function. Importantly, this is accomplished through a single interaction with the target antigen rather than in a multi-step process reliant on activated antigen presenting cells or other mechanisms. In addition, negative immune modulators such as regulatory T cells or other immunosuppressive signals from tumors or the microenvironment do not influence CAR co-stimulation, at least during the initial period of action.

Perhaps not surprisingly, third-generation CARs (Fig. 12.1d), those containing CD3 zeta and two co-stimulatory domains (e.g.: CD3z-CD28-41BB), have not been more successful than their second-generation counterparts. Despite efficient killing in short-term in vitro cytotoxicity assays, in vivo testing has not been as effective perhaps related to overactivation of CAR T cells and subsequent apoptosis [16, 17]. At the present time, most investigators are pursuing work with second-generation CARs.

Targets

Once the role of hard-wired co-stimulation had been established and CAR T cells were endowed with the signaling domains required for full T cell activation, significant toxicities were uncovered related to target antigen choice. The Surgery Branch at the National Cancer Institute opened a clinical trial of a third-generation CAR T cell targeting Her2-neu (ErbB2) in adults with Her2+ refractory disease. In contrast to modern techniques described below, the Her2-CAR T cells generated from the initial retroviral transduction and culture were subjected to a second round of stimulation in a rapid expansion protocol designed to generate more than 10^{10} cells at the lowest tested cell dose. A patient with metastatic colon cancer who received 10^{10} CAR transduced T cells after an intensive fludarabine and cyclophosphamide lymphodepleting regimen developed pulmonary edema within 15 min of cell infusion and subsequently died due to severe cytokine release syndrome likely related to low-level ErbB2 expression on lung endothelium [18]. In a second event, two patients who received a TCR-based therapy targeting the cancer testes antigen, MAGE-A3, died from necrotizing leukoencephalopathy caused by cross-reactivity of the MAGE-A3 TCR-specific T cells with MAGE-A12 expressed normally in the brain [5].

These unfortunate cases elucidate, among other factors, the importance of carefully choosing targets for T-cell based immunotherapies. In general, targets expressed in high levels on any tissue or even low levels on vital tissues should be avoided. Determining this expression is extremely difficult, however, as normal tissue protein arrays are few, incomplete and have variable sensitivity [19]. In the short term, it will be impossible to pick targets with no normal tissue expression with certainty. Therefore, alternative strategies are needed to ensure the safety of patients after CAR T cell infusion, especially as the field moves towards solid tumor targets after the initial success with hematologic malignancies.

Suicide Vectors

Suicide vectors have recently been incorporated in CARs to provide a mechanism to induce CAR T cell apoptosis should unacceptable toxicity be encountered. At the present time, there are three such vectors: inducible caspase 9, the Herpes simplex thymidine kinase (Hsv-*tk*)/ganciclovir system and rituximab therapy for CAR T cells co-expressing CD20.

The Hsv-*tk* system although efficacious has significant limitations. Hsv-*tk* is immunogenic, which potentially can result in rejection of CAR modified T cells. Furthermore, this system is activated by ganciclovir, an important anti-viral drug commonly utilized to manage CMV reactivation, for example after hematopoietic stem cell transplantation (HSCT) [20].

The CD20 co-expression system also has significant limitations. Although this strategy overcomes the immunogenicity problem, reliance on a monoclonal antibody to eliminate all circulating and tissue-resident CAR T cells in the presence of a potentially overwhelming population of normal CD20-expressing B cells is inefficient and carries the risk of suboptimal suicide response and continued CAR-mediated toxicity [21].

The most promising suicide mechanism, presently, relies on an inducible caspase 9 molecule that is incorporated in the CAR backbone. A small molecule, AP1903, administered to a patient results in dimerization of the modified caspase 9 produced by the CAR that in turn induces CAR T cell apoptosis via endogenous mechanisms. This method is extremely desirable compared to the other two methods due to the wide distribution of the drug in the body, its lack of immunogenicity and the rapid elimination of the vast majority of CAR T cells [22]. A phase I clinical trial of AP1903 has been completed [23] and the drug has been incorporated in several recent clinical trials [GD2 CAR for neuroblastoma (NCT01822652) and osteosarcoma/rhabdomyosarcoma (NCT02107963)]. Pre-clinical studies are also underway in other conditions (e.g., CD123 CAR for acute myeloid leukemia [24]; donor T cells to control graft-versus-host disease after HSCT [25]; mesenchymal stromal cells [26]).

Target Choice

Appropriate targets for CAR-based therapies are difficult to elucidate. Unlike TCR-based therapies the target must be extracellular, have limited expression on normal tissues and is not necessarily predicted by somatic mutations as in small molecule targeting of overactive kinases. Recent strategies have started with determining key differences between normal and tumor gene expression profiles, annotating which genes result in extracellular protein expression and identifying and validating a method to target this particular protein before incorporating in a CAR. As a leading example, the Pediatric Oncology Branch at the National Cancer Institute curates a database of potential extracellular targets for 12 pediatric malignancies [19].

The array of possible targets is vast and specific candidates are outside the scope of this report. In general, since CARs are not restrained by MHC, any specific extracellular molecule can be a viable target, not just proteins [27]. For example, the disialoganglioside, GD2, has been successfully targeted in neuroblastoma, osteosarcoma and some rhabdomyosarcomas using CARs [10]. It is important to note that not all CARs require an antibody to endow specificity and successful interactions. Small molecule or neurotransmitter receptors [28], cytokine receptors such as the IL-13 receptor in gliomas [29] and growth factor receptors such as EGFvIII in gliomas [30] have been targeted as well.

Manufacturing

Transduction Strategies

Three general strategies are used to express CARs in T cells: transient transfection with mRNA, transposon/transposase systems and viral-mediated transduction. Transient transfection is attractive since it is easy, relatively inexpensive and eliminates the possibility of long-term toxicity since the CAR is not stably incorporated in the cell's genome. In addition, daughter T cells produced by CAR-activated T cell parents will not express the CAR themselves as occurs in stably transduced CAR T cells. Consequently, multiple doses may be required to achieve the desired anti-tumor effects, but this may be limited by endogenous immune responses to the CAR, which has occurred in several settings [31].

The Sleeping Beauty transposon/transposase system permanently incorporates the CAR gene into non-coding regions of T cells using a plasmid-based approach. Mutational mutagenesis is largely eliminated since the transposon uses flanking inverted terminal repeats to identify sites of incorporation via the transposase. These systems can handle large cassettes (up to 10 kbp) and are less expensive to manufacture than viral systems [32, 33]. However, widespread application of this technology is currently limited by proprietary interests.

Retroviruses, both γ-retrovirus (γ-RV) and lentiviruses (LV), remain the principle mechanism of CAR transduction in T cells for most centers [34]. Despite the devastating complications related to insertional mutagenesis in stem cells observed in early gene therapy trials [35], γ-RVs have been widely used especially in terminally differentiated T cells without such complications. T cells are exposed to the replication-incompetent retrovirus ex vivo and recombination events leading to a replication-competent virus as well as insertional mutagenesis are highly unlikely [36, 37]. Retroviruses remain the most efficient method of gene transfer with transduction rates as high as 90 % in some studies [38] but the cost to establish qualified γ-RV producer lines and qualified LV viral supernatant is high and requires GMP and GCP facilities that are currently limited but expanding.

General Schema of Virally-Transduced CAR T Cell Production

CAR T cells are most commonly made from peripheral blood mononuclear cells collected from the patient or transplant donor via apheresis. T cells are activated with anti-CD3 antibody (OKT3) with or without activation of co-stimulatory molecules on the T cell (e.g., CD28) prior to exposure to the viral supernatant. This activation in coordination with cytokine support (e.g., IL-2, IL-7) facilitates efficient CAR gene transduction and results in expansion of the culture to meet cell dose requirements [39].

Some groups, especially those evaluating CAR therapy after HSCT, generate T cell lines whose endogenous TCR is specific for EBV or other viruses with the aim of reducing the risk of alloreactivity and graft-versus-host disease (GVHD) [40, 41]. This approach requires more extensive culture time and manipulation, which can dramatically increase costs, and patients with rapidly progressive disease may succumb or require initiation of therapy before the product generation is complete. Notably, there is no evidence to date to indicate that allogeneic-derived CAR T cells induce GVHD. In that regard, no children with ALL treated with an anti-CD19 CAR after a prior HSCT have developed GVHD despite the donor-derived T cells being collected directly from the patient [38, 42]. Although recruitment is ongoing, GVHD has also not been observed in any adult patients with ALL or chronic lymphocytic leukemia (CLL) after HSCT treated with CD19 CAR T cells manufactured from donor lymphocyte infusion (DLI) aliquots [43] [NCT01087294].

Proliferative Capacity

Importantly, few CAR T cells are required for significant antitumor effects. In contrast to trials employing tumor peptide-MHC specific TCRs [3, 5] where billions of cells are required for objective responses, complete responses have been seen with as few as 10^5–10^6 CD19 CAR-transduced T cells per kilogram [38, 44, 45]. Since T cells are activated ex vivo to facilitate transduction resulting in often significant proliferation, the required starting T cell numbers in culture may be quite low. This reduces manufacturing cost and facilitates ease of culture manipulations.

Such low doses of CAR T cells are possible in part due to their impressive capacity for in vivo expansion. Upon encountering antigen, CAR T cells (1) kill the target (2) produce inflammatory cytokines and (3) proliferate with each of the progeny bearing the CAR gene being fully capable of responding to antigen, killing additional target cells, producing cytokines and proliferating further. In vivo expansion in leukemia trials has been estimated to be as much as 700-fold [38, 44].

In vivo expansion is dependent at least in part on whole body target antigen burden (both malignant and non-malignant targets). Lower CAR T cell expansion is seen, for example, in ALL patients with minimal residual disease in comparison to those with morphologic evidence of leukemia in bone marrow [38]. Thus, expansion (and therefore toxicity) may be inherently linked to disease burden.

In contrast, some patients with massive disease burden who are treated at the same dose as those with lesser amounts of disease do not have clinically meaningful responses or toxicity despite higher in vivo expansion [38]. There is a finite limit to the number of generations of responsive CAR T cells that can be produced after activation as cells are driven toward exhaustive phenotypes. Some groups are pursuing upfront selection of central memory or stem cell memory T cells before CAR transduction [46, 47], but an entirely new method to determine optimal cell dose based at least on proliferative potential, disease burden and anticipated toxicities is certainly warranted.

Pre-Infusion Conditioning

Most groups pretreat patients with a lymphodepleting chemotherapy regimen prior to CAR T cell infusion. Lymphodepletion with fludarabine and/or cyclophosphamide serve to increase endogenous homeostatic cytokines (e.g., IL-7) that are thought to be supportive for the infused CAR T cells [48, 49]. Although some modeling in animals has been performed [50], the ideal lymphodepletion strategy is not known and may differ depending on the patient population. Children with ALL, for example, receive lymphodepleting chemotherapy as part of the standard of care regimen so may need a less intensive preparative regimen prior to CAR T cells. At least one clinical protocol is currently underway to address this question (NCT00924326).

CD19 CAR T Cell Therapy

Efficacy

The majority of the clinical experience with CAR T cell therapy to date has been targeting CD19 in B-lineage leukemias and lymphomas [38, 42, 43, 45]. Despite differences in receptor co-stimulation endodomains, CAR structural components and scFv's, all groups have observed impressive responses in relapsed and refractory patient populations (Table 12.1). The CD19 CAR T cell, therefore, serves as a model system on which further advances can be made towards optimizing response and persistence and mitigating toxicity.

Toxicity

In addition to uniformly impressive clinical responses, second-generation CD19 CAR T cell protocols have also been associated with severe toxicity. Cytokine release syndrome (CRS) is the dose limiting toxicity. CRS in its most benign form is characterized by low-grade fever alone, but continues as a spectrum of toxicity to include

Table 12.1 CD19 CAR trials planned or currently underway

Trial location	ClinicalTrials.gov	Age	Cell dose/kg	scFv	Costim	Preparative regimen	Published results	Notes
Baylor COM	NCT02050347	Adults and children	5×10^6 1×10^7 5×10^7	FMC63	CD28	None		Relapse after Allogeneic-HSCT; T cells from either donor or patient; Log lower dosing if unrelated donor
Baylor COM	NCT00709033[a]	≥18 years	Unknown	FMC63	CD28	+/– cyclophosphamide		CLL, NHL Determine if EBV-specific CAR T cells persist longer than non-viral specific CAR T cells
Baylor COM	NCT00840853	Pediatric and Adult	3.75×10^6 11.2×10^6 33×10^6 CD3+ cells (CAR dose varies)	FMC63	CD28	None	*Blood* 122(17): 2965–73, 2013	Donor multi-virus T cells transduced post-HSCT Not controlled for CAR transduction
Baylor COM	NCT00586391	Pediatric and Adult	$2 \times 10^7/m^2$ $1 \times 10^8/m^2$ $2 \times 10^8/m^2$	FMC63	CD28	None		May receive ipilimumab 2 weeks after cell infusion; Option for 3 additional infusions
Baylor COM	NCT01853631	≥18 years	$2 \times 10^6/m^2$ $1 \times 10^7/m^2$ $4 \times 10^7/m^2$		50 % CD28; 50 % CD28 and 41BB	+/– cyclophosphamide		Compares persistence between 2nd and 3rd generation CAR

(continued)

Table 12.1 (continued)

Trial location	ClinicalTrials.gov	Age	Cell dose/kg	scFv	Costim	Preparative regimen	Published results	Notes
Children's Hospital of Philadelphia	NCT01626495	1–24 years	Unknown	FMC63	50 % No co-stim; 50 % 4-1BB	Varies		Compares persistence between 1st and 2nd generation CAR
Chinese PLA General Hospital	NCT02081937	50–80 years	Unknown	Unknown	4-1BB	None		Mantle cell lymphoma
Chinese PLA General Hospital	NCT01864889	5–90 years	Unknown	Unknown	Mix of no co-stim and 4-1BB	None		CD19+ Leukemias, lymphomas; Track persistence of 1st and 2nd generation CARs
City of Hope	NCT02051257; NCT01815749[a]; NCT01318317[a]	≥18 years	Unknown	FMC63	CD28	Autologous PBSC transplant		NHL; Central memory-enriched CAR cells given 2–3 days after autologous PBSCT; Truncated EGFR co-expression for targeted elimination with cetuximab
City of Hope	NCT02146924[a] NCT02153580[a]	>18 years	Unknown	FMC63	CD28	Cyclophosphamide		ALL, NHL, CLL; Central memory-enriched; Truncated EGFR; Option for second infusion
Fred Hutchinson Caner Research Center	NCT01865617	≥18 years	Unknown	FMC63	CD28	Unknown		Lentivirus; Option for second infusion
Fred Hutchinson Caner Research Center	NCT01475058[a]	18–75 years	Unknown	FMC63	CD28	Allogeneic HSCT		Donor-derived, CMV or EBV-specific CD8+ central memory CAR T cells after allogeneic HSCT

Jichi Medical University, Japan	NCT02134262[a]	20–70 years	$3 \times 10^5 - 1 \times 10^7$ (split dosing)	SFG	CD28	Cyclophosphamide (1.5 g/m^2) or bendamustine (120 mg/m^2)		scFv = SFG
Memorial Sloan-Kettering Cancer Center	NCT01430390	Up to 19 years	3×10^6 1×10^7 3×10^7	FMC63	CD28	Varies		Donor EBV T cells post-HSCT
Memorial Sloan-Kettering Cancer Center	NCT01860937; NCT01044069	1-80 years	1×10^6 or 3×10^6	FMC63	CD28	Cyclophosphamide		MRD+ patients get lower dose of cells
Memorial Sloan-Kettering Cancer Center	NCT01840566	≥18 years	5×10^6 1×10^7 2×10^7	FMC63	CD28	Melphalan, cytarabine, etoposide, carmustine		NHL; Cells given after autologous HSCT
Memorial Sloan-Kettering Cancer Center	NCT00466531	≥18 years	Unknown	FMC63	50 % CD28; 50 % 41BB	+/– Cyclophosphamide or Individualized	*Sci Transl Med* 6(224):224ra25, 2014	Compares persistence between CD28 and 4-1BB co-stimulation
NCI Pediatric Oncology Branch	NCT01593696	1–30 years	1×10^6	FMC63	CD28	Fludarabine and cyclophosphamide	Pending	CNS disease eligible; MTD defined 1 x 10^6 CAR T cells/kg
NCI Surgery Branch	NCT00924326	18–68 years	1×10^6 2.5×10^6	FMC63	CD28	Randomized to +/– fludarabine/ cyclophosphamide	*Blood* 119(12): 2709–20, 2012; *Blood* 116(20): 4099–102, 2010	CLL, Mantle, Follicular, and Large cell lymphoma
NCI Experimental Transplantation and Immunology Branch	NCT01087294	18–75 years	$0.5–10 \times 10^6$	FMC63	CD28	None	*Blood* 122(25): 4129–39, 2013	Relapse after Allo-HSCT using donor for T cell source

(continued)

Table 12.1 (continued)

Trial location	ClinicalTrials.gov	Age	Cell dose/kg	scFv	Costim	Preparative regimen	Published results	Notes
Seattle Children's Hospital	NCT02028455; NCT01683279	1–26 years	Unknown 1:1 ratio of CD4:CD8 CAR T cells	FMC63	4-1BB	Individualized		Truncated EGFR
University College London	NCT01195480	Up to 18 years	$2 \times 10^8/m^2$	Unknown	None	Allogeneic HSCT		Donor EBV T cells post-HSCT followed by vaccination w/ EBV-LCL
University of Pennsylvania	NCT02030834	≥18 years	$1–5 \times 10^8$ total CAR cells	FMC63	4-1BB	Individualized		Phase IIa NHL
University of Pennsylvania	NCT02030847	≥18 years	$1–5 \times 10^8$ total CAR cells	FMC63	4-1BB	Individualized	*Sci Transl Med* 3(95):95ra73, 2011; *NEJM* 368(16): 1509–18, 2013	Phase II ALL
University of Pennsylvania	NCT01747486	≥18 years	$1–5 \times 10^7$ or $1–5 \times 10^8$ total CAR cells	FMC63	4-1BB	Individualized	*NEJM* 365(8): 725–33, 2011	Phase II CLL or SLL
University of Pennsylvania	NCT01551043[a]	≥18 years	Unknown	FMC63	4-1BB	Individualized		ALL relapsed after allo-HSCT Donor-derived T cells
University of Pennsylvania	NCT02135406	≥18 years	Unknown	FMC63	4-1BB	Autologous HSCT		Multiple myeloma relapsed after auto-HSCT
Uppsala University Hospital	NCT02132624	>18 years	Unknown	Unknown	CD28 and 4-1BB	Pending		Pending

HSCT hematopoietic stem cell transplant, *ALL* acute lymphoblastic leukemia, *CLL* chronic lymphoblastic leukemia, *SLL* small lymphocytic leukemia, *NHL* B-cell non-Hodgkin's lymphoma, *PBSCT* peripheral blood stem cell transplant, *EGFR* epidermal growth factor receptor, *MRD* minimal residual disease, *CNS* central nervous system, *MTD* maximally tolerated dose

[a]Not yet recruiting

high-grade fever, rigors, myalgia and hypotension requiring vasopressor support. In addition, significant neurotoxicities have been reported even in patients without CNS involvement of their disease including visual hallucinations, headache, ataxia and seizures [38, 45, 51]. CD19 is not expressed on neural tissues, so the leading hypothesis is that such neurotoxicity is cytokine mediated. Indeed, elevated IL-6 has been found in the CSF of patients with dysphasia or visual hallucinations [38] and is dramatically elevated (>2500 pg/mL; normal <10 pg/mL) in the plasma of patients experiencing severe CRS even without neurologic dysfunction [38, 42, 45, 51].

Toxicity Management

Cytokine Release Syndrome

Unfortunately, deaths have occurred in adults after CAR T cell infusion. An adult with ALL infused with CD19 CAR T cells in 2009 developed symptoms of severe CRS manifested by fever, hypotension and respiratory failure soon after infusion and died [52]. Two additional deaths possibly related to severe CRS were more recently reported—one from intractable seizures and the second from cardiac failure [53].

In all of these cases and others of non-lethal but severe CRS, IL-6 appears to be principally involved in the clinical syndrome. Plasma IL-6 has routinely been elevated in most responding patients and correlates with CRS severity [38, 45, 51]. Rather than making IL-6 themselves, CAR T cells appear to induce other components of the immune system, including monocytes, to release this cytokine (Fig. 12.2a). The key determinant of the severity of CRS then appears to be the amount of IL-6 that is produced and is mediated through the different affinities of the membrane-bound and soluble forms of the IL-6 receptor (IL-6R). At low levels, IL-6 mediates effector functions via the classic signaling pathway where the cytokine binds a cell-associated IL-6R that then associates with gp130 leading to STAT3 and ERK/PI3K activation (Fig. 12.2b). However, in the presence of high IL-6 concentrations, IL-6 also binds to the soluble IL-6R that can in turn crosslink gp130 without the aid of a cell-associated receptor (Fig. 12.2c*i*). Since gp130 is ubiquitously expressed, IL-6 signaling via trans activation can have profound effects. The anti-IL-6 receptor monoclonal antibody, tocilizumab, has been used to successfully stabilize many patients with severe CRS when supportive care alone is not sufficient [38, 42, 45, 51]. Tocilizumab blocks both the membrane-bound and soluble receptors (Fig. 12.2c*ii*) so that the IL-6 axis is rapidly turned off with similarly rapid improvement in clinical status. Due to the recruitment of other immune components, some patients with severe CRS may benefit from corticosteroid therapy as well.

The most important intervention for managing CRS remains aggressive supportive care initiated early [51]. Patients should be monitored closely from the time of onset of fever. If initial attempts to stabilize low blood pressure with IV fluids is unsuccessful, patients should be transferred preemptively to critical care units for intensive monitoring and support. Detailed recommendations have recently been published by a multi-institutional consortium of investigators involved in CAR clinical trials [51].

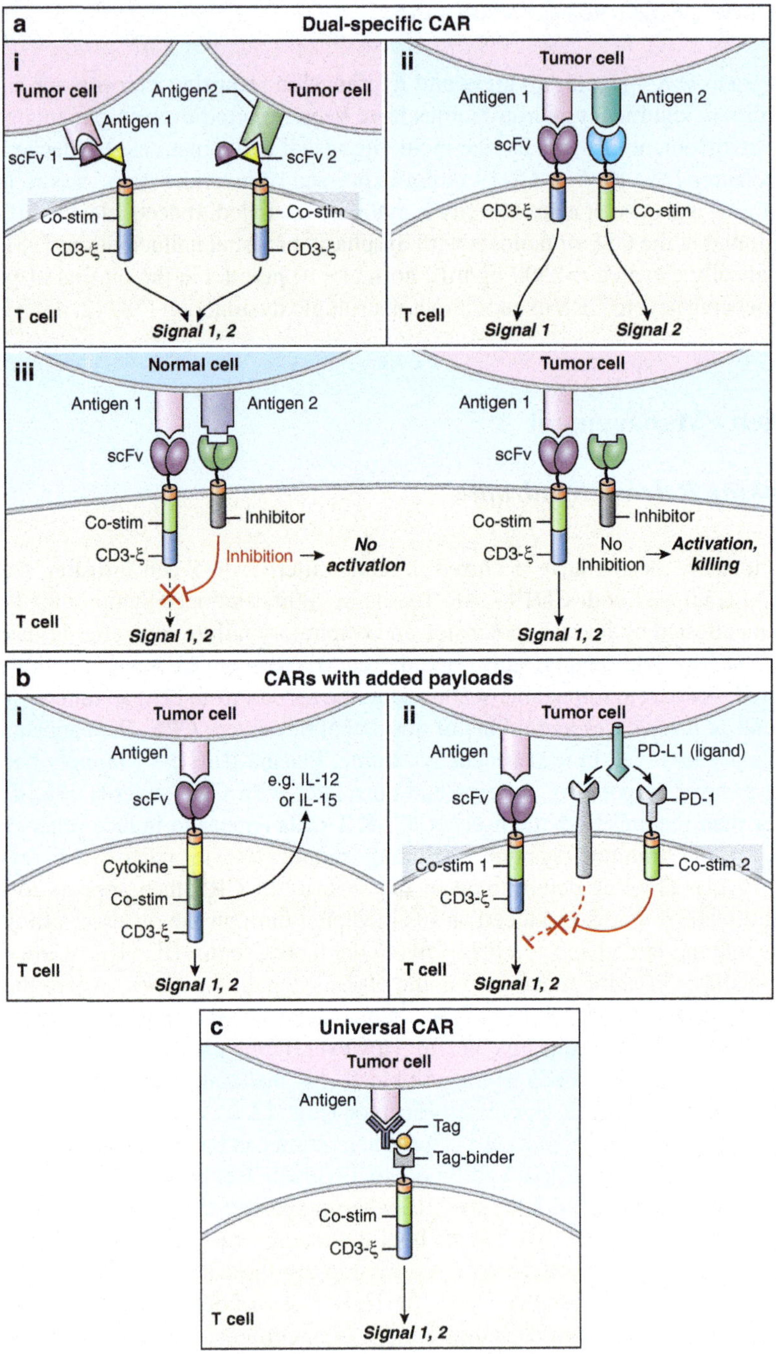

Fig. 12.2 CARs with added functionality. (**a**) Dual-specific CARs may be designed to (*i*) overcome antigenloss immune escape by incorporating two scFv's where engagement of either one results in full CAR T cell activation, (ii) impart more specificity by requiring two different antigens to be present on the same tumor cell before the CAR T cell becomes fully activated, or (iii) differentiate between normal and tumor cells. By incorporating an inhibitory signal in a second receptor specific for an antigen on normal tissues not present on tumors, CAR T cells become activated only when encountering tumor since the inhibitory signal is not produced. (**b**) CARs can also be generated to capitalize on

Importantly, if managed appropriately and anti-cytokine therapy is instituted before evidence of end organ damage, even severe CRS can be rapidly reversed with limited consequences. Whether arresting CRS with tocilizumab impacts the anti-tumor response is an area of ongoing study.

Now that guidelines have been established for managing CRS, the impact of such interventions on CAR T cell efficacy can be systematically evaluated. It is not yet clear whether grade 3–4 CRS is required for complete leukemia eradication in patients with significant disease burdens. However, several patients with ALL and low-level disease have achieved MRD negative remissions with no more than grade 1 fever [38]. While IL-6 may be the chief harbinger of toxicity, it may play little or no role in tumor cytotoxicity. If that proves to be true, one might envision a regimen whereby patients are premedicated with tocilizumab prior to CAR T cell infusion.

Additional work is also needed to evaluate the roles that other cytokines (e.g., TNFα, INFγ, GM-CSF, IL-10) play in the development of toxicity and anti-tumor response. Modulating one or more of these may provide a therapeutic advantage. Also, additional components of the innate immune system are no doubt recruited by CAR T cell activation and also contribute to toxicity and/or anti-tumor response.

Additional Toxicities

In addition to CRS, patients treated with CAR-based therapies are at risk for on-target but off-tissue cytotoxicity. Low level Her2 expression on lung epithelium was attributed as a contributing factor in the death of one patient with metastatic colon cancer who received a Her2-directed CAR [18]. Other factors such as cell dose likely also contributed and Her2 is being targeted with CARs in lower doses in sarcomas and glioblastoma [NCT00902044 and NCT01109095].

Several targets in ALL for CAR T cell therapies (e.g., CD19, CD22) also have on-target, off-tissue cytotoxicity, but the elimination of normal B cells is tolerated with appropriate supportive care. Few antigens isolated to solid tumors and not normal tissues [e.g., EGFRvIII] have been identified to date. Therefore, strategies to circumvent damage to normal, oftentimes vital tissue while maintaining potent cytotoxicity towards tumors are needed.

Dual specific CARs (Fig. 12.3a*i*) or co-expression of two CARs that transmit a full activation signal only when both receptors are engaged (Fig. 12.3a*ii*) have recently been generated. For example, cells co-transduced with an ErbB2-directed CAR containing CD3ζ and a MUC1-specific CAR containing CD28 only efficiently proliferate and kill breast cancer targets when both antigens are present [54]

Fig. 12.2 (continued) other features of immunity. (*i*) Incorporating the production of additional cytokines, such as IL-12 or IL-15, may help support CAR T cell persistence. (*ii*) Co-expressing the extracellular component of PD-1 fused with a second co-stimulatory domain may overcome the endogenous PD-1 inhibitory signal to the CAR T cell when it is engaged by its ligand, PD-L1, a potential method of immune evasion employed by tumors. (**c**) A universal CAR specific for a unique tag has been conceptualized as a means of CAR T cell activation when used in conjunction with a tagged monoclonal antibody, for example

(Fig. 12.3a*ii*). Another group demonstrated tumor-specific lysis, cytokine production, in vivo responses and persistence by T cells co-transduced with mesothelin-CD3ζ and α-folate receptor-CD28 CARs that are similar to those responses produced by second-generation CARs [55] (Fig. 12.3a*ii*).

Some groups have opted to transiently express CARs using mRNA electroporation with the aim to minimize the risk of toxicity to normal tissues. A patient with mesothelioma treated with mesothelin-specific second-generation CAR T cells using transient CAR expression had a partial response while another patient with pancreatic adenocarcinoma had stable disease [56]. Importantly, the former patient had a cardiac arrest due to anaphylaxis during repeated infusions of the CAR T cells [57], although mesothelin is expressed on pericardium [58] and could have also played a role. More studies are needed to determine if transient CAR expression improves the safety profile of CAR T cell therapy.

Another strategy to minimize the risk to normal tissues is to co-transduce T cells with two CARs targeting different antigens, one of which is present on both normal and tumor tissues and results in activation while the second is present only on normal tissues and provides an inhibitory signal preventing T cell activation (Fig. 12.3a*iii*). With this approach, the CAR T cell will only become completely activated when it encounters tumor.

CARs with Additional Functionality

CAR T cells can be engineered with additional functionality triggered by the receptor's engagement with its cognate antigen [59] (Fig. 12.3b*i*). Inducing the T cell to produce IL-7 receptor alpha, for example, has been hypothesized to better support the CAR T cell after infusion [60] and studies of this approach are ongoing. Similarly, IL-15 or IL-12 secreting CAR T cells serve to overcome the inhibitory signals received by regulatory T cells and other suppressive components of the tumor microenvironment [61–63].

Fig. 12.3 (continued) activation and recruits other immune effectors to produce IL-6. IL-6 also provides an activating signal to CAR T cells and together with other inflammatory cytokines contributes to the clinical features of CRS. Activated CAR T cells and their progeny are capable of killing additional tumor targets creating a feed-forward loop potentially leading to an even greater degree of CRS. (**b**) During periods of low plasma IL-6 levels, IL-6 signals its effectors via the classic approach by binding to membrane-bound IL-6 receptor (IL-6R) followed by association with two gp130 molecules. Formation of this complex leads to STAT3 and ERK/PI3K activation and immune activation. IL-6 in this state does not interact with soluble IL-6R. (**c**) *i*. When high levels of IL-6 are present, activation is accomplished though the classic method but also through trans since soluble IL-6R binds IL-6 the complex of which can then associate two gp130 domains without aid from membrane-bound IL-6R and result in signaling. *ii*. The anti-IL6 receptor antibody tocilizumab blocks both the membrane-bound and soluble forms of IL-6R thereby preventing IL-6 from associating with gp130 and its subsequent activation signals

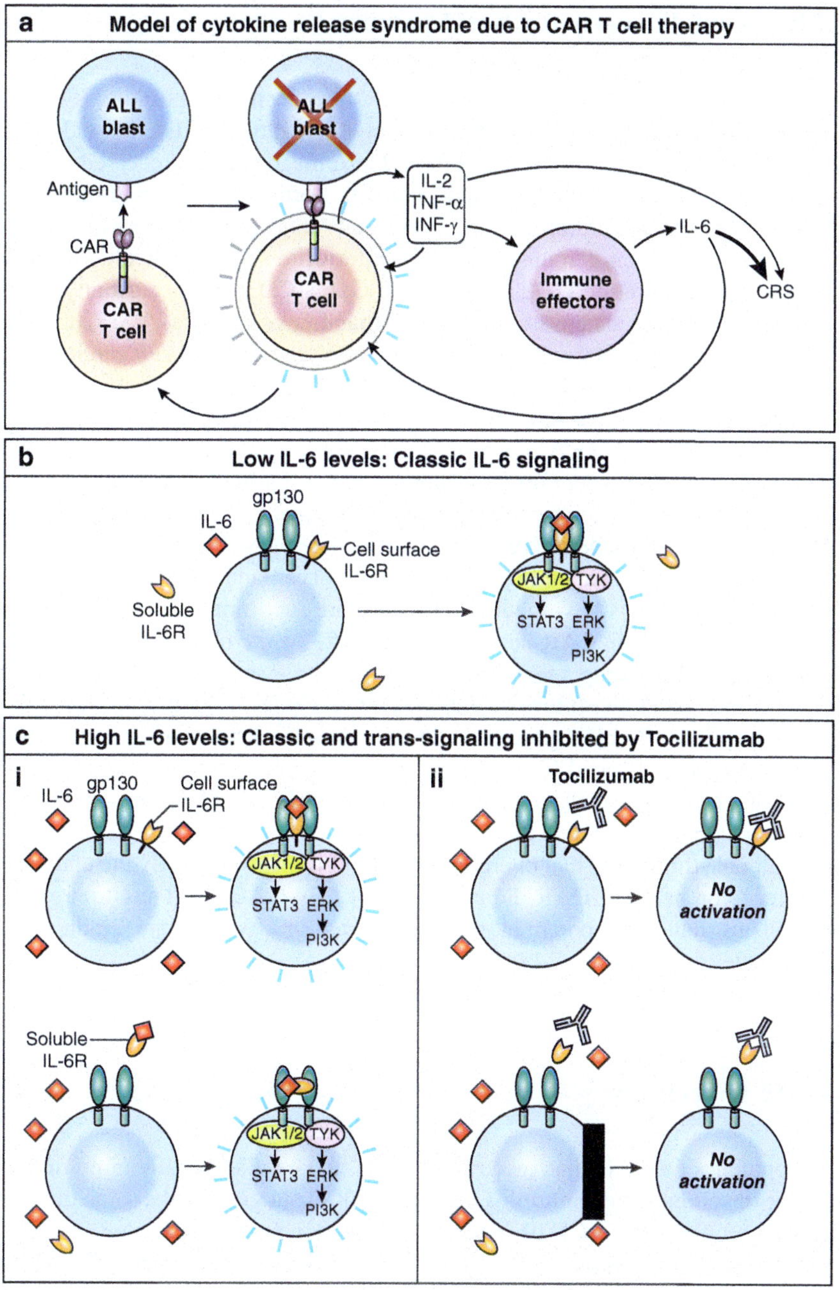

Fig. 12.3 Cytokine release syndrome (CRS) is mediated principally by IL-6 and ameliorated by tocilizumab. (**a**) Upon encountering its target, the CAR T cell becomes activated and produces inflammatory cytokines (e.g., IL-2, TNF-α, and IFN-γ) that aids in maintaining CAR T cell

CAR T Cell Therapy for Solid tumors

To date, treatment of solid tumors with CARs has been disappointing. Five of 11 children with neuroblastoma had reduction or stabilization of disease with a GD2-directed CAR [9]. In another study of neuroblastoma, 6 patients received two doses of CAR T cells targeting the L1-cell adhesion molecule (CD171) without response and limited in vivo proliferation [64]. These trials employed first-generation CARs, which have since been demonstrated to be inferior to second-generation CARs. A trial of a third-generation GD2 CAR (OX40-CD28-CD3zeta) is currently ongoing in neuroblastoma, osteosarcoma and rhabdomyosarcoma [NCT02107963; NCT01822652]. CARs targeting carbonic anhydrase IX in renal cell carcinoma also met with limited results [65, 66].

Aside from incomplete activation of first-generation CARs, one possible explanation for the lack of CAR T cell efficacy in solid tumors may be the tumor microenvironment, which is often immunosuppressive, or ineffective trafficking of CAR T cells to the tumor. A variety of approaches to augment immunotherapy (e.g., anti-CTLA-4, anti-PD-L1, chemokines) and to modulate the microenvironment in order to allow for more efficient endogenous immune surveillance of tumors are in development. Investigations have begun to incorporate such immunologic modifiers as secretable molecules in CAR-activated T cells [67] while others plan for combination therapy with existing agents [68].

Rather than block inhibitory signals with monoclonal antibodies, efforts are also directed at turning T-cell suppressive signals into stimulatory ones. One method of immune escape employed by many tumors is the upregulation of PD-L1, which suppresses activated and PD-1 expressing cytotoxic T cells. Co-expressing a second chimeric receptor of PD-1 to the CD28 signaling domain in CAR T cells provides an additional stimulation that overcomes the inhibition produced by the endogenous PD-1 receptor in these CAR T cells [69, 70] (Fig. 12.3b*ii*). Thus tumor escape mechanisms have the potential to be used as an asset in the eradication of malignancies.

Future Directions

Development of a CAR that successfully kills its target is a challenging undertaking. The development of multiple CARs for multiple targets for the same tumor will likely be required since CD19 negative escape clones have been demonstrated in leukemia patients treated with CD19-directed immunotherapies [38, 42, 71]. Targeting multiple antigens on each tumor will only add to the already complex process of developing a successful therapy.

For these reasons, an entirely different approach to targeting tumor antigens is being explored. Rather than making multiple CARs, a single, universal CAR targeting biotin, for example could be administered to patients who have been pretreated with a biotinylated monoclonal antibody to tumor antigens, in essence labeling tumor cells for targeting by CAR T cells (Fig. 12.3c). Such an approach is under development [72].

Given the early results of CD19 CAR T cell clinical trials [38, 42, 73] and the better understanding of how best to manage CRS [51] this therapy will likely receive regulatory approval. Studies are currently limited to patients with relapsed or refractory disease. So, investigations are needed to determine how best to incorporate CD19 CAR T cells in initial therapy. Although a long-term goal is to reduce chemotherapy exposure over the course of treatment for leukemia, chemotherapy will remain an important component of treatment. Importantly in this regard, it appears that CRS is more likely to be severe in patients with significant leukemia burdens than in those with minimal residual disease [38]. Further, CAR therapy will likely be more effective in patients with lower disease burden. So, for the newly diagnosed or relapsed patient intensive chemotherapy prior to CD19 CAR T cell infusion will likely be necessary, a strategy that is currently under investigation [NCT01593696].

References

1. Choudhury A et al (2006) Clinical results of vaccine therapy for cancer: learning from history for improving the future. Adv Cancer Res 95:147–202
2. Mocellin S, Pilati P, Nitti D (2009) Peptide-based anticancer vaccines: recent advances and future perspectives. Curr Med Chem 16(36):4779–4796
3. Robbins PF et al (2011) Tumor regression in patients with metastatic synovial cell sarcoma and melanoma using genetically engineered lymphocytes reactive with NY-ESO-1. J Clin Oncol 29(7):917–924
4. Chodon T et al (2014) Adoptive transfer of MART-1 T-cell receptor transgenic lymphocytes and dendritic cell vaccination in patients with metastatic melanoma. Clin Cancer Res 20(9):2457–2465
5. Morgan RA et al (2013) Cancer regression and neurological toxicity following anti-MAGE-A3 TCR gene therapy. J Immunother 36(2):133–151
6. Gross G, Waks T, Eshhar Z (1989) Expression of immunoglobulin-T-cell receptor chimeric molecules as functional receptors with antibody-type specificity. Proc Natl Acad Sci U S A 86(24):10024–10028
7. Eshhar Z, Waks T, Gross G (2014) The emergence of T-bodies/CAR T cells. Cancer J 20(2):123–126
8. Hwu P et al (1995) In vivo antitumor activity of T cells redirected with chimeric antibody/T-cell receptor genes. Cancer Res 55(15):3369–3373
9. Pule MA et al (2008) Virus-specific T cells engineered to coexpress tumor-specific receptors: persistence and antitumor activity in individuals with neuroblastoma. Nat Med 14(11):1264–1270
10. Louis CU et al (2011) Antitumor activity and long-term fate of chimeric antigen receptor-positive T cells in patients with neuroblastoma. Blood 118(23):6050–6056
11. Pallasch CP et al (2014) Sensitizing protective tumor microenvironments to antibody-mediated therapy. Cell 156(3):590–602
12. Gribben JG et al (1997) Biologic response modifiers in acute lymphoblastic leukemia. Leukemia 11(Suppl 4):S31–S33
13. Vinay DS, Kwon BS (1998) Role of 4-1BB in immune responses. Semin Immunol 10(6):481–489

14. Croft M (2003) Co-stimulatory members of the TNFR family: keys to effective T-cell immunity? Nat Rev Immunol 3(8):609–620
15. Watts TH (2005) TNF/TNFR family members in costimulation of T cell responses. Annu Rev Immunol 23:23–68
16. Haso W et al (2013) Anti-CD22-chimeric antigen receptors targeting B-cell precursor acute lymphoblastic leukemia. Blood 121(7):1165–1174
17. Hombach AA, Rappl G, Abken H (2013) Arming cytokine-induced killer cells with chimeric antigen receptors: CD28 outperforms combined CD28-OX40 "super-stimulation". Mol Ther 21(12):2268–2277
18. Morgan RA et al (2010) Case report of a serious adverse event following the administration of T cells transduced with a chimeric antigen receptor recognizing ERBB2. Mol Ther 18(4):843–851
19. Orentas RJ et al (2012) Identification of cell surface proteins as potential immunotherapy targets in 12 pediatric cancers. Front Oncol 2:194
20. Bonini C et al (1997) HSV-TK gene transfer into donor lymphocytes for control of allogeneic graft-versus-leukemia. Science 276(5319):1719–1724
21. Introna M et al (2000) Genetic modification of human T cells with CD20: a strategy to purify and lyse transduced cells with anti-CD20 antibodies. Hum Gene Ther 11(4):611–620
22. Straathof KC et al (2005) An inducible caspase 9 safety switch for T-cell therapy. Blood 105(11):4247–4254
23. Iuliucci JD et al (2001) Intravenous safety and pharmacokinetics of a novel dimerizer drug, AP1903, in healthy volunteers. J Clin Pharmacol 41(8):870–879
24. Mardiros A et al (2013) T cells expressing CD123-specific chimeric antigen receptors exhibit specific cytolytic effector functions and antitumor effects against human acute myeloid leukemia. Blood 122(18):3138–3148
25. Di Stasi A et al (2011) Inducible apoptosis as a safety switch for adoptive cell therapy. N Engl J Med 365(18):1673–1683
26. Ramos CA et al (2010) An inducible caspase 9 suicide gene to improve the safety of mesenchymal stromal cell therapies. Stem Cells 28(6):1107–1115
27. Orentas RJ, Lee DW, Mackall C (2012) Immunotherapy targets in pediatric cancer. Front Oncol 2:3
28. Gattenlohner S et al (2006) Rhabdomyosarcoma lysis by T cells expressing a human autoantibody-based chimeric receptor targeting the fetal acetylcholine receptor. Cancer Res 66(1):24–28
29. Krebs S et al (2014) T cells redirected to interleukin-13Ralpha2 with interleukin-13 mutein-chimeric antigen receptors have anti-glioma activity but also recognize interleukin-13Ralpha1. Cytotherapy 16(8):1121–31
30. Sampson JH et al (2014) EGFRvIII mCAR-modified T-cell therapy cures mice with established intracerebral glioma and generates host immunity against tumor-antigen loss. Clin Cancer Res 20(4):972–984
31. Jensen MC et al (2010) Antitransgene rejection responses contribute to attenuated persistence of adoptively transferred CD20/CD19-specific chimeric antigen receptor redirected T cells in humans. Biol Blood Marrow Transplant 16(9):1245–1256
32. Hackett PB et al (2013) Evaluating risks of insertional mutagenesis by DNA transposons in gene therapy. Transl Res 161(4):265–283
33. Hackett PB, Largaespada DA, Cooper LJ (2010) A transposon and transposase system for human application. Mol Ther 18(4):674–683
34. Fry TJ, Mackall CL (2013) T-cell adoptive immunotherapy for acute lymphoblastic leukemia. Hematology Am Soc Hematol Educ Program 2013:348–353
35. Hacein-Bey-Abina S et al (2003) LMO2-associated clonal T cell proliferation in two patients after gene therapy for SCID-X1. Science 302(5644):415–419
36. Zhang L, Thrasher AJ, Gaspar HB (2013) Current progress on gene therapy for primary immunodeficiencies. Gene Ther 20(10):963–969
37. Matrai J, Chuah MK, VandenDriessche T (2010) Recent advances in lentiviral vector development and applications. Mol Ther 18(3):477–490

38. Lee DW, Kochenderfer JN, Stetler-Stevenson M, Cui YK, Delbrook C, Feldman SA, Fry TJ, Orentas R, Sabatino M, Shah NN, Steinberg SM, Stroncek D, Tschernia N, Yuan C, Zhang H, Zhang L, Rosenberg SA, Wayne AS, Mackall CL (2015) T Cells expressing CD19 chimeric antigen receptors for acute lymphoblastic leukemia in children and young adults: a phase 1 dose-escalation trial. Lancet 385(9967):517–528
39. Tumaini B et al (2013) Simplified process for the production of anti-CD19-CAR-engineered T cells. Cytotherapy 15(11):1406–1415
40. Cruz CR et al (2013) Infusion of donor-derived CD19-redirected virus-specific T cells for B-cell malignancies relapsed after allogeneic stem cell transplant: a phase 1 study. Blood 122(17):2965–2973
41. Terakura S et al (2012) Generation of CD19-chimeric antigen receptor modified CD8+ T cells derived from virus-specific central memory T cells. Blood 119(1):72–82
42. Grupp SA et al (2013) Chimeric antigen receptor-modified T cells for acute lymphoid leukemia. N Engl J Med 368(16):1509–1518
43. Kochenderfer JN et al (2013) Donor-derived CD19-targeted T cells cause regression of malignancy persisting after allogeneic hematopoietic stem cell transplantation. Blood 122(25):4129–4139
44. Porter DL et al (2011) Chimeric antigen receptor-modified T cells in chronic lymphoid leukemia. N Engl J Med 365(8):725–733
45. Davila ML et al (2014) Efficacy and toxicity management of 19-28z CAR T cell therapy in B cell acute lymphoblastic leukemia. Sci Transl Med 6(224):224ra25
46. Wang X et al (2012) Phenotypic and functional attributes of lentivirus-modified CD19-specific human CD8+ central memory T cells manufactured at clinical scale. J Immunother 35(9):689–701
47. Lugli E et al (2013) Identification, isolation and in vitro expansion of human and nonhuman primate T stem cell memory cells. Nat Protoc 8(1):33–42
48. Laurent J et al (2010) Impact of 3 different short-term chemotherapy regimens on lymphocyte-depletion and reconstitution in melanoma patients. J Immunother 33(7):723–734
49. Cui Y et al (2009) Harnessing the physiology of lymphopenia to support adoptive immunotherapy in lymphoreplete hosts. Blood 114(18):3831–3840
50. Wrzesinski C et al (2010) Increased intensity lymphodepletion enhances tumor treatment efficacy of adoptively transferred tumor-specific T cells. J Immunother 33(1):1–7
51. Lee DW et al (2014) How I treat: current concepts in the diagnosis and management of cytokine release syndrome. Blood 124:188–195
52. Brentjens R et al (2010) Treatment of chronic lymphocytic leukemia with genetically targeted autologous T cells: case report of an unforeseen adverse event in a phase I clinical trial. Mol Ther 18(4):666–668
53. Carroll J (2014) Two deaths force MSK to hit the brakes on engineered T cell cancer study, in FierceBiotech (http://www.fiercebiotech.com/story/memorial-sloan-kettering-hits-brakes-engineered-t-cell-cancer-study/2014-04-06)
54. Wilkie S et al (2012) Dual targeting of ErbB2 and MUC1 in breast cancer using chimeric antigen receptors engineered to provide complementary signaling. J Clin Immunol 32(5):1059–1070
55. Lanitis E et al (2013) Chimeric antigen receptor T Cells with dissociated signaling domains exhibit focused antitumor activity with reduced potential for toxicity in vivo. Cancer Immunol Res 1(1):43–53
56. Beatty GL et al (2014) Mesothelin-specific chimeric antigen receptor mRNA-engineered T cells induce anti-tumor activity in solid malignancies. Cancer Immunol Res 2(2):112–120
57. Maus MV et al (2013) T cells expressing chimeric antigen receptors can cause anaphylaxis in humans. Cancer Immunol Res 1(1):26–31
58. Hassan R, Bera T, Pastan I (2004) Mesothelin: a new target for immunotherapy. Clin Cancer Res 10(12 Pt 1):3937–3942
59. Dotti G et al (2014) Design and development of therapies using chimeric antigen receptor-expressing T cells. Immunol Rev 257(1):107–126

60. Perna SK et al (2014) Interleukin-7 mediates selective expansion of tumor-redirected cytotoxic T lymphocytes (CTLs) without enhancement of regulatory T-cell inhibition. Clin Cancer Res 20(1):131–139
61. Chmielewski M, Abken H (2012) CAR T cells transform to trucks: chimeric antigen receptor-redirected T cells engineered to deliver inducible IL-12 modulate the tumour stroma to combat cancer. Cancer Immunol Immunother 61(8):1269–1277
62. Quintarelli C et al (2007) Co-expression of cytokine and suicide genes to enhance the activity and safety of tumor-specific cytotoxic T lymphocytes. Blood 110(8):2793–2802
63. Hoyos V et al (2010) Engineering CD19-specific T lymphocytes with interleukin-15 and a suicide gene to enhance their anti-lymphoma/leukemia effects and safety. Leukemia 24(6):1160–1170
64. Park JR et al (2007) Adoptive transfer of chimeric antigen receptor re-directed cytolytic T lymphocyte clones in patients with neuroblastoma. Mol Ther 15(4):825–833
65. Lamers CH et al (2006) Treatment of metastatic renal cell carcinoma with autologous T-lymphocytes genetically retargeted against carbonic anhydrase IX: first clinical experience. J Clin Oncol 24(13):e20–e22
66. Lamers CH et al (2013) Treatment of metastatic renal cell carcinoma with CAIX CAR-engineered T cells: clinical evaluation and management of on-target toxicity. Mol Ther 21(4):904–912
67. Craddock JA et al (2010) Enhanced tumor trafficking of GD2 chimeric antigen receptor T cells by expression of the chemokine receptor CCR2b. J Immunother 33(8):780–788
68. John LB et al (2013) Anti-PD-1 antibody therapy potently enhances the eradication of established tumors by gene-modified T cells. Clin Cancer Res 19(20):5636–5646
69. Prosser ME et al (2012) Tumor PD-L1 co-stimulates primary human CD8(+) cytotoxic T cells modified to express a PD1:CD28 chimeric receptor. Mol Immunol 51(3–4):263–272
70. Ankri C et al (2013) Human T cells engineered to express a programmed death 1/28 costimulatory retargeting molecule display enhanced antitumor activity. J Immunol 191(8):4121–4129
71. Topp MS et al (2011) Targeted therapy with the T-cell-engaging antibody blinatumomab of chemotherapy-refractory minimal residual disease in B-lineage acute lymphoblastic leukemia patients results in high response rate and prolonged leukemia-free survival. J Clin Oncol 29(18):2493–2498
72. Urbanska K et al (2012) A universal strategy for adoptive immunotherapy of cancer through use of a novel T-cell antigen receptor. Cancer Res 72(7):1844–1852
73. Kochenderfer JN et al (2012) B-cell depletion and remissions of malignancy along with cytokine-associated toxicity in a clinical trial of anti-CD19 chimeric-antigen-receptor-transduced T cells. Blood 119(12):2709–2720

Chapter 13
Rapamycin-Resistant T Cells and Pentostatin-Based Immuno-Selective Conditioning for the Allogeneic T Cell Therapy of Cancer

Daniel H. Fowler and David C. Halverson

Abstract Allogeneic hematopoietic stem cell transplantation (HSCT) is the most common T cell therapy procedure, yet little is known about how best to harness the graft-versus-tumor (GVT) effect that underlies this curative modality. Importantly, it has proven difficult to clinically dissect GVT effects away from severe GVHD. In this review, we summarize the often times competing roles of key donor and host cellular players in the transplant setting; in particular, it is important to consider the influence of T cells and antigen-presenting-cells (APC) on the fate of host tumor cells. Going forward, it will be critical to evaluate whether anti-tumor effects can be harnessed in the mixed chimeric state, which inherently poses less risk for GVHD. We summarize the history of allogeneic HSCT, with particular attention to its reliance upon myeloablation, which can limit T cell therapy directly by causing conditioning-related toxicity or indirectly through potentiation of GVHD. Novel approaches to the safe and more effective delivery of allogeneic T cell therapy are clearly needed, especially in settings of advanced, chemotherapy refractory malignancy. Our laboratory and clinical trial research has focused on two primary strategies to address these obstacles, namely: (1) minimization of host conditioning intensity through use of pentostatin-based chemotherapy; and (2) adoptive transfer of ex vivo manufactured, rapamycin-resistant allogeneic T cells. Through use of these two approaches, the anti-tumor mechanism of transplantation is dramatically shifted away from chemotherapy and towards immunotherapy, thereby exposing the relative merits and limitations of a given T cell product. Continued advances in allogeneic T cell therapy will rely upon improvements in host conditioning and enhancement of allogeneic T cell products.

Keywords Reduced-intensity allogeneic stem cell transplantation • Rapamycin • Pentostatin • Th1/Th2 • CD4+ T cells • Non-Hodgkin's lymphoma • Graft rejection • Graft-versus-host disease • Graft-versus-tumor effect • Co-stimulation

D.H. Fowler, M.D. (✉) • D.C. Halverson, M.D.
Experimental Transplantation and Immunology Branch, National Institutes of Health, National Cancer Institute, 10 Center Drive; CRC; Room 3-3330, Bethesda, MD 20892, USA
e-mail: dhfowler@helix.nih.gov; david.halverson@nih.gov

P.A. Ascierto et al. (eds.), *Developments in T Cell Based Cancer Immunotherapies*, Cancer Drug Discovery and Development,
DOI 10.1007/978-3-319-21167-1_13

Shared Biology of Graft-Versus-Host Disease and Graft-Versus-Tumor Effects

Murine models have provided a framework for understanding the pathogenesis of graft-versus-host disease (GVHD). Initially, it was discovered that both donor $CD4^+$ and $CD8^+$ T cells contribute to GVHD through recognition of host alloantigen as major histocompatibility class II and class I differences, respectively [1]. Subsequently, a landmark study identified donor T cell IFN-γ as being critical for experimental GVHD; specifically, IFN-γ primed the monocyte/macrophage lineage for response to intestinal-derived bacterial lipopolysaccharide (LPS), which resulted in lethal levels of TNF-α systemically [2]. As such, by 1992, key aspects of GVHD pathogenesis were firmly rooted, namely, identification of a role for: (1) both donor $CD4^+$ and $CD8^+$ T cells, in particular subsets capable of IFN-γ secretion; (2) gut damage with subsequent systemic LPS exposure; and (3) a distal inflammatory cytokine cascade. This fundamental understanding of GVHD in large part still grounds our current translational efforts, which focus on minimization of host tissue damage (in particular, reducing neutropenia-associated gut damage that can lead to systemic LPS exposure) in the context of adoptive transfer of allogeneic T cells of increased function.

An understanding of the role of cytokine-secreting subsets of donor $CD4^+$ T cells in GVHD has increased dramatically since this early linkage of T cell IFN-γ secretion to the distal inflammatory state. In an initial experimental murine study [3], we found that allograft augmentation with IL-4 primed donor T cells of T-helper 2 (Th2) cytokine phenotype restricted IFN-γ driven alloreactivity and prevented LPS-driven, TNF-α mediated GVHD. As such, when one considers the GVHD potential of a T cell inoculum, it is important to consider not only T cell dose and relative CD4 vs. CD8 subset composition, but also T cell cytokine phenotype. Other investigations have similarly shown a counter-regulatory role of donor T cells of Th2-type: for example, infusion of donor T cells deficient in the production of multiple Th2 cytokines increased experimental GVHD [4].

On the other hand, other studies have identified a role for both Th1- and Th2-type T cells in experimental GVHD, including a model whereby cytokine-secreting cells were selectively deleted in vivo [5]. Indeed, there exists a form of reciprocity in terms of T cell subsets in GVHD: for example, reduction in the Th1 component can increase Th2- or Th17-mediated GVHD [6]. Of note, GVHD mediated by a given inflammatory subset (Th1, Th2, or Th17) induced a differential pattern of target tissue involvement. Finally, CD4 cells can also exist in a fourth subset, regulatory T cells (T-Regs), which inhibit GVHD [7] and uniquely have not been shown to mediate target tissue damage. A previous review has compared and contrasted the biology of the two T-helper subsets known to reduce GVHD, the Th2 and T-Reg subsets [8]. It should be noted that the Th1-dominant biology of GVHD is to a great extent illustrated in mouse models of acute GVHD, and it is less clear whether this biology is operational in chronic GVHD [which is oftentimes primarily summarized as a Th2-driven process [9]] or in GVHD in humans. Nonetheless, clinical chronic

GVHD, at least in some target organs, can be associated with infiltrating T cells of Th1 phenotype [10]. In sum, T cell cytokine phenotype helps determine the pattern of GVHD, the net balance of which can be influenced by a reciprocal balance between functional T cell subsets.

In addition to GVHD, it is important to consider the effect of donor T cell subsets on graft rejection, which is mediated by a host-versus-graft (HVG) rejection response that is primarily T cell-driven. This consideration is particularly relevant in settings of lower-intensity host conditioning, as an increased number of host T cells remain that can mediate an HVG reaction. In an initial experimental murine model that utilized sub-lethal host irradiation, we found that allograft augmentation with donor Th2-type cells reduced GVHD while eliminating graft rejection [11]. This method of preventing graft rejection relies largely on the alloresponse (GVHD), and therefore has a relatively narrow therapeutic window. In a subsequent model that involved an F_1-into-parent strain combination that was thereby devoid of alloreactivity in the GVH direction, allograft augmentation with ex vivo manufactured donor $CD4^+$Th2/$CD8^+$Tc2 cells prevented graft rejection [12]; this result thus offered a potential strategy for use of T cell subsets in the prevention of rejection with reduced GVHD. More recently [13], using potent donor $CD4^+$ Th2 cells rendered rapamycin-resistant by ex vivo manufacturing, we identified a new mechanism of rejection abrogation whereby donor Th2 cells secrete IL-4 which then conditions host T cells to adopt Th2-type differentiation; importantly, such host Th2-type cells were greatly reduced in their capacity to mediate allograft rejection. This newly discovered IL-4 cytokine-mediated mechanism, which involves donor influence on host elements, differed from prior approaches of rejection abrogation that have been characterized primarily as cytolytic in mechanism [14]. Given the inhibitory role of T cell-derived IL-4 in preventing [15] or treating [16] GVHD, allograft augmentation with donor rapamycin-resistant $CD4^+$ T cells enriched for a Th2 component represents a new strategy to simultaneously prevent graft rejection while reducing GVHD. This approach is particularly applicable to transplants involving minimal host conditioning, where both graft rejection and GVHD are clinical concerns.

There exists significant overlap in terms of the molecular mechanisms and cellular players that contribute to GVHD and GVT effects [see review [17]]. This shared biology between GVHD and GVT effects poses an ongoing obstacle to allogeneic transplantation, particularly for patients with bulky or chemotherapy-refractory malignancy. In our own initial studies using a murine leukemia model, we found that recipients of Th2- and Tc2-augmented allografts had reduced GVHD that occurred at the expense of increased death due to leukemia [18, 19]. Subsequently, using a murine model involving metastatic breast cancer cells, we found that the Tc2 population mediated greatly reduced GVHD relative to Tc1 cells; however, the Tc1 population was far superior in terms of GVT potential [20].

The murine literature is certainly replete with examples of interventions that can "separate" GVHD from GVT effects. In a recent example, the cytolytic molecule TRAIL mediated anti-tumor effects while reducing experimental murine GVHD [21]. The TRAIL cytolytic pathway thus appears to differ from the donor T cell perforin and fas ligand effector mechanisms, as each of these contributes to the early stages of

GVHD [22] while also mediating distinct GVT effects [23]. However, the extent to which such "separation" strategies can be successfully translated to the clinic is largely unknown. As such, we have elected to pursue a strategy that seeks to provide a distinct component associated with known anti-tumor mediation (Th1-type cells) in combination with a distinct component for GVHD modulation (Th2-type cells).

When one evaluates adoptive T cell therapy in either allogeneic or autologous settings, one must consider the differentiation status of the infused T cell product (namely, the mix of naïve vs. T central memory [T_{CM}] vs. T effector memory [T_{EM}] subsets). In murine models, T_{EM} cells (operationally defined by their genetic lack of expression of the lymph node homing molecule CD62L) mediated reduced GVHD [24]. In other studies using a CD8-driven model of murine GVHD, the T_{CM} subset mediated increased GVHD relative to the T_{EM} subset [25]. In further studies, a "stem cell like" subset of $CD8^+$ T cells expressing the SCA^+ marker efficiently propagated GVHD upon secondary transfer [26]; such stem-like $CD8^+$ T cells have similarly been characterized in models of autologous T cell therapy [27]. Similarly, in further studies, naïve $CD8^+$ T cells mediated increased GVHD relative to T_{CM} cells [28]; still yet, it was observed that adoptive transfer of T_{EM} cells alone generated an allo-response that was unsustained, thereby leading to a blunted GVHD response [29]. In a more recent study, it was the $CD8^+$ T_{EM} subset (rather than the $CD4^+$ T_{EM} subset) that was capable of mediating a substantial GVL response with reduced GVHD [30]. Human clinical results seem to corroborate these murine data, as allografts enriched for $CD4^+CCR7^+$ T_{CM} cells had increased GVHD potential [31].

In sum, these data demonstrate the increased in vivo potential of allogeneic T cell products of reduced differentiation status. Although the examples above focused on GVHD potential, the same principle applies to GVHD counter-regulatory populations; that is, the $CD62L^+$ subset of T-Reg cells had an increased capacity to prevent GVHD [32]. And, in our own studies, rapamycin-resistant Th2 cells enriched for the T_{CM} markers CD62L and CCR7 were enhanced in their ability to down-regulate Th1-driven alloresponses [33]. As such, it is critical to assess T cell differentiation status in the broader context of the cytokine and regulatory phenotype.

In any type of adoptive T cell therapy, it is important to consider the role of antigen-presenting-cells (APC); this consideration is particularly relevant in the allogeneic transplant setting, as both host and donor APC contribute to both GVHD and GVT effects. In an initial landmark study [34], the critical role of host APC in the initiation of experimental acute murine GVHD was established; more recently, non-hematopoietic host APC have also been shown to initiate GVHD [35]. It is also now known that all host APC subsets are not identical in this regard, as a subset of host $CD8^+$ APC can potentiate GVT effects without exacerbating GVHD [36].

However, the necessity of host APC for GVHD induction is not absolute, as donor APC alone can drive GVHD in a sex-mismatched, H-Y antigen driven murine model [37]. Importantly, host conditioning contributes substantially to donor APC-driven GVHD through release of host antigen that is subsequently "cross-presented" by donor APC [38]. Because donor APC are critical for maximal GVHD but not maximal GVL effects [39], we reason that attempts can be made to improve the GVHD-to-GVL balance by reducing donor APC-driven GVHD; specifically, reduction in host conditioning can limit antigen cross-presentation and also delay the onset of donor myeloid cell engraftment, thereby reducing donor myeloid APC populations post-transplant. It is further interesting to note that the source of APC may drive antigen-specific GVHD manifestations, namely: maximal chronic GVHD of the gut required donor APC whereas either donor or host APC were sufficient to drive skin GVHD [40]. In sum, these data indicate that consideration of APC biology (both donor and host) is critical for experimental allogeneic T cell therapy efforts.

Finally, when one considers the allogeneic T cell therapy of cancer, it will be increasingly important to consider specific aspects of the tumor biology at-hand. That is, although an initial observation in 1979 associated GVHD itself with clinical anti-leukemia effects [41], further progress in the field will require an improved transplantation therapeutic index (increased anti-tumor effects/reduced GVHD). A recent NCI relapse conference was convened to summarize the current state-of-the-field in terms of tumor biology as it relates to transplantation therapies [42]. It may be important to know tumor sensitivity to various cytolytic pathways (TRAIL vs. fas ligand vs. perforin) or various cytokine mechanisms (Th1 vs. Th2 vs. Th17); an understanding of this tumor biology may guide the selection of allogeneic T cell populations to supply the appropriate mix of effector function for a particular tumor. And from a second angle, it will be important to understand the extent to which the initial, broader allogeneic T cell response might be dissected from the tumor antigen specific T cell response that would presumably associate with less GVHD. The generation of alloantigen-independent $CD8^+$ T cell anti-tumor responses that arise during allogeneic transplantation was nicely illustrated in a solid tumor model [43] and more recently, in a clinical trial of WT-1 specific T cells in leukemia patients [44]. In sum, these data indicate that it will become increasingly important to consider both T cell and tumor cell phenotype for attempts to optimize allogeneic transplant therapy.

It is critical to evaluate new approaches to allogeneic transplantation, as it has been difficult to separate GVHD from GVT effects at a clinical level. This difficulty remains a sobering reality: for example, in a recent study, the existence of GVL effects was positively correlated with the occurrence of clinical GVHD by the new NIH consensus criteria [45]. Current efforts in the field are laudable and perhaps transformative, but do not necessarily solve the GVHD/GVL imbalance of allogeneic transplantation. As one example, the use of post-transplant cyclophosphamide allows for the safer implementation of haplo-identical allogeneic transplantation [46]; however, malignant disease relapse remains the primary therapeutic obstacle with this approach [47]. As another example, enforced expression of a novel, caspase-activating suicide gene in allogeneic T cells represents a new method to control ongoing GVHD [48]; however, elimination of the entire repertoire of in vivo activated allogeneic T cells might theoretically blunt an ongoing anti-tumor memory response. And third, it is exciting to observe the rapid development in the field of genetic T cell modulation with chimeric antigen receptors (CAR), which offers a method to dial-in tumor antigen specificity, even in the context of allogeneic transplantation [49]; it is important to note that, in this latter study, CAR-expressing allogeneic T cell therapy was not associated with clinical GVHD. However, such efforts are somewhat tangential to the GVHD/GVL balance dilemma, as potent

CAR-modulated T cell responses can occur in the autologous setting [see recent review [50]]; in other words, with further developments in the CAR field, it is not clear how the allo-setting will contribute to this new form of immune therapy. Distinct from the above efforts and therapeutic approaches, we will summarize our translational efforts to improve the GVHD/GVL balance after allogeneic HCT. Specifically, we will highlight our efforts to minimize host conditioning (through pentostatin-based therapy) and to maximize donor T cell intensity (through adoptive transfer of rapamycin-resistant donor T cells).

Pentostatin-Based, Immuno-Selective Host Conditioning

Myeloablative host conditioning, such as use of high-dose cyclophosphamide in combination with high-dose total body irradiation (TBI), was a central component of initial efforts in allogeneic bone marrow transplantation. This clinical practice was motivated by an understanding that high-dose therapy could mediate leukemia cure [51]. In addition, there was a realization that elimination of host cellular elements reduced the risk of graft rejection; indeed, subsequent experimental models demonstrated that increased doses of TBI could eradicate the host alloreactive T cells accountable for graft rejection [52]. However, the morbidity and mortality associated with myeloablative conditioning occurs in young patients and these effects increase with age [53]; as such, myeloablation is not a suitable therapy for the majority of patients due to the advanced age at most cancer diagnoses. These toxicity concerns, combined with the realization that a component of the curative effect of transplantation is due to a graft-versus-leukemia effect [54] rather than solely conditioning-mediated, led investigators to explore allogeneic transplantation after "non-myeloablative" or "reduced-intensity" conditions [see review of current progress [55]]. Here, we discuss how this movement towards less-intensive conditioning is incomplete, and outline how interventions using pentostatin-based, selective immune modulation offer an avenue toward a more complete therapeutic transition. It is in this context of minimally-intensive conditioning that we are evaluating rapamycin-resistant allogeneic T cell subsets, which will be discussed later in this chapter.

When one considers the overall contribution of conditioning to the morbidity and mortality of allogeneic transplantation, it is important to consider both direct effects (neutropenia-related infection, acute and chronic end-organ damage) and indirect effects (most notably, potentiation of GVHD). In murine models, it has long been known that the intensity of conditioning correlates with GVHD severity [56]. Conditioning can result in elevated levels of pro-inflammatory cytokines such as TNF-α, which contributes to GVHD pathogenesis [57]. Recent studies have further elucidated the link between conditioning and GVHD, namely: conditioning-mediated tissue damage increases extra-cellular ATP levels, which in turn activate the inflammasome that is critical for initiation of complex, down-stream inflammatory processes such as GVHD [58]. Our therapeutic approach is predicated on the adoptive transfer of potent allogeneic T cell products, such as the rapamycin-resistant subsets, which are capable of causing tissue injury with subsequent activation of inflammatory cascades. Therefore, we reason that if one is to optimize the T cell component of allogeneic transplantation it will be necessary to develop less-intensive, less-inflammatory approaches to conditioning.

As delineated in a recent review [55], reduced-intensity conditioning regimens come in a substantial variety of intensities, but each published regimen can be considered to cause significant host inflammation. That is, published regimens utilize sub-lethal TBI [which, in experimental models, causes intestinal crypt cell apoptosis [59]] and/or doses of chemotherapy that induce neutropenia and can cause inflammation [see recent review of the role of conditioning-related inflammation [60]]. In our previous studies, we utilized a reduced-intensity regimen of fludarabine combined with cyclophosphamide at a total dose of 4800 mg/m^2. This regimen, although 'reduced' relative to myeloablative regimens, still required intensive inpatient monitoring, was associated with an approximate 2-week period of neutropenia, and resulted in acute GVHD in approximately two-thirds of patients [61]; of interest, GVHD primarily involved the gut and was associated with increased inflammatory cytokines [62]. Thus, although substantial graft-versus-lymphoma effects could be generated with this regimen [63], acute GVHD was a clear constraining factor that motivated us to subsequently pursue conditioning regimens that were 75 % reduced in alkylator dose (Cy total dose, 1200 mg/m^2). Other approaches that are less intensive than the approach we initially evaluated, such as the combination of low-dose TBI with fludarabine, have also been associated with substantial GVHD, in particular late acute GVHD of the gut [64]. It is interesting to speculate that the low-dose TBI component of the conditioning packet is a particular driver of GVHD, as even low doses of TBI are pro-inflammatory in experimental models [65]. In sum, these clinical data indicate that further steps are clearly required to evaluate whether further alterations (reductions) in conditioning intensity can translate into reduced GVHD. Nonetheless, on a more positive note, GVHD that occurs after reduced-intensity conditioning has an improved therapeutic index relative to GVHD that occurs after myeloablative conditioning (safer mediation of a GVL effect) [66].

One key component of our translational research is to harness the drug pentostatin in an attempt to limit inflammation associated with host conditioning. At an operational level, we define this simply as an ability to reduce the host absolute lymphocyte count (ALC; measure of efficacy in terms of reducing the HVG response) without substantial reduction in the host absolute neutrophil count (ANC; measure of toxicity in terms of damage to non-lymphoid cells). In 1980, a landmark article demonstrated the link between adenosine deaminase (ADA) deficiency and lymphoid cell depletion [67], with ADA deficiency now recognized as a common molecular basis for severe combined immunodeficiency disease. Over the next 35 years, there has been a dramatic increase in our understanding of the complexities of adenosine biology, particularly as it resides at the interface between inflammation, immunity, and cancer [see recent review [68]].

Pentostatin, which is a specific inhibitor of the ADA enzyme [69], appears to be relatively specific for lymphocytes; this specificity appears to be due to down-stream pathway alterations rather than to differences in ADA expression [see review [70]].

That is, lymphocytes inherently have high deoxycytidine kinase activity, which converts deoxyadenosine (dAdo) to deoxyadenosine monophosphate (dAMP); and, lymphocytes have intrinsically low 5′ nucleotidase activity, which degrades dAMP to dAdo. This lymphocyte combination of enzyme activity allows the rapid accumulation of initial dAMP and subsequent deoxyadenosine diphosphate (dADP) and deoxyadenosine triphosphate (dATP). As such, inhibition of ADA activity by pentostatin in lymphocytes represents a sensitive pathway to yield increased levels of dATP, which is a key molecule that mediates cellular cytotoxicity.

Thus, it is important to emphasize that the specificity of pentostatin for lymphocyte modulation is not absolute; this recognition is important both when one considers the potential toxicities and potential anti-tumor effects of pentostatin-based conditioning. Patients with genetic deficiency of ADA have severe immune deficiency, but also have dysfunction in multiple organs, most notably brain and bone [67]. Unfortunately, a paucity of information exists pertaining to whether pharmacologic blockade of ADA similarly results in tissue damage, which may then prime for GVHD; however, the observation that addition of pentostatin to a high-dose cyclophosphamide containing preparative regimen resulted in lethal cardiac failure indicates the relevancy of this consideration [71]. From a practical standpoint, it is important to note that pentostatin is a drug whose half-life is critically influenced by renal function and therefore must be dose reduced in patients with impaired creatinine clearance [72].

Given the relative lymphocyte specificity of pentostatin, it is not surprising that initial phase I clinical trials and current clinical usage of pentostatin is restricted primarily to lymphoid malignancy, most notable hairy cell leukemia, T cell lymphoma, and chronic lymphocytic leukemia (CLL) [see review [73]]. However, just as one should maintain concern about the non-lymphocyte toxicity of pentostatin, one should not rule out the possibility that pentostatin, particularly when used in combination with other agents, might yield anti-tumor effects in non-lymphoid malignancy; of note, to our knowledge, combination pentostatin regimens have not been evaluated in a systematic manner in non-lymphoid malignancy. The combination regimen of pentostatin (4 mg/m^2) plus cyclophosphamide (600 mg/m^2) has been evaluated extensively for therapy of CLL [74]. This regimen, although associated with anti-tumor effects, is associated with grade 3 or 4 neutropenia in 35 % of patients. As such, we were motivated to evaluate alternative methods of delivering combination pentostatin plus cyclophosphamide therapy that would avoid neutropenia and thereby be more compatible with a non-inflammatory conditioning platform prior to allogeneic T cell therapy.

To develop alternative approaches to pentostatin-based conditioning, we turned to an experimental model of allogeneic bone marrow graft rejection, which we previously used to test the combination of fludarabine plus cyclophosphamide [75]. In this effort, we found that the combination of pentostatin plus cyclophosphamide (PC) was more effective than the combination of fludarabine plus cyclophosphamide (FC) in preventing marrow graft rejection [76]. Several key findings from this study have helped inform our subsequent translational efforts, including: (1) the immune-depleting (defined by decreased T cell numbers) and immune-suppressing (defined by decreased T cell effector cytokine secretion) effects of pentostatin were modest but highly synergistic with daily, low-dose therapy with the DNA alkylator, cyclophosphamide (Cy); (2) a schedule of intermittent pentostatin plus daily, low-dose Cy could be employed to attain severe host T cell depletion without severe neutropenia; and (3) the immunosuppressive effects of the PC regimen lasted several days, thereby suggesting a prolonged biologic half-life.

In a second experimental model, we evaluated whether the PC regimen could be used to prevent host capacity to produce neutralizing antibody to a foreign protein, namely the anti-mesothelin immunotoxin SS1P, which represents a potential therapeutic modality for a wide variety of mesothelin-expressing tumors. However, previous clinical trials of SS1P did not result in anti-tumor responses, at least in part due to neutralizing antibody formation, which typically limited the duration of SS1P therapy to one treatment cycle [77]. In our murine model [78], we developed a PC regimen that extensively depleted host immune T and B cells without substantial depletion of myeloid elements and with no apparent morbidity and no mortality. Importantly, this immune-depleting, non-myelosuppressive PC treatment regimen permitted the dosing of six serial injections of SS1P immunotoxin without neutralizing antibody formation. These results were then translated into a clinical trial (clinicaltrials.gov, NSC# 01362790), which demonstrated that a non-neutropenia inducing, yet immune-depleting PC regimen could: delay the formation of neutralizing antibody to SS1P; result in high serum immunotoxin concentrations; and, for the first time, produce responses against refractory malignant mesothelioma [79]. Further studies will be required to determine the mechanism(s) of anti-tumor action of this new treatment regimen (direct immunotoxin effect vs. direct pentostatin effect on tumor vs. indirect effect of pentostatin on immunity). Nonetheless, this pilot study established the principle that an immune-depleting yet non-myelosuppressive PC regimen could be safely translated to the clinic for therapeutic gain. Figure 13.1 summarizes the potential effects of pentostatin inhibition of ADA when one considers the seemingly disparate transplantation variables of host conditioning, T cell biology, and tumor cell biology.

In addition to clinical use of the PC regimen in the setting of immunotoxin therapy, we have tested the PC regimen as an immune-depleting, non-neutropenia inducing host conditioning regimen prior to allogeneic HSCT (clinicaltrials.gov, NSC# 00923845). Pentostatin has previously been administered as a component of conditioning regimens for allogeneic HSCT; however, each of these regimens was combined with other chemotherapy agents [80] or with TBI [81, 82] at doses that resulted in substantial neutropenia. In our new protocol effort, 3 weekly infusions of pentostatin (each at a dose of 4 mg/m^2) are combined with low-dose daily Cy (200 mg flat dose per day) and administered over a 21-day interval. The results of this clinical trial have not been published. However, analysis of data (Principal Investigator, Daniel Fowler) from the first ten patients treated on this trial indicates that the PC regimen: (1) safely induced immune T cell depletion and suppression without neutropenia; (2) resulted in an alloengraftment pattern marked by prominent mixed chimerism in both lymphoid and myeloid lineages; and (3) was associated with a virtual absence of clinical acute GVHD.

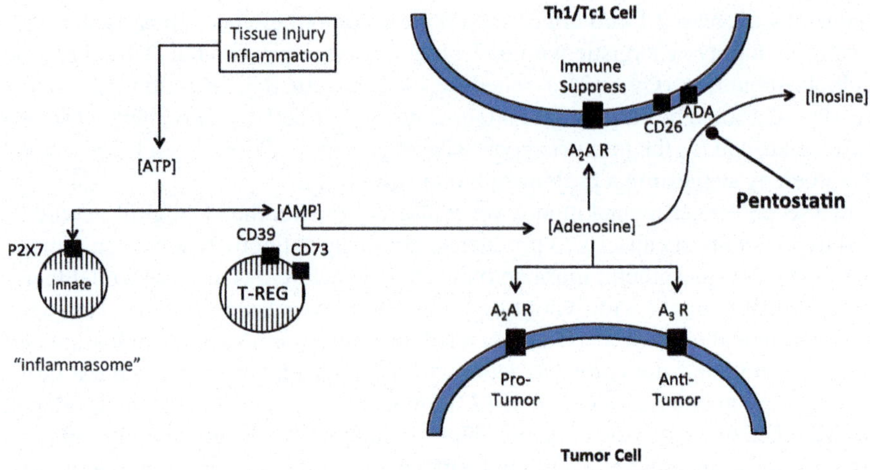

Fig. 13.1 Pentostatin at the interface of conditioning-related inflammation, T cell suppression, and tumor cell biology. Transplant preparative regimens cause host tissue injury, which can cause extracellular release of ATP. Such ATP can promote inflammation via binding to P2X7 receptors and subsequent activation of the inflammasome on innate immune cells (see reference [58]). Alternatively, in the presence of T-Reg cells, ATP can be processed into AMP and then to adenosine through CD39 and CD73 cell surface molecules; pentostatin inhibits the ADA enzyme, which has CD26 as co-factor. In the presence of pentostatin, adenosine levels increase because ADA acts to convert adenosine to inosine. Elevated levels of adenosine, acting largely through the adenosine A_2A receptors, can suppress immune Th1/Tc1 cells and can have a pro-tumor effect; however, adenosine can mediate anti-tumor effects in some models, for example, through signaling of the alternative A_3 receptor (reference [68])

In this study, which enrolled patients with metastatic renal cell carcinoma, we used high-dose sirolimus post-transplant therapy for GVHD prophylaxis, with the hypothesis that sirolimus may cause a direct anti-tumor effect [similar to the documented effect of the rapalog temsirolimus [83]] while permitting an allogeneic GVT effect [which can occur in at least 20 % of appropriately selected patients with metastatic renal cell cancer [84, 85]]. However, we observed that adoptive transfer of donor rapamycin-resistant T cells combined with T cell-replete, HLA-matched peripheral blood stem cell allografts was associated with persistent mixed chimerism, limited T cell effector function post-transplant, and an absence of clinical anti-tumor responses. The absence of anti-tumor effects with this regimen may have been due to: selection of patients with adverse prognostic characteristics; tolerizing effects of post-transplant therapy with high-dose sirolimus [which has been observed in the setting of allogeneic therapy of sickle cell anemia [86]]; or to the non-inflammatory nature of conditioning using the PC regimen (absence of neutropenia).

In summary, using two distinct murine models, we have demonstrated that the strategy of selective immune modulation with the PC regimen can be used for either the prevention of graft rejection or the prevention of host neutralization of foreign protein therapeutics. And, the PC regimen can be safely translated to the clinic, as it can cause immune depletion and immune suppression without neutropenia or other significant end-organ damage. In future experimental models and clinical trials, it will be essential to further evaluate: (1) the extent to which the PC regimen is accountable for reduced GVHD on the basis of lymphocyte selectivity; and (2) the extent to which the PC regimen might be harnessed to directly mediate anti-tumor responses through modulation of adenosine pathways in the tumor-T cell micro-environment.

Allogeneic hematopoietic stem cell transplantation using a highly selective immune-depleting and immune-suppressive pentostatin plus cyclophosphamide regimen represents a more complete transition towards allogeneic T cell therapy that is further dissected from the toxic effects of chemotherapy and total body irradiation. This shift will hopefully pay dividends in terms of reducing the direct toxicity of conditioning and through an indirect effect (reduced priming of GVHD). At the same time, however, minimization of host conditioning through the PC regimen places a higher anti-tumor burden upon the allograft immunotherapeutic component. It is our contention that this represents an obstacle that can be overcome, as: (1) the current, typical patient enters transplant in a partial or complete remission; and (2) for the other subset of patients that enter transplant with chemotherapy refractory disease, increases in the intensity of host conditioning are unlikely to represent the key element to a curative intervention. Indeed, it is the chemotherapy refractory patients who are currently typically underserved by allo-transplantation (they are either not referred for transplant or receive transplantation without efficacy); therefore, this subgroup of patients is likely to benefit from novel strategies that enhance allogeneic T cell efficacy. To this end, we have evaluated the effect of allogeneic T cells with potential for increased in vivo activity, namely, subsets of rapamycin-resistant T cells.

Adoptive Transfer of Rapamycin-Resistant T Cells

mTOR, now known as the mechanistic target of rapamycin, was discovered in 1994 as the target of the rapamycin-FKBP12 complex [87]. Since that time, there has been a rapid advancement of research and a broadened understanding regarding the role of mTOR in health and disease [see review [88]]. It should be noted that rapamycin initially inhibits only the mTORC1 complex via an allosteric effect; however, mTORC2 is sequentially inhibited in an indirect manner with prolonged drug exposure [89]. This indirect and incomplete mechanism whereby rapamycin inhibits mTOR represents only one of several mechanisms of T cell or tumor cell acquisition of rapamycin resistance [see review [90]]. It is important to differentiate rapamycin from the newly described direct, competitive inhibitors of the ATP-binding site of mTOR that potently and simultaneously inhibit both mTORC1 and mTORC2 [91]. Of note, compared to rapamycin, these direct mTOR inhibitors preferentially lead to cellular death rather than the attainment of resistance [92]. One must also differentiate this pharmacologic analysis from genetic approaches, which have characterized and cleanly dissected the relative roles of mTORC1 and mTORC2 on T cell differentiation [93, 94].

In initial studies, rapamycin was found to induce T cell anergy, even in conditions of T cell co-stimulation [95]. More recently, ex vivo rapamycin combined with relatively weak co-stimulation and the absence of effector cytokine polarization promoted the generation of T-Reg cells [96]. This predisposition towards regulatory T cell differentiation in the presence of rapamycin is due in part to the T-Reg cell transcription factor Foxp3, which up-regulates pim2 as an alternative kinase signaling pathway to the blocked, mTOR pathway [97]. In vivo therapy with rapamycin after allo-transplant preferentially inhibits effector T cells while sparing T-Reg cells, with consequent reduction in experimental GVHD [98]. It is critical to note that experimental models that have found an association between rapamycin and T-Reg cell differentiation have been performed in the absence of T-helper cell polarizing cytokines such as IL-4 or IL-12. Indeed, in other studies, IL-4 [99] or IL-12 [100] promote GATA-3 or T-bet transcription factors, respectively, while restricting Foxp3 expression. Based on these data, we have concluded that the association of rapamycin with a certain pattern of T-helper subset differentiation is contextual and in part determined by the absence of polarizing cytokines (preferential T-Reg induction) or the presence of polarizing cytokines (Th1- or Th2-permissive).

In our initial studies of T cell subsets in experimental transplantation, we utilized an ex vivo manufacturing method whereby donor T cells were stimulated with host APC in the presence of polarizing cytokines to generate Th1- or Th2-type alloreactive T cells [18]. However, manufacturing of Th1- or Th2-type antigen-specific T cells can lack feasibility and reproducibility; and, theoretically, ex vivo clonal expansion to alloantigen expressed on APC may deviate an immune response away from anti-tumor specificities. Given these considerations, we began to evaluate whether the processes of cytokine polarization and antigen-specificity might be dissociated from one another, namely: to achieve cytokine polarization ex vivo in a polyclonal manner via APC-free co-stimulation [101] and then to allow clonal expansion to occur in vivo post-transplant. Indeed, in our studies, we found that donor T cell cytokine polarity could be achieved ex vivo using polyclonal co-stimulation, with subsequent alloantigen-driven clonal expansion occurring in vivo. Importantly, the in vivo derived antigen-specific T cells maintained the cytokine polarity of the input polyclonal T cells, thereby demonstrating that the processes of cytokine polarity and antigen-driven clonal expansion could be dissected from one another [20].

In subsequent experiments, we evaluated the effect of rapamycin in our method of ex vivo T cell co-stimulation and cytokine polarization. We found that rapamycin-resistant $CD4^+$Th1/$CD8^+$Tc1 or $CD4^+$Th2/$CD8^+$Tc2 cells could be generated in the presence of rapamycin depending on whether IL-12 or IL-4 was added to the culture, respectively [33]. These results stood somewhat in contrast to prior in vivo findings, where it was found that rapamycin inhibited T cells of Th1 cytokine phenotype when administered after experimental allogeneic BMT [102]. It is possible that these seemingly disparate results relate in part to rapamycin influence on APC populations, which is not a factor in our ex vivo method; that is, rapamycin can inhibit APC function in vivo for inhibition of Th1-mediated immunity [103]. In this way, the T cell manufacturing method that we have established allows for the isolation of rapamycin effects on T cells without a competing drug effect at the level of APC populations.

We found that a combination of high-dose rapamycin, co-stimulation, and IL-4/IL-2 polarization generated $CD4^+$ T cells enriched in their capacity to prevent experimental murine acute GVHD [15]. Rapamycin-resistant Th2 cells had minimal capacity for effector cytokine secretion at the time of adoptive transfer; however, such recipients had greatly increased type II cytokine secretion in vivo, a marked drop in IFN-γ secretion in vivo, and superior protection against GVHD. These immune biology and clinical effects were largely abrogated if the Th2 cells were generated from IL-4 deficient donors, thus confirming a Th2-type mechanism. And, we found that a sequential infusion of unmanipulated T cells capable of Th1-type differentiation followed by Th2 cell therapy yielded the best balance of GVT effects to GVHD. In a separate project, we also found that Th2 cell secretion of IL-10 and the consumption of exogenous IL-2 were important for reduction of GVHD by Th2 cell therapy [16]. This latter observation suggests that the Th2 cells manufactured in rapamycin may outcompete other T cell populations for limiting growth factors. In sum, these results provide evidence for a multi-faceted mechanism whereby rapamycin-resistant Th2 cells can regulate ongoing GVHD.

In an initial clinical trial, we evaluated the effect of allograft augmentation with co-stimulated $CD4^+$ T cells that were polarized with IL-4 but not manufactured in rapamycin [61]. The T cells were propagated in culture for approximately 20-days using 2-rounds of co-stimulation and secreted high levels of a mixed Th2/Th1 pattern of cytokines. Of note, in this trial, acute GVHD was observed in approximately two-thirds of patients independent of whether patients received the additional, IL-4 polarized T cell population. As a result, we concluded that this first-generation, highly differentiated T cell population was not efficacious and that further studies should evaluate an alternative method of T cell manufacturing (that is, inclusion of rapamycin).

In light of our murine data indicating an improved ability of rapamycin-resistant Th2 cells to both reduce GVHD and prevent graft rejection, we developed a new protocol (www.cancer.gov/clinicaltrials; #NCT 00077480) that incorporated rapamycin during ex vivo manufacturing. In addition, we modified our transplant platform in an attempt to improve the safety of allogeneic T cell therapy. First, we reduced the transplant preparative regimen intensity (decrease in cyclophosphamide dose by 75 %; total dose, 1200 mg/m^2) to a level that could be safely administered in an outpatient setting. Second, we added a short-course of sirolimus therapy in the first 14 days post-transplant to complement standard cyclosporine GVHD prophylaxis. Third, and perhaps most importantly, we delayed the infusion of the experimental T cells from day 0 of transplant to day 14 post-transplant in an attempt to: (1) separate potential side effects of the manufactured T cells from side effects attributable to the conditioning or to the unmanipulated T cells contained in the initial allograft; (2) establish and confirm a state of mixed chimerism at day 14 post-transplant prior to the experimental donor lymphocyte infusion (DLI); and (3) avoid the transfer of T-Rapa cells during concurrent sirolimus drug administration, which might inhibit the manufactured T cells in vivo.

In this clinical protocol, n=40 patients were initially treated with pre-emptive DLI at day 14 post-transplant with rapamycin-resistant donor T cells [104]. The T cells were generated by a 12-day culture method using a single round of co-stimulation. The resultant T cells were comprised of a mix of CD4⁺ Th2 and Th1 cells, as determined by transcription factor analysis and cytokine secretion; the T cell products were minimally differentiated, as defined by capacity for only nominal magnitudes of cytokine secretion. By gene expression micro-array analysis, there was minimal variability in T cell phenotype and a consistent pattern of alteration in several gene families, including: up-regulation of cell cycle and stress response genes; and, down-regulation of apoptosis and inflammation genes.

In this trial, as anticipated given the low-intensity conditioning, we found that patients had varying levels of mixed donor/host chimerism at day 14 post-transplant (at the time of pre-emptive DLI). Then, 2 weeks after DLI (at day 28 post-transplant), we observed a relatively marked increase in donor engraftment in both lymphoid and myeloid lineages. Several findings suggest that the manufactured T cell DLI were safe and effective in vivo, including: absence of engraftment syndrome (cytokine-related toxicity that causes rash, fever, fluid third-spacing, and pulmonary infiltrates), low rate and severity of GVHD, and absence of transplant-related mortality; preferential post-transplant expansion of donor $CD4^+$ T cells relative to donor $CD8^+$ T cells; and promotion of a balanced pattern of Th2- and Th1-type of cytokines post-transplant. Importantly, complete remissions were attained, including in some patients with chemotherapy-refractory disease or high-risk hematologic malignancy diagnoses. In sum, these results indicate that the infusion of rapamycin-resistant donor $CD4^+$ T cells on this new, low-intensity transplant platform resulted in a potentially favorable balance between GVL effects and GVHD.

Nonetheless, approximately one-half of patients on this study eventually died from their malignancy, and as such, we established the goal of developing and evaluating modified populations of rapamycin-resistant T cells that might mediate increased anti-tumor effects. Towards this end, we reasoned that truncation of the T cell manufacturing interval might generate a unique cell product with increased in vivo function; in an analogous effort, reduction in culture time of tumor-infiltrating-lymphocytes (TIL) products was effective in the treatment of metastatic melanoma [105]. In our current efforts, we are evaluating rapamycin-resistant T cells manufactured for 6-days (T-$Rapa_6$ cells) as an alternative to the 12-day method in our initial study (T-$Rapa_{12}$ cells).

We have now found that the T-$Rapa_6$ and T-$Rapa_{12}$ cell products were very similar in terms of standard assays (such as cytokine secretion profile) but differed significantly by gene expression micro-array analysis [106]. Using the exact same transplant platform as we have recently described [104], we have now finished accruing to a sequential phase II study whereby patients receive the T-$Rapa_6$ cell product instead of the T-$Rapa_{12}$ cell product at the day 14 post-transplant time point. The initial results of this trial have been presented at the 2014 Meeting of the American Society of Blood and Marrow Transplantation [107]. Remarkably, T-$Rapa_6$ cell recipients had an increase in the rate of classical acute GVHD relative to the T-$Rapa_{12}$ cell recipients (increase from 10 to 40.5 %); T-$Rapa_6$ cell recipients also had an earlier time to onset and an increase in severity of acute GVHD. However, overall survival was not decreased in T-$Rapa_6$ cell recipients and deaths due to

Table 13.1 Sequential NCI clinical trials evaluating IL-4 polarized donor T cells: minimizing host conditioning intensity while maximizing T cell intensity

Preparative regimen[a]	PBSCT[b]	IL-4 polarized T cells[c]	Reference number[d]
Flu/Cy (4800 mg/m^2)	√	Th2 (20 days)	# 63
Flu/Cy (1200 mg/m^2)	√	Th2.R (12 days)	# 104
Flu/Cy (1200 mg/m^2)	√	Th2.R (6 days)	# 107
No Flu/Cy	√	Th2.R (6 days)	Study in progress

[a]Flu, fludarabine; Cy, cyclophosphamide; dose represent total Cy dosing
[b]PBSCT, indicates T cell-replete peripheral blood stem cell allograft
[c]Type of manufactured T cell population; Th2.R indicates cells that are both IL-4 polarized and rendered rapamycin-resistant. The number in parenthesis indicates total number of days of ex vivo culture
[d]Further details of clinical trial results described in the listed references

malignancy appears to have been reduced; as such, it is possible that the T-Rapa$_6$ cell population mediated both increased GVHD and increased GVL effects. In sum, these results indicate that a seemingly subtle change in ex vivo T cell manufacturing (reduction in culture time) can result in substantial changes in T cell function that appear to influence in vivo transplantation outcome.

Current research seeks to better understanding the molecular and pathway differences that exist between these two T cell products, as this information may provide leads for the further development of T cell products with increased function. And, from a translational angle, current efforts are directed towards altering the transplant platform to more safely accommodate a T cell product that exerts increased in vivo function (the T-Rapa$_6$ product). Specifically, given the known inter-relationship between conditioning intensity and T cell-mediated GVHD, we hypothesize that the acute GVHD associated with the T-Rapa$_6$ cell product will be minimized if further reductions in host conditioning were used prior to adoptive T cell transfer. To address this hypothesis, we will make use of the immune-selective, pentostatin-based conditioning regimens that we have detailed and avoid the use of regimens that can cause neutropenia, even on a limited level (such as the low-dose Flu/Cy regimen). The success of such an endeavor would represent a further step away from myeloablative transplant and towards a transplant approach that employs maximal T cell intensity with minimal conditioning intensity (see Table 13.1, which summarizes our sequential efforts in this endeavor).

References

1. Korngold R (1992) Lethal graft-versus-host disease in mice directed to multiple minor histocompatibility antigens: features of CD8+ and CD4+ T cell responses. Bone Marrow Transplant 9:355–364
2. Nestel FP, Price KS, Seemayer TA, Lapp WS (1992) Macrophage priming and lipopolysaccharide-triggered release of tumor necrosis factor alpha during graft-versus-host disease. J Exp Med 175:405–413

3. Fowler DH, Kurasawa K, Husebekk A, Cohen PA, Gress RE (1994) Cells of Th2 cytokine phenotype prevent LPS-induced lethality during murine graft-versus-host reaction. Regulation of cytokines and CD8+ lymphoid engraftment. J Immunol 152:1004–1013
4. Tawara I, Maeda Y, Sun Y, Lowler KP, Liu C, Toubai T, Mckenzie AN, Reddy P (2008) Combined Th2 cytokine deficiency in donor T cells aggravates experimental acute graft-vs-host disease. Exp Hematol 36:988–996
5. Liu J, Anderson BE, Robert ME, Mcniff JM, Emerson SG, Shlomchik WD, Shlomchik MJ (2001) Selective T-cell subset ablation demonstrates a role for T1 and T2 cells in ongoing acute graft-versus-host disease: a model system for the reversal of disease. Blood 98: 3367–3375
6. Yi T, Chen Y, Wang L, Du G, Huang D, Zhao D, Johnston H, Young J, Todorov I, Umetsu DT, Chen L, Iwakura Y, Kandeel F, Forman S, Zeng D (2009) Reciprocal differentiation and tissue-specific pathogenesis of Th1, Th2, and Th17 cells in graft-versus-host disease. Blood 114:3101–3112
7. Taylor PA, Lees CJ, Blazar BR (2002) The infusion of ex vivo activated and expanded CD4(+)CD25(+) immune regulatory cells inhibits graft-versus-host disease lethality. Blood 99:3493–3499
8. Fowler D, Hoffmann P, Edinger M (2009) Functionally defined T cell subsets in transplantation biology and therapy: regulatory T cells and Th2 cells. Cancer Treat Res 144:155–186
9. Blazar B, White ES, Couriel D (2012) Understanding chronic GVHD from different angles. Biol Blood Marrow Transplant 18:S184–S188
10. Imanguli MM, Swaim WD, League SC, Gress RE, Pavletic SZ, Hakim FT (2009) Increased T-bet+cytotoxic effectors and type I interferon-mediated processes in chronic graft-versus-host disease of the oral mucosa. Blood 113:3620–3630
11. Fowler DH, Kurasawa K, Smith R, Eckhaus MA, Gress RE (1994) Donor CD4-enriched cells of Th2 cytokine phenotype regulate graft-versus-host disease without impairing allogeneic engraftment in sublethally irradiated mice. Blood 84:3540–3549
12. Erdmann AA, Jung U, Foley JE, Toda Y, Fowler DH (2004) Co-stimulated/Tc2 cells abrogate murine marrow graft rejection. Biol Blood Marrow Transplant 10:604–613
13. Mariotti J, Foley J, Ryan K, Buxhoeveden N, Kapoor V, Amarnath S, Fowler DH (2008) Graft rejection as a Th1-type process amenable to regulation by donor Th2-type cells through an interleukin-4/STAT6 pathway. Blood 112:4765–4775
14. Martin PJ, Akatsuka Y, Hahne M, Sale G (1998) Involvement of donor T-cell cytotoxic effector mechanisms in preventing allogeneic marrow graft rejection. Blood 92:2177–2181
15. Foley JE, Jung U, Miera A, Borenstein T, Mariotti J, Eckhaus M, Bierer BE, Fowler DH (2005) Ex vivo rapamycin generates donor Th2 cells that potently inhibit graft-versus-host disease and graft-versus-tumor effects via an IL-4-dependent mechanism. J Immunol 175:5732–5743
16. Foley JE, Mariotti J, Ryan K, Eckhaus M, Fowler DH (2008) Th2 cell therapy of established acute graft-versus-host disease requires IL-4 and IL-10 and is abrogated by IL-2 or host-type antigen-presenting cells. Biol Blood Marrow Transplant 14:959–972
17. Fowler DH (2006) Shared biology of GVHD and GVT effects: potential methods of separation. Crit Rev Oncol Hematol 57:225–244
18. Fowler DH, Breglio J, Nagel G, Hirose C, Gress RE (1996) Allospecific CD4+, Th1/Th2 and CD8+, Tc1/Tc2 populations in murine GVL: type I cells generate GVL and type II cells abrogate GVL. Biol Blood Marrow Transplant 2:118–125
19. Fowler DH, Breglio J, Nagel G, Eckhaus MA, Gress RE (1996) Allospecific CD8+ Tc1 and Tc2 populations in graft-versus-leukemia effect and graft-versus-host disease. J Immunol 157:4811–4821
20. Jung U, Foley JE, Erdmann AA, Eckhaus MA, Fowler DH (2003) CD3/CD28-costimulated T1 and T2 subsets: differential in vivo allosensitization generates distinct GVT and GVHD effects. Blood 102:3439–3446

21. Ghosh A, Dogan Y, Moroz M, Holland AM, Yim NL, Rao UK, Young LF, Tannenbaum D, Masih D, Velardi E, Tsai JJ, Jenq RR, Penack O, Hanash AM, Smith OM, Piersanti K, Lezcano C, Murphy GF, Liu C, Palomba ML, Sauer MG, Sadelain M, Ponomarev V, van den Brink MR (2013) Adoptively transferred TRAIL+ T cells suppress GVHD and augment anti-tumor activity. J Clin Invest 123:2654–2662
22. Jiang Z, Podack E, Levy RB (2001) Major histocompatibility complex-mismatched allogeneic bone marrow transplantation using perforin and/or Fas ligand double-defective CD4(+) donor T cells: involvement of cytotoxic function by donor lymphocytes prior to graft-versus-host disease pathogenesis. Blood 98:390–397
23. Schmaltz C, Alpdogan O, Horndasch KJ, Muriglan SJ, Kappel BJ, Teshima T, Ferrara JL, Burakoff SJ, van den Brink MR (2001) Differential use of Fas ligand and perforin cytotoxic pathways by donor T cells in graft-versus-host disease and graft-versus-leukemia effect. Blood 97:2886–2895
24. Chen BJ, Cui X, Sempowski GD, Liu C, Chao NJ (2004) Transfer of allogeneic CD62L-memory T cells without graft-versus-host disease. Blood 103:1534–1541
25. Zhang Y, Joe G, Hexner E, Zhu J, Emerson SG (2005) Alloreactive memory T cells are responsible for the persistence of graft-versus-host disease. J Immunol 174:3051–3058
26. Zhang Y, Joe G, Hexner E, Zhu J, Emerson SG (2005) Host-reactive CD8+ memory stem cells in graft-versus-host disease. Nat Med 11:1299–1305
27. Gattinoni L, Zhong XS, Palmer DC, Ji Y, Hinrichs CS, Yu Z, Wrzesinski C, Boni A, Cassard L, Garvin LM, Paulos CM, Muranski P, Restifo NP (2009) Wnt signaling arrests effector T cell differentiation and generates CD8+ memory stem cells. Nat Med 15:808–813
28. Zheng H, Matte-Martone C, Jain D, Mcniff J, Shlomchik WD (2009) Central memory CD8+ T cells induce graft-versus-host disease and mediate graft-versus-leukemia. J Immunol 182:5938–5948
29. Chen BJ, Deoliveira D, Cui X, Le NT, Son J, Whitesides JF, Chao NJ (2007) Inability of memory T cells to induce graft-versus-host disease is a result of an abortive alloresponse. Blood 109:3115–3123
30. Dutt S, Baker J, Kohrt HE, Kambham N, Sanyal M, Negrin RS, Strober S (2011) CD8+CD44(hi) but not CD4+CD44(hi) memory T cells mediate potent graft antilymphoma activity without GVHD. Blood 117:3230–3239
31. Yakoub-Agha I, Saule P, Depil S, Micol JB, Grutzmacher C, Boulanger-Villard F, Bauters F, Jouet JP, Dessaint JP, Labalette M (2006) A high proportion of donor CD4+ T cells expressing the lymph node-homing chemokine receptor CCR7 increases incidence and severity of acute graft-versus-host disease in patients undergoing allogeneic stem cell transplantation for hematological malignancy. Leukemia 20:1557–1565
32. Taylor PA, Panoskaltsis-Mortari A, Swedin JM, Lucas PJ, Gress RE, Levine BL, June CH, Serody JS, Blazar BR (2004) L-Selectin(hi) but not the L-selectin(lo) CD4+25+ T-regulatory cells are potent inhibitors of GVHD and BM graft rejection. Blood 104:3804–3812
33. Jung U, Foley JE, Erdmann AA, Toda Y, Borenstein T, Mariotti J, Fowler DH (2006) Ex vivo rapamycin generates Th1/Tc1 or Th2/Tc2 Effector T cells with enhanced in vivo function and differential sensitivity to post-transplant rapamycin therapy. Biol Blood Marrow Transplant 12:905–918
34. Shlomchik WD, Couzens MS, Tang CB, Mcniff J, Robert ME, Liu J, Shlomchik MJ, Emerson SG (1999) Prevention of graft versus host disease by inactivation of host antigen-presenting cells. Science 285:412–415
35. Koyama M, Kuns RD, Olver SD, Raffelt NC, Wilson YA, Don AL, Lineburg KE, Cheong M, Robb RJ, Markey KA, Varelias A, Malissen B, Hammerling GJ, Clouston AD, Engwerda CR, Bhat P, Macdonald KP, Hill GR (2012) Recipient nonhematopoietic antigen-presenting cells are sufficient to induce lethal acute graft-versus-host disease. Nat Med 18:135–142
36. Toubai T, Sun Y, Luker G, Liu J, Luker KE, Tawara I, Evers R, Liu C, Mathewson N, Malter C, Nieves E, Choi S, Murphy KM, Reddy P (2013) Host-derived CD8+ dendritic cells are required for induction of optimal graft-versus-tumor responses after experimental allogeneic bone marrow transplantation. Blood 121:4231–4241

37. Toubai T, Tawara I, Sun Y, Liu C, Nieves E, Evers R, Friedman T, Korngold R, Reddy P (2012) Induction of acute GVHD by sex-mismatched H-Y antigens in the absence of functional radiosensitive host hematopoietic-derived antigen-presenting cells. Blood 119: 3844–3853
38. Wang X, Li H, Matte-Martone C, Cui W, Li N, Tan HS, Roopenian D, Shlomchik WD (2011) Mechanisms of antigen presentation to T cells in murine graft-versus-host disease: cross-presentation and the appearance of cross-presentation. Blood 118:6426–6437
39. Matte CC, Liu J, Cormier J, Anderson BE, Athanasiadis I, Jain D, Mcniff J, Shlomchik WD (2004) Donor APCs are required for maximal GVHD but not for GVL. Nat Med 10:987–992
40. Anderson BE, Mcniff JM, Jain D, Blazar BR, Shlomchik WD, Shlomchik MJ (2005) Distinct roles for donor- and host-derived antigen-presenting cells and costimulatory molecules in murine chronic graft-versus-host disease: requirements depend on target organ. Blood 105:2227–2234
41. Weiden PL, Flournoy N, Thomas ED, Prentice R, Fefer A, Buckner CD, Storb R (1979) Antileukemic effect of graft-versus-host disease in human recipients of allogeneic-marrow grafts. N Engl J Med 300:1068–1073
42. Miller JS, Warren EH, van den Brink MR, Ritz J, Shlomchik WD, Murphy WJ, Barrett AJ, Kolb HJ, Giralt S, Bishop MR, Blazar BR, Falkenburg JH (2010) NCI First International Workshop on the biology, prevention, and treatment of relapse after allogeneic hematopoietic stem cell transplantation: report from the committee on the biology underlying recurrence of malignant disease following allogeneic HSCT: graft-versus-tumor/leukemia reaction. Biol Blood Marrow Transplant 16:565–586
43. Stelljes M, Strothotte R, Pauels HG, Poremba C, Milse M, Specht C, Albring J, Bisping G, Scheffold C, Kammertoens T, Oelmann E, Silling G, Berdel WE, Kienast J (2004) Graft-versus-host disease after allogeneic hematopoietic stem cell transplantation induces a CD8+ T cell-mediated graft-versus-tumor effect that is independent of the recognition of alloantigenic tumor targets. Blood 104:1210–1216
44. Chapuis AG, Ragnarsson GB, Nguyen HN, Chaney CN, Pufnock JS, Schmitt TM, Duerkopp N, Roberts IM, Pogosov GL, Ho WY, Ochsenreither S, Wolfl M, Bar M, Radich JP, Yee C, Greenberg PD (2013) Transferred WT1-reactive CD8+ T cells can mediate antileukemic activity and persist in post-transplant patients. Sci Transl Med 5:174ra27
45. Thepot S, Zhou J, Perrot A, Robin M, Xhaard A, de Latour RP, Ades L, Ribaud P, Petropoulou AD, Porcher R, Socie G (2010) The graft-versus-leukemia effect is mainly restricted to NIH-defined chronic graft-versus-host disease after reduced intensity conditioning before allogeneic stem cell transplantation. Leukemia 24:1852–1858
46. Luznik L, O'donnell PV, Symons HJ, Chen AR, Leffell MS, Zahurak M, Gooley TA, Piantadosi S, Kaup M, Ambinder RF, Huff CA, Matsui W, Bolanos-Meade J, Borrello I, Powell JD, Harrington E, Warnock S, Flowers M, Brodsky RA, Sandmaier BM, Storb RF, Jones RJ, Fuchs EJ (2008) HLA-haploidentical bone marrow transplantation for hematologic malignancies using nonmyeloablative conditioning and high-dose, posttransplantation cyclophosphamide. Biol Blood Marrow Transplant 14:641–650
47. Kanakry JA, Kasamon YL, Gocke CD, Tsai HL, Davis-Sproul J, Ghosh N, Symons h, Bolanos-Meade J, Gladstone DE, Swinnen LJ, Luznik L, Fuchs EJ, Jones RJ, Ambinder RF (2013) Outcomes of related donor HLA-identical or HLA-haploidentical allogeneic blood or marrow transplantation for peripheral T cell lymphoma. Biol Blood Marrow Transplant 19:602–606
48. di Stasi A, Tey SK, Dotti G, Fujita Y, Kennedy-Nasser A, Martinez C, Straathof K, Liu E, Durett AG, Grilley B, Liu H, CRUZ CR, Savoldo B, Gee AP, Schindler J, Krance RA, Heslop HE, Spencer DM, Rooney CM, Brenner MK (2011) Inducible apoptosis as a safety switch for adoptive cell therapy. N Engl J Med 365:1673–1683
49. Kochenderfer JN, Dudley ME, Carpenter RO, Kassim SH, Rose JJ, Telford WG, Hakim FT, Halverson DC, Fowler DH, Hardy NM, Mato AR, Hickstein DD, Gea-Banacloche JC, Pavletic SZ, Sportes C, Maric I, Feldman SA, Hansen BG, Wilder JS, Blacklock-Schuver B,

Jena B, Bishop MR, Gress RE, Rosenberg SA (2013) Donor-derived CD19-targeted T cells cause regression of malignancy persisting after allogeneic hematopoietic stem cell transplantation. Blood 122:4129–4139
50. Maus MV, Grupp SA, Porter DL, June CH (2014) Antibody modified T cells: CARs take the front seat for hematologic malignancies. Blood 123:2625–2635
51. Thomas ED (1982) The use and potential of bone marrow allograft and whole-body irradiation in the treatment of leukemia. Cancer 50:1449–1454
52. Schwartz E, Lapidot T, Gozes D, Singer TS, Reisner Y (1987) Abrogation of bone marrow allograft resistance in mice by increased total body irradiation correlates with eradication of host clonable T cells and alloreactive cytotoxic precursors. J Immunol 138:460–465
53. Gupta T, Kannan s, Dantkale V, Laskar S (2011) Cyclophosphamide plus total body irradiation compared with busulfan plus cyclophosphamide as a conditioning regimen prior to hematopoietic stem cell transplantation in patients with leukemia: a systematic review and meta-analysis. Hematol Oncol Stem Cell Ther 4:17–29
54. Bortin MM, Rimm AA, Saltzstein EC (1973) Graft versus leukemia: quantification of adoptive immunotherapy in murine leukemia. Science 179:811–813
55. Gill S, Porter DL (2013) Reduced-intensity hematopoietic stem cell transplants for malignancies: harnessing the graft-versus-tumor effect. Annu Rev Med 64:101–117
56. Truitt RL, Atasoylu AA (1991) Impact of pretransplant conditioning and donor T cells on chimerism, graft-versus-host disease, graft-versus-leukemia reactivity, and tolerance after bone marrow transplantation. Blood 77:2515–2523
57. Holler E, Kolb HJ, Mittermuller J, Kaul M, Ledderose G, Duell T, Seeber B, Schleuning M, Hintermeier-Knabe R, Ertl B et al (1995) Modulation of acute graft-versus-host-disease after allogeneic bone marrow transplantation by tumor necrosis factor alpha (TNF alpha) release in the course of pretransplant conditioning: role of conditioning regimens and prophylactic application of a monoclonal antibody neutralizing human TNF alpha (MAK 195F). Blood 86:890–899
58. Wilhelm K, Ganesan J, Muller T, Durr C, Grimm M, Beilhack A, Krempl CD, Sorichter S, Gerlach UV, Juttner E, Zerweck A, Gartner F, Pellegatti P, di Virgilio F, Ferrari D, Kambham N, Fisch P, Finke J, Idzko M, Zeiser R (2010) Graft-versus-host disease is enhanced by extracellular ATP activating P2X7R. Nat Med 16:1434–1438
59. Ch'ang HJ, Maj JG, Paris F, Xing HR, Zhang J, Truman JP, Cardon-Cardo C, Haimovitz-Friedman A, Kolesnick R, Fuks Z (2005) ATM regulates target switching to escalating doses of radiation in the intestines. Nat Med 11:484–490
60. Sjoqvist C, Snarski E (2013) Inflammatory markers in patients after hematopoietic stem cell transplantation. Arch Immunol Ther Exp (Warsz) 61:301–307
61. Fowler DH, Odom J, Steinberg SM, Chow CK, Foley J, Kogan Y, Hou J, Gea-Banacloche J, Sportes C, Pavletic S, Leitman S, Read EJ, Carter C, Kolstad A, Fox R, Beatty GL, Vonderheide RH, Levine BL, June CH, Gress RE, Bishop MR (2006) Phase I clinical trial of costimulated, IL-4 polarized donor CD4+ T cells as augmentation of allogeneic hematopoietic cell transplantation. Biol Blood Marrow Transplant 12:1150–1160
62. Fowler DH, Foley J, Whit-Shan Hou J, Odom J, Castro K, Steinberg SM, Gea-Banacloche J, Kasten-Sportes C, Gress RE, Bishop MR (2004) Clinical "cytokine storm" as revealed by monocyte intracellular flow cytometry: correlation of tumor necrosis factor alpha with severe gut graft-versus-host disease. Clin Gastroenterol Hepatol 2:237–245
63. Bishop MR, Hou JW, Wilson WH, Steinberg SM, Odom J, Castro K, Kasten-Sportes C, Gea-Banacloche J, Marchigiani D, Gress R, Fowler DH (2003) Establishment of early donor engraftment after reduced-intensity allogeneic hematopoietic stem cell transplantation to potentiate the graft-versus-lymphoma effect against refractory lymphomas. Biol Blood Marrow Transplant 9:162–169
64. Mielcarek M, Martin PJ, Leisenring W, Flowers ME, Maloney DG, Sandmaier BM, Maris MB, Storb R (2003) Graft-versus-host disease after nonmyeloablative versus conventional hematopoietic stem cell transplantation. Blood 102:756–762

65. Nowosielska EM, Cheda A, Wrembel-Wargocka J, Janiak MK (2011) Anti-neoplastic and immunostimulatory effects of low-dose X-ray fractions in mice. Int J Radiat Biol 87:202–212
66. Weisdorf D, Zhang MJ, Arora M, Horowitz MM, Rizzo JD, Eapen M (2012) Graft-versus-host disease induced graft-versus-leukemia effect: greater impact on relapse and disease-free survival after reduced intensity conditioning. Biol Blood Marrow Transplant 18:1727–1733
67. Mitchell BS, Kelley WN (1980) Purinogenic immunodeficiency diseases: clinical features and molecular mechanisms. Ann Intern Med 92:826–831
68. Antonioli L, Blandizzi C, Pacher P, Hasko G (2013) Immunity, inflammation and cancer: a leading role for adenosine. Nat Rev Cancer 13:842–857
69. Kraut EH, Neff JC, Bouroncle BA, Gochnour D, Grever MR (1990) Immunosuppressive effects of pentostatin. J Clin Oncol 8:848–855
70. Johnston JB (2011) Mechanism of action of pentostatin and cladribine in hairy cell leukemia. Leuk Lymphoma 52(Suppl 2):43–45
71. Gryn J, Gordon R, Bapat A, Goldman N, Goldberg J (1993) Pentostatin increases the acute toxicity of high dose cyclophosphamide. Bone Marrow Transplant 12:217–220
72. Lathia C, Fleming GF, Meyer M, Ratain MJ, Whitfield L (2002) Pentostatin pharmacokinetics and dosing recommendations in patients with mild renal impairment. Cancer Chemother Pharmacol 50:121–126
73. Ho AD, Hensel M (2006) Pentostatin for the treatment of indolent lymphoproliferative disorders. Semin Hematol 43:S2–S10
74. Weiss MA, Maslak PG, Jurcic JG, Scheinberg DA, Aliff TB, Lamanna N, Frankel SR, Kossman SE, Horgan D (2003) Pentostatin and cyclophosphamide: an effective new regimen in previously treated patients with chronic lymphocytic leukemia. J Clin Oncol 21: 1278–1284
75. Petrus MJ, Williams JF, Eckhaus MA, Gress RE, Fowler DH (2000) An immunoablative regimen of fludarabine and cyclophosphamide prevents fully MHC-mismatched murine marrow graft rejection independent of GVHD. Biol Blood Marrow Transplant 6:182–189
76. Mariotti J, Taylor J, Massey PR, Ryan K, Foley J, Buxhoeveden N, Felizardo TC, Amarnath S, Mossoba ME, Fowler DH (2011) The pentostatin plus cyclophosphamide nonmyeloablative regimen induces durable host T cell functional deficits and prevents murine marrow allograft rejection. Biol Blood Marrow Transplant 17:620–631
77. Hassan R, Bullock S, Premkumar A, Kreitman RJ, Kindler H, Willingham MC, Pastan I (2007) Phase I study of SS1P, a recombinant anti-mesothelin immunotoxin given as a bolus I.V. infusion to patients with mesothelin-expressing mesothelioma, ovarian, and pancreatic cancers. Clin Cancer Res 13:5144–5149
78. Mossoba ME, Onda M, Taylor J, Massey PR, Treadwell S, Sharon E, Hassan R, Pastan I, Fowler DH (2011) Pentostatin plus cyclophosphamide safely and effectively prevents immunotoxin immunogenicity in murine hosts. Clin Cancer Res 17:3697–3705
79. Hassan R, Miller AC, Sharon E, Thomas A, Reynolds JC, Ling A, Kreitman RJ, Miettinen MM, Steinberg SM, Fowler DH, Pastan I (2013) Major cancer regressions in mesothelioma after treatment with an anti-mesothelin immunotoxin and immune suppression. Sci Transl Med 5:208ra147
80. Kharfan-Dabaja MA, Anasetti C, Fernandez HF, Perkins J, Ochoa-Bayona JL, Pidala J, Perez LE, Ayala E, Field T, Alsina M, Nishihori T, Locke F, Pinilla-Ibarz J, Tomblyn M (2013) Phase II study of CD4+–guided pentostatin lymphodepletion and pharmacokinetically targeted busulfan as conditioning for hematopoietic cell allografting. Biol Blood Marrow Transplant 19:1087–1093
81. Miller KB, Roberts TF, Chan G, Schenkein DP, Lawrence D, Sprague K, Gorgun G, Relias V, Grodman H, Mahajan A, Foss FM (2004) A novel reduced intensity regimen for allogeneic hematopoietic stem cell transplantation associated with a reduced incidence of graft-versus-host disease. Bone Marrow Transplant 33:881–889
82. Pavletic SZ, Bociek RG, Foran JM, Rubocki RJ, Kuszynski CA, Wisecarver JL, Hatcher L, Lucas DM, Byrd JC, Grever MR, Joshi SS, Hardiman P, Smith LM, Mcguire TR, Bierman PJ,

Vose JM, Armitage JO, Talmadge JE (2003) Lymphodepleting effects and safety of pentostatin for nonmyeloablative allogeneic stem-cell transplantation. Transplantation 76:877–881
83. Hudes G, Carducci M, Tomczak P, Dutcher J, Figlin R, Kapoor A, Staroslawska E, Sosman J, Mcdermott D, Bodrogi I, Kovacevic Z, Lesovoy V, Schmidt-Wolf IG, Barbarash O, Gokmen E, O'toole T, Lustgarten S, Moore L, Motzer RJ (2007) Temsirolimus, interferon alfa, or both for advanced renal-cell carcinoma. N Engl J Med 356:2271–2281
84. Childs R, Chernoff A, Contentin N, Bahceci E, Schrump D, Leitman S, Read EJ, Tisdale J, Dunbar C, Linehan WM, Young NS, Barrett AJ (2000) Regression of metastatic renal-cell carcinoma after nonmyeloablative allogeneic peripheral-blood stem-cell transplantation. N Engl J Med 343:750–758
85. Barkholt L, Bregni M, Remberger M, Blaise D, Peccatori J, Massenkeil G, Pedrazzoli P, Zambelli A, Bay JO, Francois S, Martino R, Bengala C, Brune M, Lenhoff S, Porcellini A, Falda M, Siena S, Demirer T, Niederwieser D, Ringden O (2006) Allogeneic haematopoietic stem cell transplantation for metastatic renal carcinoma in Europe. Ann Oncol 17:1134–1140
86. Hsieh MM, Kang EM, Fitzhugh CD, Link MB, Bolan CD, Kurlander R, Childs RW, Rodgers GP, Powell JD, Tisdale JF (2009) Allogeneic hematopoietic stem-cell transplantation for sickle cell disease. N Engl J Med 361:2309–2317
87. Brown EJ, Albers MW, Shin TB, Ichikawa K, Keith CT, Lane WS, Schreiber SL (1994) A mammalian protein targeted by G1-arresting rapamycin-receptor complex. Nature 369:756–758
88. Laplante M, Sabatini DM (2012) mTOR signaling in growth control and disease. Cell 149:274–293
89. Sarbassov DD, Ali SM, Sengupta S, Sheen JH, Hsu PP, Bagley AF, Markhard AL, Sabatini DM (2006) Prolonged rapamycin treatment inhibits mTORC2 assembly and Akt/PKB. Mol Cell 22:159–168
90. Fowler DH (2014) Rapamycin-resistant effector T-cell therapy. Immunol Rev 257:210–225
91. Chresta CM, Davies BR, Hickson I, Harding T, Cosulich S, Critchlow SE, Vincent JP, Ellston R, Jones D, Sini P, James D, Howard Z, Dudley P, Hughes G, Smith L, Maguire S, Hummersone M, Malagu K, Menear K, Jenkins R, Jacobsen M, Smith GC, Guichard S, Pass M (2010) AZD8055 is a potent, selective, and orally bioavailable ATP-competitive mammalian target of rapamycin kinase inhibitor with in vitro and in vivo antitumor activity. Cancer Res 70:288–298
92. Sini P, James D, Chresta C, Guichard S (2010) Simultaneous inhibition of mTORC1 and mTORC2 by mTOR kinase inhibitor AZD8055 induces autophagy and cell death in cancer cells. Autophagy 6:553–554
93. Chi H (2012) Regulation and function of mTOR signalling in T cell fate decisions. Nat Rev Immunol 12:325–338
94. Delgoffe GM, Pollizzi KN, Waickman AT, Heikamp E, Meyers DJ, Horton MR, Xiao B, Worley PF, Powell JD (2011) The kinase mTOR regulates the differentiation of helper T cells through the selective activation of signaling by mTORC1 and mTORC2. Nat Immunol 12:295–303
95. Powell JD, Lerner CG, Schwartz RH (1999) Inhibition of cell cycle progression by rapamycin induces T cell clonal anergy even in the presence of costimulation. J Immunol 162:2775–2784
96. Battaglia M, Stabilini A, Roncarolo MG (2005) Rapamycin selectively expands CD4+CD25+FoxP3+ regulatory T cells. Blood 105:4743–4748
97. Basu S, Golovina T, Mikheeva T, June CH, Riley JL (2008) Cutting edge: Foxp3-mediated induction of pim 2 allows human T regulatory cells to preferentially expand in rapamycin. J Immunol 180:5794–5798
98. Zeiser R, Leveson-Gower DB, Zambricki EA, Kambham N, Beilhack A, Loh J, Hou JZ, Negrin RS (2008) Differential impact of mammalian target of rapamycin inhibition on CD4+CD25+Foxp3+ regulatory T cells compared with conventional CD4+ T cells. Blood 111:453–462

99. Wang Y, Souabni A, Flavell RA, Wan YY (2010) An intrinsic mechanism predisposes Foxp3-expressing regulatory T cells to Th2 conversion in vivo. J Immunol 185:5983–5992
100. Zhao J, Zhao J, Perlman S (2012) Differential effects of IL-12 on Tregs and non-Treg T cells: roles of IFN-gamma, IL-2 and IL-2R. PLoS One 7, e46241
101. Levine BL, Bernstein WB, Connors M, Craighead N, Lindsten T, Thompson CB, June CH (1997) Effects of CD28 costimulation on long-term proliferation of CD4+ T cells in the absence of exogenous feeder cells. J Immunol 159:5921–5930
102. Blazar BR, Taylor PA, Panoskaltsis-Mortari A, Vallera DA (1998) Rapamycin inhibits the generation of graft-versus-host disease- and graft-versus-leukemia-causing T cells by interfering with the production of Th1 or Th1 cytotoxic cytokines. J Immunol 160:5355–5365
103. Reichardt W, Durr C, von Elverfeldt D, Juttner E, Gerlach UV, Yamada M, Smith B, Negrin RS, Zeiser R (2008) Impact of mammalian target of rapamycin inhibition on lymphoid homing and tolerogenic function of nanoparticle-labeled dendritic cells following allogeneic hematopoietic cell transplantation. J Immunol 181:4770–4779
104. Fowler DH, Mossoba ME, Steinberg SM, Halverson DC, Stroncek D, Khuu HM, Hakim FT, Castiello L, Sabatino M, Leitman SF, Mariotti J, Gea-Banacloche JC, Sportes C, Hardy NM, Hickstein DD, Pavletic SZ, Rowley S, Goy A, Donato M, Korngold R, Pecora A, Levine BL, June CH, Gress RE, Bishop MR (2013) Phase 2 clinical trial of rapamycin-resistant donor CD4+ Th2/Th1 (T-Rapa) cells after low-intensity allogeneic hematopoietic cell transplantation. Blood 121:2864–2874
105. Dudley ME, Gross CA, Langhan MM, Garcia MR, Sherry RM, Yang JC, Phan GQ, Kammula US, Hughes MS, Citrin DE, Restifo NP, Wunderlich JR, Prieto PA, Hong JJ, Langan RC, Zlott DA, Morton KE, White DE, Laurencot CM, Rosenberg SA (2010) CD8+ enriched "young" tumor infiltrating lymphocytes can mediate regression of metastatic melanoma. Clin Cancer Res 16:6122–6131
106. Castiello L, Mossoba M, Viterbo A, Sabatino M, Fellowes V, Foley JE, Winterton M, Halverson DC, Civini S, Jin P, Fowler DH, Stroncek DF (2013) Differential gene expression profile of first-generation and second-generation rapamycin-resistant allogeneic T cells. Cytotherapy 15:598–609
107. Halverson DC, Rowley S, Hansen BG, Schuver BB, Mossoba M, Steinberg SM, Hakim F, Kurlander R, Gea-Banacloche J, Sportes C, Hardy NM, Hickstein DD, Pavletic S, Khuu HM, Castiello L, Sabatino M, Stroncek D, Levine B, June CH, Donato M, Goy A, Korngold R, Pecora A, Bishop MR, Gress R, Fowler D (2014) T-Rapa6 and T-Rapa12 cells differentially mediate acute GVHD after low-intensity allogeneic HCT. Biol Blood Marrow Transplant 20:S130–S131

ERRATUM

Developments in T Cell Based Cancer Immunotherapies

Paolo A. Ascierto, David F. Stroncek, and Ena Wang

P.A. Ascierto et al. (eds.), *Developments in T Cell Based Cancer Immunotherapies*, Cancer Drug Discovery and Development,
DOI 10.1007/978-3-319-21167-1

DOI 10.1007/978-3-319-21167-1_14

The publisher regrets that the affiliation of the Book editor Ena Wang was incorrect in the print and online versions of this book. The correct affiliation is given below.

Ena Wang
Sidra Medical and Research Center,
Research Branch, Doha, Qatar

The updated original online version of the original book can be found at
http://dx.doi.org/10.1007/978-3-319-21167-1

P.A. Ascierto et al. (eds.), *Developments in T Cell Based Cancer Immunotherapies*, Cancer Drug Discovery and Development,
DOI 10.1007/978-3-319-21167-1_14

Index

A

Abscopal effect, 211–224
Adoptive cell therapy/adoptive T cell therapy (ACT), 76, 118, 152–165, 167, 168, 170, 250–251, 254, 284
Anergy, 55, 56, 58–60, 65, 66, 233, 254, 292

B

Biomarkers, 15, 157, 169, 170, 220, 222, 240, 248, 252
B7-H1 (PD-L1), 55, 57, 59, 61, 65, 105, 167, 170, 231, 233, 240

C

Cancer, 2–9, 11–17, 30–32, 34, 36, 41–43, 54, 57, 59, 61–64, 66, 76–78, 83, 85–87, 100–105, 107, 109, 110, 112, 113, 116, 118, 130, 144, 152–155, 157–159, 164–166, 168–170, 183, 192, 194, 196, 199, 213, 217–220, 222–224, 231, 233–235, 237, 238, 246–255, 259, 262, 273, 283, 285–287, 290, 293
 immunotherapy, 4, 17, 42, 101, 103, 112, 113, 118, 185, 247, 249, 252–254
 vaccines, 2, 4, 7, 43, 100, 219, 221, 246, 248
CD4+ T cells, 7, 39, 81, 83, 84, 105, 144, 160, 161, 165, 189, 261, 282, 283, 293, 294
CD8, 2, 4, 6, 7, 9, 11–13, 15, 16, 34–40, 59, 61–64, 78, 79, 81, 83, 85, 100, 102–105, 108, 110, 114, 140, 144, 153, 157–161, 163–171, 185, 188–194, 196, 198–200, 221, 223, 232, 238, 246, 248, 261, 268, 270, 282–285, 292, 294
Cell culture, 131, 133, 134, 139
Cell processing, 132, 133
Cell selection, 140
Checkpoint inhibitor, 14–16, 41–43, 104, 106, 213, 221, 222, 224
Chimeric antigen, 76, 130, 159, 184, 251, 260, 285
Clinical trials, 4, 6–8, 11–13, 15, 31, 32, 36, 42, 43, 55, 61, 65, 78, 83, 85, 86, 100, 103, 131, 132, 145, 146, 155, 156, 161, 162, 165, 166, 168, 170, 194, 196, 198, 200, 217–219, 221, 238, 249, 250, 252, 253, 262, 263, 271, 277, 285, 288, 289, 291, 293, 295
Combination therapy, 13, 17, 221, 276
Co-stimulation, 2, 15, 58, 59, 159, 251, 261, 262, 266, 269, 292–294
Cytokines, 2, 11, 17, 30, 36, 38, 39, 41, 58, 61, 64, 84, 100, 132, 141, 143, 146, 152–154, 158, 162, 163, 167, 187, 189, 191, 193, 195, 198, 215, 222, 246, 249, 265, 272–273, 275–274, 282, 286, 287, 292–294
 prime-boost regimen, 17
Cytotoxic T-lymphocyte antigen 4 (CTLA-4), 15, 30, 41–43, 55–57, 60, 61, 64, 100, 112, 114, 115, 169, 213, 217, 218, 220, 222, 231–233, 236, 238–240, 250, 252, 276

P.A. Ascierto et al. (eds.), *Developments in T Cell Based Cancer Immunotherapies*, Cancer Drug Discovery and Development,
DOI 10.1007/978-3-319-21167-1

E

Epitope enhancement, 30, 31, 34
Exhaustion, 55, 61, 62, 65, 66, 164, 197, 238, 239

G

Gene modified T cells, 77, 78, 80, 81, 84–87
Gene therapy, 132, 133, 137, 249, 264
Genetically engineered T cells, 87
Good manufacturing practices, 130
Graft rejection, 283, 286, 288, 290, 293
Graft-versus-host disease, 187, 263, 265, 282
Graft-versus-tumor effect (GVT), 283–285, 290, 293

H

Homeostatic cytokines, 155, 250, 266

I

Immune checkpoint blockade, 103–104, 217, 231
Immune metabolism, 191
Immune regulation, 14, 58
Immunomodulatory antibodies, 246, 252
Immunotherapy, 3, 5, 6, 9, 17, 32, 41, 43, 76, 77, 100–109, 111–113115–118, 154, 155, 168–170, 192–194, 197–199, 215, 219, 224, 240, 246, 247, 250, 260, 276
Induced pluripotent stem (iPS) cells, 195, 197
Interleukin-2 (IL-2), 35, 41, 64, 103–105, 108, 114, 116–118, 132, 137, 142, 144, 154, –162, 164, 167, 170, 172, 185, 189, 192–196, 198, 200, 217, 233, 238, 250, 261, 265, 275, 293

M

Melanoma, 2, 6, 11–13, 15, 16, 36, 42, 57, 61, 62, 65, 76, 78–81, 84–86, 100, 102–105, 108–110, 112, 114, 116, 117, 130, 144, 154–156, 159–161, 163, 165, 167–170, 184, 192, 193, 197, 200, 217–219, 221–224, 231, 233, 234, 236–240, 246, 247, 249, 250, 252, 253, 259, 294
Memory T cell, 63, 64, 112, 143, 164, 167, 185–193, 196, 199, 232, 248, 249, 266

N

NKT cells, 16, 30, 38–41, 153
Non-Hodgkin's lymphoma, 199, 252, 270

P

PD-L1, 14, 15, 41, 59, 104, 105, 108, 170, 217, 218, 220, 231–238, 240, 253, 255, 272–273, 276
Pentostatin, 286–291
Peptide cancer vaccines, 1–17
Polyfunctionality, 64–66
Programmed cell death protein 1 (PD-1), 15, 16, 30, 41–43, 55, 56, 60, 61, 65, 100, 105, 112, 153, 167–168, 171, 193, 217, 218, 222, 231–240, 252, 253, 272–273, 276

R

Rapamycin, 191, 192, 196, 283, 284, 286, 290–295
Reduced-intensity allogeneic stem cell transplantation, 286
Reprogramming, 195, 197, 198

S

Small molecules, 192, 195, 196
Stemness, 63–66
Stereotactic ablative radiotherapy (SABR), 213–216, 218, 220–222, 224

T

T cell, 4–7, 11, 13–17, 30–32, 34–42, 54–66, 76–87, 100–103, 105, 106, 108, 110, 112, 114, 118, 130–132, 134, 136–142, 144–147, 153, 154, 156–161, 163–172, 183–200, 213, 215, 217, 219, 221, 223, 224, 231–233, 238, 239, 246–254, 259–271, 273, 275–274, 276, 277, 282–286, 288–295
T cell avidity, 34, 35, 38
T cell differentiation, 63, 152, 163–165, 185, 188–198, 284, 291, 292

T cell receptor (TCR), 5, 31, 59, 76, 105, 130, 159, 184, 222, 250, 259
T memory stem cells, 63, 187
TCR affinity, 79, 81–84, 86, 130, 250–251
Th1/Th2, 282, 283, 285, 292–294
Th17, 56, 63–66, 110, 282, 285
Transcription factors, 4, 54, 60, 102, 188, 195, 197, 198, 292, 294
Tumor, 2, 32, 53–66, 76, 100, 130, 153, 184, 214, 231, 246, 260, 285
Tumor immunity, 39–41
Tumor-infiltrating leukocytes/lymphocyte (TIL), 16, 76, 154–157, 168, 183, 184, 250, 294

V

Vaccination, 6, 9, 10, 12, 32, 33, 43, 64, 103, 108, 113, 116–118, 196, 216, 219–221, 224, 238, 247, 248, 251, 254, 270

The manufacturer's authorised representative in the EU is Springer Nature Customer Service Centre GmbH, Europaplatz 3, 69115 Heidelberg, Germany. If you have any concerns regarding our products, please contact ProductSafety@springernature.com

Printed and bound by CPI Group (UK) Ltd, Croydon, CR0 4YY
28/09/2026
02224079-0002